Organic Reactions

Organic Reactions

VOLUME 29

JOHN WILEY & SONS, INC.

New York · Chichester · Brisbane · Toronto · Singapore

Published by John Wiley & Sons, Inc.

Library of Congress Catalog Card Number 42-20265

ISBN 0-471-87490-6

Printed in the United States of America

10 9 8 7 6 5 4 3 2 1

PREFACE TO THE SERIES

In the course of nearly every program of research in organic chemistry the investigator finds it necessary to use several of the better-known synthetic reactions. To discover the optimum conditions for the application of even the most familiar one to a compound not previously subjected to the reaction often requires an extensive search of the literature; even then a series of experiments may be necessary. When the results of the investigation are published, the synthesis, which may have required months of work, is usually described without comment. The background of knowledge and experience gained in the literature search and experimentation is thus lost to those who subsequently have occasion to apply the general method. The student of preparative organic chemistry faces similar difficulties. The textbooks and laboratory manuals furnish numerous examples of the application of various syntheses, but only rarely do they convey an accurate conception of the scope and usefulness of the processes.

For many years American organic chemists have discussed these problems. The plan of compiling critical discussions of the more important reactions thus was evolved. The volumes of *Organic Reactions* are collections of chapters each devoted to a single reaction, or a definite phase of a reaction, of wide applicability. The authors have had experience with the processes surveyed. The subjects are presented from the preparative viewpoint, and particular attention is given to limitations, interfering influences, effects of structure, and the selection of experimental techniques. Each chapter includes several detailed procedures illustrating the significant modifications of the method. Most of these procedures have been found satisfactory by the author or one of the editors, but unlike those in *Organic Syntheses* they have not been subjected to careful testing in two or more laboratories.

Each chapter contains tables that include all the examples of the reaction under consideration that the author has been able to find. It is inevitable, however, that in the search of the literature some examples will be missed, especially when the reaction is used as one step in an extended synthesis. Nevertheless, the investigator will be able to use the tables and their accompanying bibliographies in place of most or all of the literature search so often required.

Because of the systematic arrangement of the material in the chapters and the entries in the tables, users of the books will be able to find information desired by reference to the table of contents of the appropriate chapter. In the interest of economy the entries in the indices have been kept to a minimum, and, in particular, the compounds listed in the tables are not repeated in the indices.

The success of this publication, which will appear periodically, depends upon the cooperation of organic chemists and their willingness to devote time and effort to the preparation of the chapters. They have manifested their interest already by the almost unanimous acceptance of invitations to contribute to the work. The editors will welcome their continued interest and their suggestions for improvements in *Organic Reactions*.

Chemists who are considering the preparation of a manuscript for submission to *Organic Reactions* are urged to write either secretary before they begin work.

CONTENTS

Organic Reactions

CHAPTER 1

REPLACEMENT OF ALCOHOLIC HYDROXYL GROUPS BY HALOGENS AND OTHER NUCLEOPHILES VIA OXYPHOSPHONIUM INTERMEDIATES

BERTRAND R. CASTRO

Ecole Nationale Superieure de Chimie de Montpellier
Montpellier, France

CONTENTS

ACKNOWLEDGMENT

I gratefully acknowledge the expert extensive assistance of Dr. Robert M. Joyce in preparing the manuscript for publication.

INTRODUCTION

The Michaelis–Arbuzov reaction involves the formation of dialkyl alkyl-phosphonates by heating alkyl halides with trialkyl phosphites, via an intermediate phosphonium salt $\mathbf{1}$.[1,2]

$$(RO)_3P + R'X \xrightarrow{\quad\quad} (RO)_3P^+R', X^- \xrightarrow{\text{Heat}} (RO)_2P(O)R' + RX$$

$$\mathbf{1}$$

$$\text{(Eq. 1)}$$

In recent years it has been found that phosphonium salts analogous to $\mathbf{1}$ can be formed by reaction of a variety of trivalent phosphorus compounds with oxidizing electrophiles, and that the resulting phosphonium intermediates can be trapped with an alcohol to form alkoxyphosphonium cations. These cations can then undergo nucleophilic displacements, either with their own counterions or with added nucleophiles (Eqs. 2 and 3).

$$Y_3P + XE \longrightarrow Y_3P^+E, X^- \xrightarrow{ROH} Y_3P^+OR \qquad \text{(Eq. 2)}$$

$$Y_3P^+OR \quad
\begin{array}{l}
\xrightarrow{\text{(a) } E^-} Y_3PO + RE \\
\xrightarrow{\text{(b) } X^-} Y_3PO + RX \\
\xrightarrow[\text{(c)}]{A^-} Y_3PO + RA
\end{array}
\qquad \text{(Eq. 3)}$$

In Eq. 2, XE is the oxidizing electrophile. The path followed in Eq. 3 depends on the reagents and the reaction conditions.

Although the electrophilic phosphorus species Y_3P^+E, X^- in Eq. 2 are formally equivalents of phosphorus pentachloride, this review is limited to species formed *in situ* from trivalent phosphorus compounds and various oxidants. The most commonly used trivalent phosphorus compounds are triphenyl phosphite, triphenylphosphine, and tris(dimethylamino)phosphine. The most commonly used oxidants are molecular halogens, alkyl halides, carbon tetrahalides, N-haloamides, N-haloamines, and diethyl azodicarboxylate (DEAD). Other oxidants that have been used include mercuric salts, peroxides, dithioesters, and sulfenamides. The more common combinations of phosphorus(III) compounds and oxidants are summarized in Table 1.

The replacement of hydroxyl groups by halogen, the most common use of oxyphosphonium intermediates, is the principal subject of this chapter. However, the replacement of hydroxyl groups by oxygen, nitrogen, sulfur, and carbon nucleophiles is also discussed. This review is restricted to alcoholic hydroxyl groups, and does not include a discussion of carboxylic hydroxyl groups.

TABLE 1. COMBINATIONS OF PHOSPHORUS(III)
COMPOUNDS AND OXIDANTS FOR REPLACEMENT OF
HYDROXYL GROUPS BY HALOGENS (X) OR OTHER
NUCLEOPHILES (O, N, S, C)

Oxidant	$P(OC_6H_5)_3$	$P(C_6H_5)_3$	$P[N(CH_3)_2]_3$
		PY$_3$	
X_2	X	X	
RX	X		
CX_4		X, N	X, N, S, O, C
R_2NX		X	X, N, S, O, C
DEAD		O, N, C, X	

The principal combinations covered are triphenylphosphine–diethyl azodicarboxylate, triphenyl phosphite-methiodide, triphenylphosphine-halogen, triphenylphosphine–tetrahalomethane, triphenylphosphine–N-haloimide, and tris(dimethylamino)phosphine–tetrahalomethane. Some of these combinations have been discussed in recent reviews.[3-7]

MECHANISM

The general course of the replacement of alcoholic hydroxyl groups via oxyphosphonium intermediates is shown in Eqs. 2 and 3. Some insight into the details of the mechanism has been obtained from studies of the triphenylphosphine–diethyl azodicarboxylate, triphenylphosphine–carbon tetrachloride, and tris(dimethylamino)phosphine–carbon tetrachloride systems.

Triphenylphosphine–Diethyl Azodicarboxylate

The structure of the initial reaction product of a trivalent phosphorus compound with DEAD depends on the nature of the groups attached to phosphorus. With triphenylphosphine, both of the betaine intermediates $2a$[8] and $2b$[9,10] have been postulated, and evidence in support of a betaine intermediate is found in the ^{31}P NMR spectrum of the adduct, in which the phosphorus chemical shift is 43 ppm downfield from phosphoric acid. On the other hand ^{31}P NMR spectra indicate that a pentacovalent phosphorane 3 is the intermediate when RO or R_2N groups are attached to phosphorus.[11,12] These intermediates may be in dynamic equilibrium, as shown in Eq. 4.

In the presence of an alcohol R′OH the R′O group of the alcohol attaches to phosphorus to form an alkoxyphosphonium intermediate 4 or a phosphorane intermediate 5 (Eq. 5); the former appears more probable. In the absence of another nucleophile a displacement occurs to form a substituted hydrazino ester 6 (Eq. 6). However, in the presence of another nucleophile nucleophilic substitution takes place at the R′ group of the alcohol (Eq. 7).

$$\text{(Eq. 4)}$$

$$\xrightarrow{\text{R'OH}} \quad \begin{array}{c} R_3P^+OR' \\[4pt] C_2H_5O_2C\overset{-}{N}\text{---}NHCO_2C_2H_5 \\ \mathbf{4} \end{array} \quad \rightleftharpoons \quad \begin{array}{c} \overset{OR'}{\underset{}{R_3P}} \\ \diagdown \\ N\text{---}NHCO_2C_2H_5 \\ | \\ CO_2C_2H_5 \\ \mathbf{5} \end{array}$$

$$\text{(Eq. 5)}$$

$$\begin{array}{c} R_3\overset{+}{P}O\text{---}R' \\ C_2H_5O_2C\overset{-}{N}\text{---}NHCO_2C_2H_5 \\ \mathbf{4} \end{array} \quad \longrightarrow \quad R_3PO + \begin{array}{c} R'N \diagup ^{CO_2C_2H_5} \\ \diagdown _{NHCO_2C_2H_5} \\ \mathbf{6} \end{array}$$

$$\text{(Eq. 6)}$$

$$\begin{array}{c} R_3P^+OR' \\ C_2H_5O_2C\overset{-}{N}\text{---}NHCO_2C_2H_5 \\ \mathbf{4} \end{array} \quad \xrightarrow{\text{HY}} \quad R_3\overset{+}{P}O\text{---}R' \ Y^-$$

$$+$$

$$C_2H_5O_2CNHNHCO_2C_2H_5$$

$$\longrightarrow R_3PO + R'Y \qquad \text{(Eq. 7)}$$

The nucleophilic substitution shown in Eq. 7 generally proceeds by an S_N2 mechanism with inversion of configuration, even with allylic substrates[13–15] and with substrates that can undergo neighboring group participation.[16] This characteristic inversion of configuration is one of the useful features of this reaction. On the other hand mixtures of products are usually found in substitutions at the anomeric position of sugars, and in reactions of cholesterol,[17] where an S_N1 mechanism appears to be involved.

Triphenylphosphine–Carbon Tetrachloride

Pure triphenylphosphine and pure carbon tetrachloride react only very slowly, even at reflux temperature.[18,19] A small amount of a polar substance such as water, an alcohol, a ketone, or a solvent such as methylene chloride or acetonitrile is necessary to initiate the reaction. Furthermore, the polarity of the solvent plays an important role in determining the reaction rate, which is 10^4 and 10^6 times higher in methylene chloride and acetonitrile, respectively, than it is in carbon tetrachloride alone.

It has been shown that in the absence of an alcohol the reaction proceeds as shown in Eqs. 8–11.[20,21]

$$R_3P + CCl_4 \longrightarrow R_3P^+CCl_3, Cl^- \qquad \text{(Eq. 8)}$$
$$\mathbf{7}$$

$$R_3P^+CCl_3, Cl^- + R_3P \longrightarrow R_3P{=}CCl_2 + R_3PCl_2 \qquad \text{(Eq. 9)}$$

$$R_3P{=}CCl_2 + R_3PCl_2 \longrightarrow R_3P^+CCl_2P^+R_3, 2Cl^- \qquad \text{(Eq. 10)}$$

$$R_3P^+CCl_2P^+R_3, 2Cl^- + R_3P \longrightarrow R_3P^+CCl{=}PR_3, Cl^- + R_3PCl_2$$

$$\text{(Eq. 11)}$$

There is no accumulation of any of the intermediates in these four steps, and the reaction appears to proceed directly as shown in Eq. 12.

$$3R_3P + CCl_4 \longrightarrow R_3P^+CCl{=}PR_3, Cl^- + R_3PCl_2 \quad \text{(Eq. 12)}$$

The intermediates shown in Eqs. 8–11 have been isolated and characterized. However, since the first step of the reaction does not occur spontaneously but must be catalyzed by a substrate or a solvent, charge-transfer species such as **8** and **9** appear to be involved.

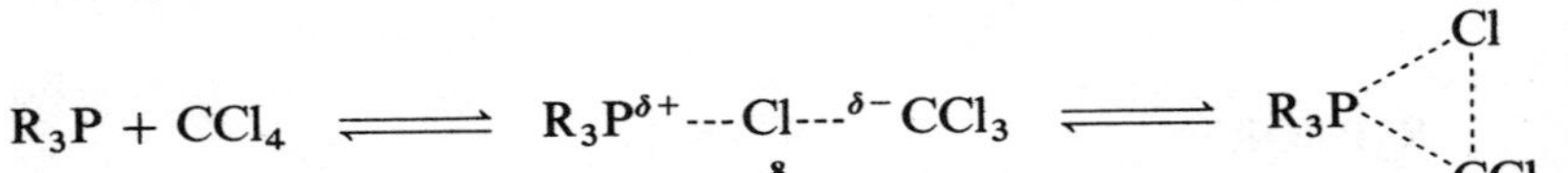

$$R_3P + CCl_4 \rightleftharpoons \underset{8}{R_3P^{\delta+}\text{---}Cl\text{---}^{\delta-}CCl_3} \rightleftharpoons \underset{9}{R_3P\overset{Cl}{\underset{CCl_3}{\diagdown}}}$$

$$\longrightarrow \underset{7}{R_3P^+CCl_3, Cl^-}$$

When the reaction is carried out in the presence of an alcohol, oxyphosphonium intermediates (**10**) can be formed by two different routes:

1. In the *chloroform route* (Eq. 13) the alcohol reacts directly with the charge-transfer species, liberating chloroform.

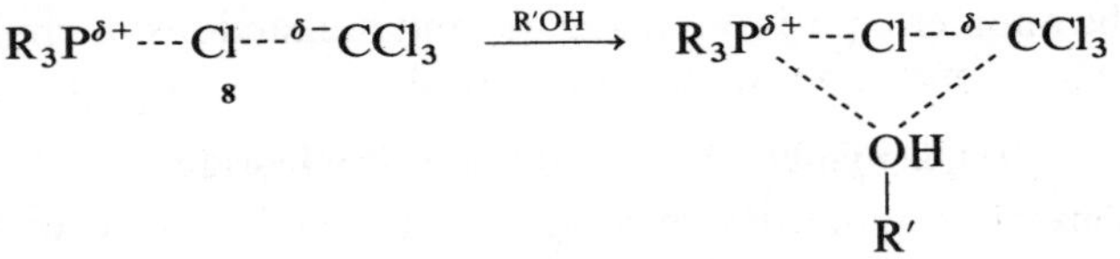

$$\underset{8}{R_3P^{\delta+}\text{---}Cl\text{---}^{\delta-}CCl_3} \xrightarrow{R'OH} R_3P^{\delta+}\text{---}Cl\text{---}^{\delta-}CCl_3$$
$$\underset{\underset{R'}{\overset{|}{OH}}}{}$$

$$\longrightarrow \underset{10}{R_3P^+OR', Cl^- + CHCl_3} \quad \text{(Eq. 13)}$$

2. In the *ylid route* (Eq. 14) the reaction of the phosphine and the halide proceeds by way of Eqs. 8 and 9, and the oxyphosphonium ion **10** is formed by trapping the basic ylid and the chloromethylphosphonium chloride **11**.

$$R_3P{=}CCl_2 + R_3PCl_2 + R'OH$$
$$\longrightarrow \underset{11}{R_3P^+CHCl_2}, Cl^- + \underset{10}{R_3P^+OR'}, Cl^- \quad \text{(Eq. 14)}$$

When the reaction proceeds by way of Eq. 14, the intermediate phosphonium salt **11** is dehalogenated by triphenylphosphine to give an ylid **12** and a triphenylphosphine–chlorine complex **13** that can be trapped by another molecule of alcohol (Eq. 15).

$$R_3P^+CHCl_2, Cl^- + R_3P \longrightarrow R_3P=CHCl + R_3PCl_2$$

11   **12**   **13**

$$\xrightarrow{R'OH} R_3P^+CH_2Cl, Cl^- + R_3P^+OR', Cl^-$$

10

(Eq. 15)

Reactions that proceed by the ylid route (Eqs. 14 and 15) are characterized by the facts that (1) little or no chloroform is evolved, and (2) more triphenylphosphine is required to complete the reaction. The latter point is important to the achievement of high yields.

The polarity of the reaction solvent appears to influence which of the two routes is followed. When the reaction is carried out with benzyl alcohol in the moderately polar methylene chloride, 46% of the theoretical amount of chloroform is evolved, and the balance of the reaction proceeds by the ylid route.[20] On the other hand, when the reaction is carried out with cyclohexanol in the less polar carbon tetrachloride, less than 5% of the theoretical amount of chloroform is evolved, and the chloromethylphosphonium chloride **11** is isolated in 95% yield.[22]

The oxyphosphonium intermediate **10** is generally written in the ionic form first postulated by Downie and co-workers.[23]

$$(C_6H_5)_3P^+OR, Cl^-$$

10

However, later investigators favor a molecular pentacovalent intermediate, and such intermediates have been isolated from reactions of hindered alcohols such as neopentyl alcohol,[24,25] isocholesterol,[26] norbornanols,[24] and benzo-bicyclooctadienols.[27]

Formation of intermediate salt **10** is generally fast and largely independent of steric hindrance.[25] On the other hand the rates of decay of the intermediate salts **10** to the products are much more sensitive to steric hindrance: primary intermediates give halides very quickly at room temperature, whereas secondary and neopentylic intermediates require prolonged heating.

This feature often allows selective reaction of primary and secondary alcoholic hydroxyl groups; even if both hydroxyls are activated in the first step of the reaction, careful optimization of reaction time and temperature can allow decay of the primary intermediate and hydrolysis of the secondary one during workup.

The intermediate salt **10** usually exhibits a strong tendency to undergo an S_N2 displacement with inversion of configuration, and neopentylic rearrangement does not take place.

$$(S)\text{-}t\text{-}C_4H_9CHDOH \xrightarrow{(C_6H_5)_3P/CBr_4} (R)\text{-}t\text{-}C_4H_9CHDBr$$
$$\textbf{14}$$

The optical rotation of the deuteriobromide **14** is 3.5 times higher than that of **14** obtained by displacement of the corresponding alcohol tosylate with lithium bromide.[28]

Alcohols that have a tendency to undergo S_N1 reactions generally give rearrangement products if the reaction cannot be performed at low temperature; such rearrangements are often observed with allylic substrates.[29,30] Reactions at the anomeric position of sugars generally proceed by an S_N1 mechanism.[31] Skeletal rearrangements are usually observed with hindered structures bearing groups that can participate in the reaction.[24,27]

Tris(dimethylamino)phosphine–Carbon Tetrachloride

This reaction appears to proceed by the mechanism shown in Eq. 16.

$$[(CH_3)_2N]_3P + ClCCl_3 \xrightarrow{\text{Fast}} [(CH_3)_2N]_3P^+Cl, CCl_3^- \xrightarrow{\text{ROH}}$$
$$\textbf{15}$$

$$HCCl_3 + [(CH_3)_2N]_3P^+OR, Cl^-$$
$$\textbf{16}$$

$$\xrightarrow{\text{Slow}} [(CH_3)_2N]_3PO + RCl$$

$$\xrightarrow[\text{H}_2\text{O}]{\text{NH}_4\text{PF}_6} [(CH_3)_2N]_3P^+OR, PF_6^-$$

$$\text{(Eq. 16)}$$

The high nucleophilicity of the aminophosphine allows complete charge transfer to the trichloromethyl anion, whose existence has been demonstrated by trapping experiments with aldehydes, ketones, and acid anhydrides.[32–34] The existence of the oxyphosphonium salt **16** is established by running the reaction at low temperatures, since the salt does not break down to the alkyl chloride at temperatures below 0°. The formation of salt **16** is very rapid with primary alcohols. However, with bulky or secondary alcohols the aminophosphine must be added to the carbon tetrachloride solution of the alcohol slowly to obtain high yields; too rapid an addition is signaled by a darkening of the solution. This effect can be explained by the equilibrium shown in Eq. 17. When the formation of the intermediate **17** is slow because of the bulk of the R group, the alternative reactions shown in Eqs. 18 and 19 become prominent.

$$[(CH_3)_2N]_3P^+Cl, CCl_3^- + ROH \rightleftharpoons CHCl_3 + [(CH_3)_2N]_3P^+Cl, RO^-$$
$$\textbf{15} \qquad\qquad\qquad\qquad\qquad\qquad\qquad\qquad \textbf{17}$$

$$\longrightarrow [(CH_3)_2N]_3P^+OR, Cl^- \quad \text{(Eq. 17)}$$
$$\textbf{16}$$

$$[(CH_3)_2N]_3P^+Cl, CCl_3^- \xrightarrow{\hspace{2cm}} \begin{array}{l} CCl_2 + [(CH_3)_2N]_3P^+Cl, Cl^- \quad \text{(Eq. 18)} \\[1em] [(CH_3)_2N]_3P^+CCl_3, Cl^- \quad \text{(Eq. 19)} \end{array}$$

15

Replacement of carbon tetrachloride with N-chlorodiisopropylamine gives a more stable conjugate base than the trichloromethide ion, allowing the preparation of oxyphosphonium salts in high yields from secondary alcohols.[35]

The reaction is faster in nonpolar solvents, where the ionic partners form a tight ion pair, than in polar solvents, where the ions are dissociated.[36] The hexamethylaminophosphoryl group appears to be a poorer leaving group than triphenylphosphoryl in the triphenylphosphine–carbon tetrachloride system, and there is less tendency for an S_N1 displacement in the former case. Hexafluorophosphate salts derived from anomeric hydroxyl groups of carbohydrates[37–40] and from substituted allylic alcohols[41] are not easily isolated at room temperature because they are hydrolyzed to the starting alcohols. They are detectable at low temperatures by their characteristic ^{31}P NMR signal, which is 35 ppm downfield from phosphoric acid.

SCOPE AND LIMITATIONS

Each of the phosphorus(III) systems covered in this review has its own characteristics and area of utility; accordingly, the scope and limitations of each are discussed separately in this section.

Triphenylphosphine–Diethyl Azodicarboxylate and Related Systems

The first studies of the reactions of azodicarboxylic esters with trivalent phosphites and phosphines appeared from 1958 to 1960.[42,43] It was later shown that the triphenylphosphine–diethyl azodicarboxylate combination can activate the alcoholic hydroxyl group, leading to formation of substituted hydrazino esters containing the R group of the alcohol.[8] Addition of various nucleophiles to this combination has provided routes for conversion of alcohols to a variety of carbon–nitrogen compounds, esters, ethers, and alcohol dehydration products. This system has become the most versatile and synthetically useful of all the phosphorus(III) systems for replacement of alcoholic hydroxyl groups.

Formation of Carbon–Nitrogen Bonds.* The reaction of an alcohol, DEAD, and triphenylphosphine generally proceeds in less than 1 hour at room temperature to give an N-alkyl-N,N'-dicarbethoxyhydrazine (**6**, Eqs. 4–6).[8,44–46] A tricarbethoxy compound $RN(CO_2C_2H_5)N(CO_2C_2H_5)_2$ is often obtained as a by-product; the reason for its formation is not clear.[44] Another side

---

* For a review of this reaction see ref. 7.

reaction is the formation of an ethyl alkyl carbonate, particularly when tris(dimethylamino)phosphine is used in place of triphenylphosphine.[47] Reaction at the anomeric position of sugars gives glycosylhydrazines.[45]

In four-component systems, where the alcohol is used with another nucleophile, the triphenylphosphine–diethyl azodicarboxylate combination has the role of a water acceptor.

Condensations with imides, particularly phthalimides, can be used to form N-alkylphthalimides, which are useful precursors of alkylamines; this reaction is a modification of the Gabriel reaction.[48–52] Phthalimidation of secondary alcohols proceeds with inversion of configuration.

$$(S)\text{-}(+)\text{-}C_6H_{13}CHOHCH_3 + HN(CO)_2C_6H_4$$

$$\longrightarrow \quad (R)\text{-}(-)\text{-}C_6H_{13}CH(CH_3)N(CO)_2C_6H_4$$

Reaction of allylic alcohols in the unsaturated sugar series can proceed by S_N2 (inversion, Eq. 20), S_N2' (retention, Eq. 21), or with elimination (Eq. 22), depending on the structure of the substrate.

$$(Eq. 20)$$

$$(Eq. 21)$$

$$(Eq. 22)$$

Phthalimidation of the anomeric position of sugars usually gives a mixture of α and β anomers.

Other imides, such as benzyloxycarbonyl benzamide[53–55] and saccharin,[56] can be used. Mixtures of N- and O-alkylation products are generally obtained (Eq. 23).

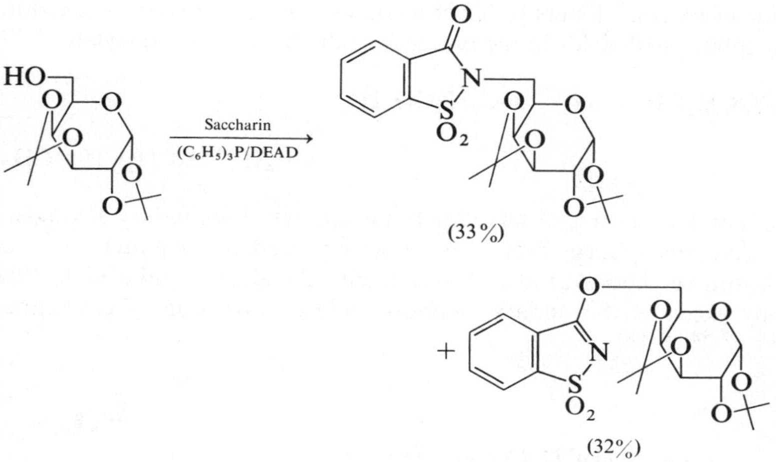

(33%)

$+$ (Eq. 23)

(32%)

Nucleosides have been made by reacting sugars with 6-chloropurine in the presence of methyldiphenylphosphine and diethyl azodicarboxylate. The anomeric selectivity has been poor, apparently because the substrates have no participating or orienting groups.[57]

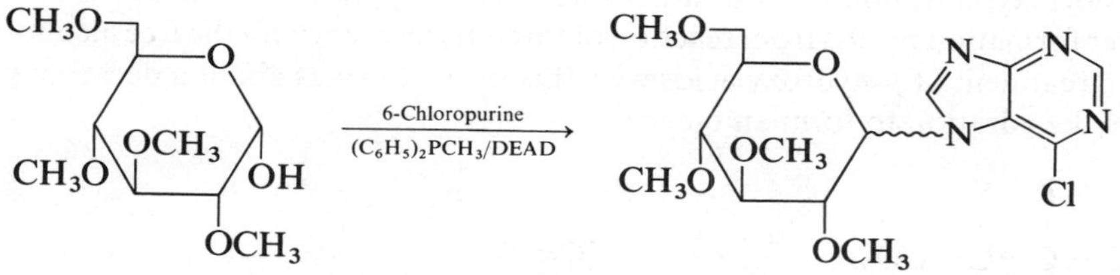

Use of hydrazoic acid as the nucleophile provides a very useful method for converting alcoholic hydroxyl groups into azides,[11,58-64] and can lead to formation of a carbon–nitrogen bond in reactions in which phthalimide simply dehydrates the alcohol.

$C_2H_5O_2CCHOHCH_2CO_2C_2H_5$

Phthalimide → $C_2H_5O_2CCH=CHCO_2C_2H_5$
(trans) (80%)

$(C_6H_5)_3P$
DEAD

HN_3 → $C_2H_5O_2CCH(N_3)CH_2CO_2C_2H_5$
(75%)

Diphenylphosphoryl azide has been used as a source of azide ion.[65]

3-β-Cholestanol $\xrightarrow[\text{(C}_6\text{H}_5\text{O)}_2\text{P(O)N}_3]{\text{(C}_6\text{H}_5\text{)}_3\text{P/DEAD}}$ 3-α-Azidocholestane

Formation of Esters. Esters of lower alcohols can be prepared by reaction of alkyl phosphites with acids in the presence of diethyl azodicarboxylate.[10,66]

$$(C_2H_5O)_2POCH_2CH=CH_2 + n\text{-}C_4H_9CO_2H$$

$$\xrightarrow{\text{DEAD}} n\text{-}C_4H_9CO_2CH_2CH=CH_2$$

Inversion of configuration is observed with phosphites of secondary alcohols.

The trivalent phosphorus system most widely used for forming esters is triphenylphosphine–diethyl azodicarboxylate plus the alcohol and acid. Yields are generally excellent. Secondary alcohols undergo inversion of configuration.[11,13,15–17,58,66–69]

$$+ \; C_6H_5CH_2CO_2H \xrightarrow[\text{DEAD}]{(C_6H_5)_3P}$$

When highly hindered alcohols are used, only low yields of esters are obtained, and with retention of configuration.[70] Here it appears that the intermediate is an acyloxyphosphonium ion rather than an alkoxyphosphonium ion, and that ester formation results from reaction of this intermediate with the free alcohol.[71]

Treatment of β-hydroxy acids with this reagent brings about a decarboxylative dehydration to form an olefin.[72]

$$+ \; CO_2 + (C_6H_5)_3PO + (C_2H_5O_2CNH)_2$$

This reaction can be used for stereoselective preparation of (Z) and (E) alkenes.[73,74] For example, the *threo*-β-hydroxy acid **18** is converted into the (Z) alkene **19**, with a small amount of the *cis*-β-lactone **20** as a by-product.

However, if R^1 and R^2 are bulky, an abnormal carboxyl activation occurs, resulting in exclusive formation of a *trans*-β-lactone (21).

This reaction reflects the fact that hindered hydroxyl groups react more sluggishly with the triphenylphosphine–diethyl azodicarboxylate betaine intermediate, allowing time for activation of the carboxyl group. The effect seems closely related to the retention of configuration observed in the esterification of diacetone glucose.[70,71]

Substitution of allylic sugar alcohols leads to inverted, nontransposed products.[14]

Long-chain ω-hydroxy acids can be converted into macrocyclic diolide lactones.[75] This type of cyclization has proven useful in syntheses of natural products. Thus double lactonizations of 22a and 22b to the carbonyl-protected derivatives of the diolides norpyrenophorin and pyrenophorin (23a and 23b) have been accomplished in 60–75% yields (Eq. 24).[76] A significantly lower yield (24%) was obtained in converting 22c to the vermiculine derivative 23c. The deleterious effect of an unprotected carbonyl group in reactions with this reagent is manifested by the low yield (15%) in an earlier synthesis of vermiculine (24), which involved only a single lactonization (Eq. 25).[77] Another example of a low yield in the lactonization of a compound containing an open carbonyl group is the synthesis of the antibiotic 26 (designated A26771B).[78] Treatment of 25 with triphenylphosphine and diethyl azodicarboxylate gave a 1% yield of **26**; when the betaine **27**, preformed from triphenylphosphine and diethyl azodicarboxylate, was used as the reagent, the yield of **26** was 8% (Eq. 26).

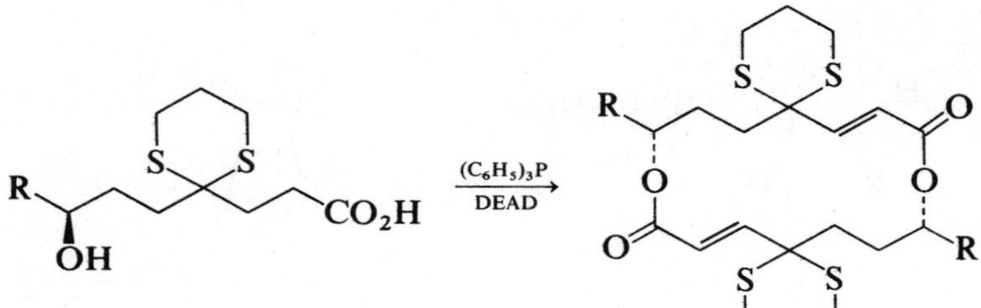

(Eq. 24)

22

a   R = H
b   R = CH$_3$
c   R = CH$_2$COCH$_3$

23

a   (60%)
b   (75%)
c   (24%)

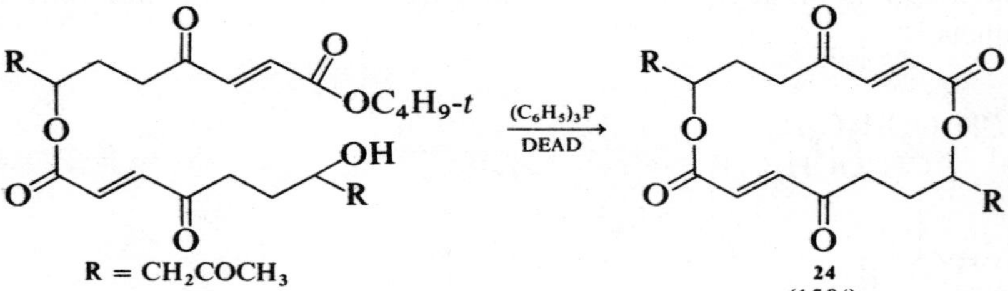

R = CH$_2$COCH$_3$

24
(15%)

(Eq. 25)

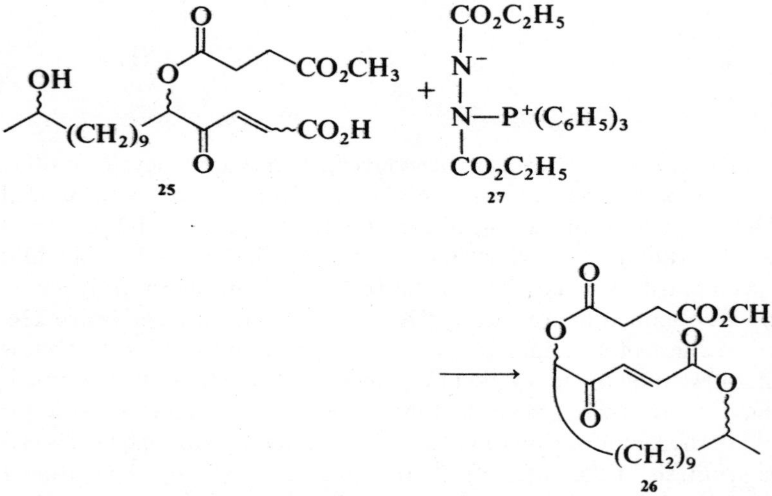

25   **27**

(Eq. 26)

26
(8%)

Stereospecific synthesis of the nine-membered lactone ring of griseoviridine (29) has been reported;[79] the chiral center in the hydroxy acid 28 controls the stereochemistry at the methyl group of lactone 29 during ring closure.

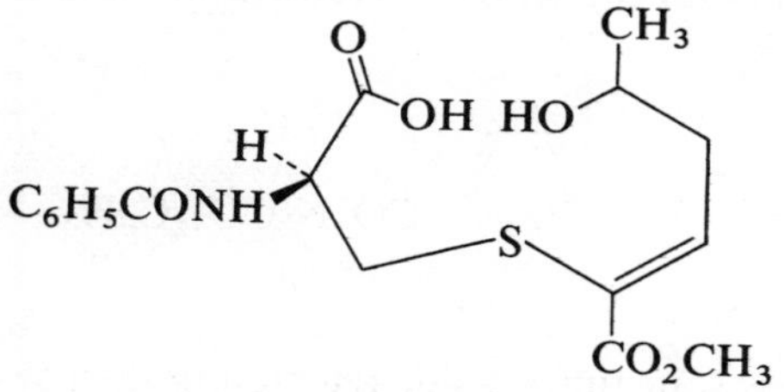

28

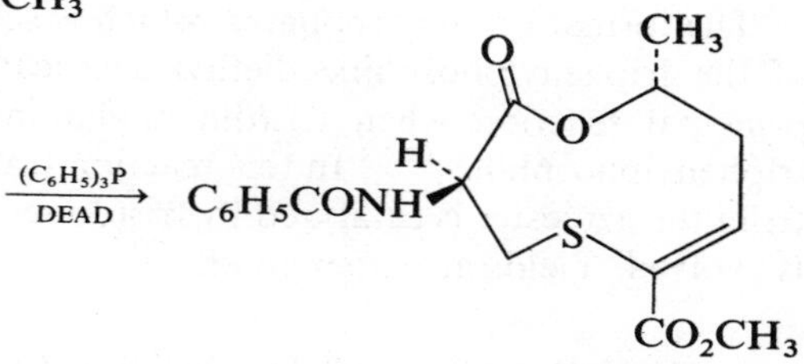

$$\xrightarrow[\text{DEAD}]{(C_6H_5)_3P}$$

29
(47%)

Benzoylation of aliphatic glycols generally gives mixtures of mono- and disubstituted esters, even where primary or secondary glycols are involved.

$$CH_3CHOHCH_2CH_2OH$$

$$\xrightarrow[C_6H_5CO_2H]{(C_6H_5)_3P/DEAD} CH_3CHOHCH_2CH_2O_2CC_6H_5 + CH_3CHCH_2CH_2O_2CC_6H_5$$

$$(70\%) \qquad\qquad\qquad\qquad\qquad O_2CC_6H_5$$

$$(7\%)$$

Selective esterification of the less hindered of two secondary hydroxyl groups in a steroid has been observed (Eq. 27).[69] The primary hydroxyl group of sugars is esterified selectively (Eq. 28).[13,68,80]

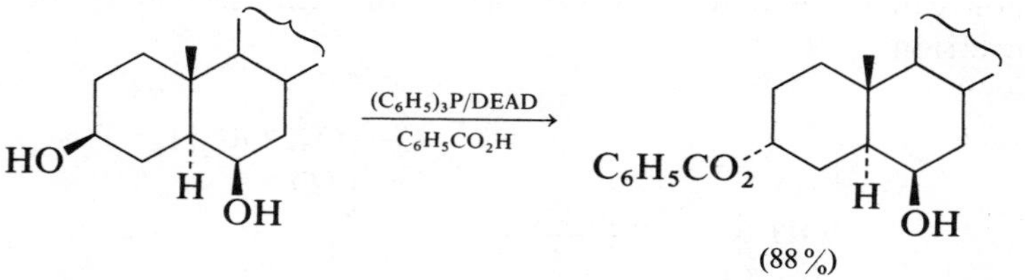

(Eq. 27)

(88%)

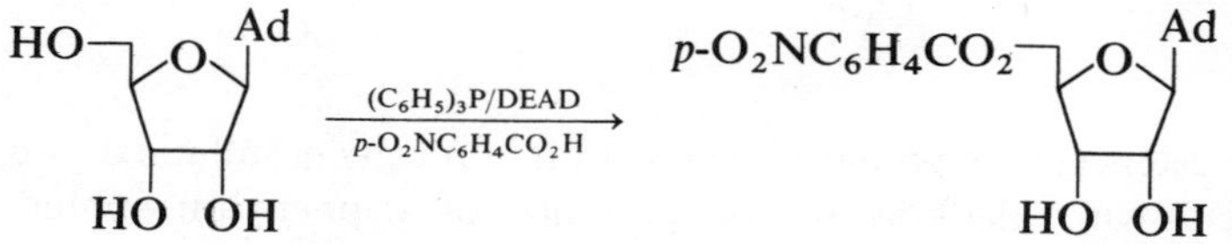

Ad = 9-Adenyl

(Eq. 28)

(63%)

Esters of N-hydroxysuccinimide and N-hydroxyphthalimide can be prepared with this reagent.[81] The reaction appears to involve formation of an oxyphosphonium derivative of the N-hydroxyimide, which is attacked by the acid to give a transient acyloxyphosphonium cation that breaks down to the ester.[82]

$$\text{(furyl)}-CO_2H + HON(CO)_2C_6H_4 \longrightarrow \text{(furyl)}-CO_2N(CO)_2C_6H_4$$
$$(95\%)$$

The formation of carbonates, which is sometimes a side reaction in the use of the triphenylphosphine–diethyl azodicarboxylate reagent,[45] becomes the principal reaction when tris(dimethylamino)phosphine is used in place of triphenylphosphine.[47,83] In this reaction transfer of the alkoxycarbonyl group from the azo ester is catalyzed by tris(dimethylamino)phosphine, and nitrogen is evolved. Yields are often good.

$$(C_2H_5O_2CN)_2 + [(CH_3)_2N]_3P + ROH$$

$$\longrightarrow C_2H_5O_2CNNHCO_2C_2H_5$$
$$\overset{|}{{}^+P[N(CH_3)_2]_3}, \ RO^-$$

$$\longrightarrow C_2H_5O_2CN-NH-\overset{\overset{O^-}{|}}{C}(OC_2H_5)OR$$
$$\overset{|}{{}^+P[N(CH_3)_2]_3}$$

$$\longrightarrow [(CH_3)_2N]_3P + C_2H_5OCO_2R + C_2H_5O_2CN{=}NH$$

$$\longrightarrow C_2H_5O_2CH + N_2$$

The reaction can be used to form carbonates of sugar secondary alcohols without inversion.

Sulfinic esters can be prepared by reaction of p-toluenesulfinic or n-dodecylsulfinic acid with alcohols in the presence of triphenylphosphine–diethyl azodicarboxylate.[84]

The triphenylphosphine–diethyl azodicarboxylate reagent can be used to prepare trisubstituted phosphates from disubstituted phosphates.[10,53,66,85]

$$(C_6H_5CH_2O)_2P(O)OH + ROH \xrightarrow[\text{DEAD}]{(C_6H_5)_3P} (C_6H_5CH_2O)_2P(O)OR$$

Such esterification of di-*t*-butyl phosphate gives excellent yields of triesters that can be readily converted to monoalkyl phosphates by cleavage with trifluoroacetic acid.[86] This reaction can be used for selective phosphorylation of the 5′ position of nucleosides.[85]

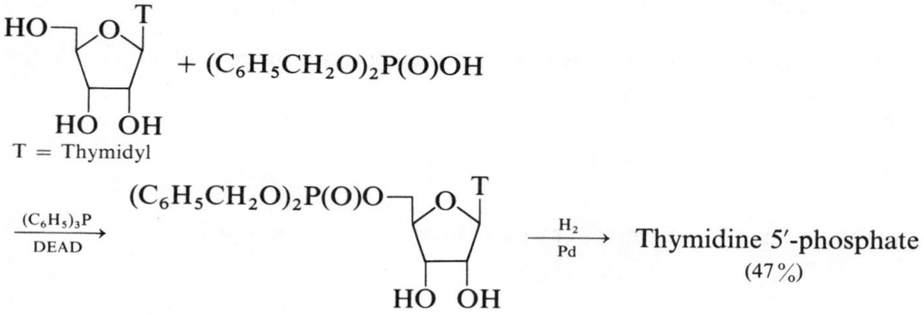

T = Thymidyl

Thymidine 5′-phosphate (47%)

Formation of Ethers. N-Hydroxyimides can be reacted with alcohols to give N-alkoxyimides;[87-89] the latter can be converted to O-alkylhydroxylamines by cleavage with hydrazine.

$$C_6H_5CH_2OH + HON(CO)_2C_6H_4 \xrightarrow[\text{DEAD}]{(C_6H_5)_3P} C_6H_5CH_2ON(CO)_2C_6H_4$$

A similar reaction has been described with N-hydroxybenzotriazole.[90] Reaction of N-hydroxyphthalimide with phenols produces aryl N-aryloxycarbonyl-anthranilates.[91]

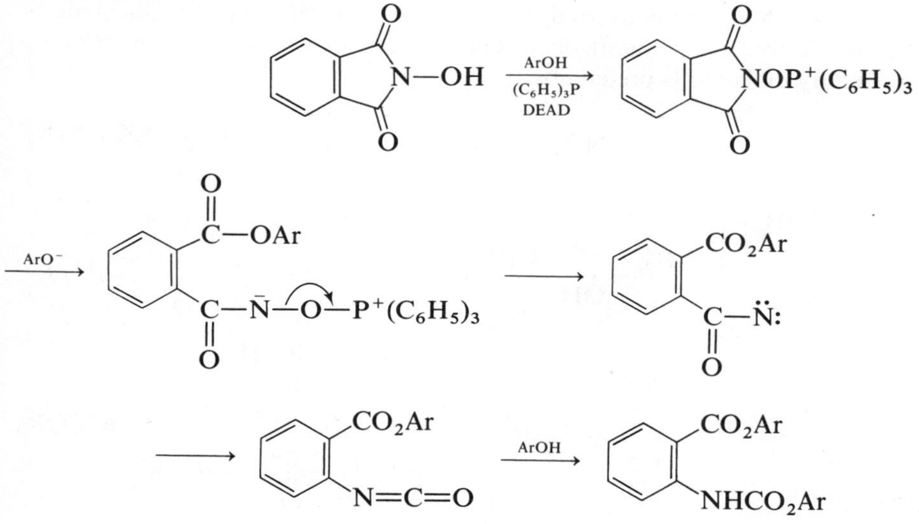

The triphenylphosphine–diethyl azodicarboxylate reagent is useful for alkylation of phenolic hydroxyl groups.[92,93] These two references differ as to whether nitrophenols and tertiary alcohols can be used in this reaction.

α,ω-Glycols can be cyclized into ethers containing 3–8 members in the ring.[94]

$$HO(CH_2)_nOH \xrightarrow[\text{DEAD}]{(C_6H_5)_3P} \quad O(CH_2)_n$$

$$n = 2–7$$

Free sugars can be converted into aryl glycosides in good yields.[95,96]

(80%)

Glycosides of aliphatic alcohols can be made if mercuric chloride is added to the reaction mixture; it is likely that the reaction proceeds through formation of the glycosyl chloride. Anomeric mixtures are generally formed.[97,98]

Reaction of alcohols with 2,6-di-*t*-butyl-4-nitrophenol produces *aci*-nitro ethers, which can be converted by treatment with base into the aldehyde or ketone derived from the alcohol. The primary hydroxyl group of primary-secondary diols reacts preferentially.[99]

Alcohol Dehydration.   Dehydration of alcohols to alkenes is often observed as a side reaction in substitutions.[58] Acyl derivatives of serine and threonine can be converted to the corresponding vinyl amino acids by a simple β-elimination.[100]

$$
\underset{\substack{| \\ H}}{\overset{\substack{CO_2CH_3 \\ |}}{RCO_2NHCCH_2OH}} \quad \xrightarrow[\text{DEAD}]{(C_6H_5)_3P} \quad \overset{\substack{CO_2CH_3 \\ |}}{RCO_2NHC=CH_2}
$$

An unusual dehydration with formation of a thioether bond has been accomplished in the cyclization of a penam to a β-lactam.[101]

Other Applications.   Alcoholic hydroxyl groups can be converted into halides by using alkyl halides as the source of halide ion.[11] Treatment of diacetone glucose with triphenylphosphine–diethyl azodicarboxylate in the presence of methyl iodide gives an oxyphosphonium iodide that decomposes in boiling toluene to the inverted iodide.[102] The same replacement, also with inversion, can be achieved with triphenylphosphine, iodine, and imidazole or with triphenylphosphine and 2,4,5-triiodoimidazole.[103] In contrast, treatment of diacetone glucose with triphenylphosphine–carbon tetrachloride, triphenyl phosphite methiodide, or triphenyl phosphite dibromide effects rearrangement of the 5,6-isopropylidene acetal to the 3,5 positions and halogenation at C_6.[104,105]

There are a few reports of the use of triphenylphosphine–diethyl azodicarboxylate to form carbon–carbon bonds. Thus ethyl cyanoacetate can be alkylated with alcohols (Eqs. 29 and 30).[54,75] The alcoholic hydroxyl group in a steroid has been replaced with a cyano group by carrying out the reaction in the presence of hydrogen cyanide (Eq. 31).[11]

$$C_2H_5O_2CCH_2CN + n\text{-}C_3H_7OH \xrightarrow[\text{DEAD}]{(C_6H_5)_3P} C_2H_5O_2CCH(CN)C_3H_7\text{-}n$$
(52%)

(Eq. 29)

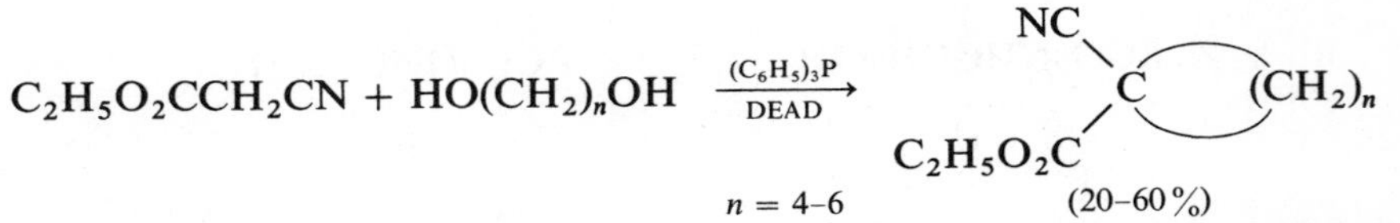

$$C_2H_5O_2CCH_2CN + HO(CH_2)_nOH \xrightarrow[\text{DEAD}]{(C_6H_5)_3P}$$

$n = 4\text{-}6$ (20–60%)

(Eq. 30)

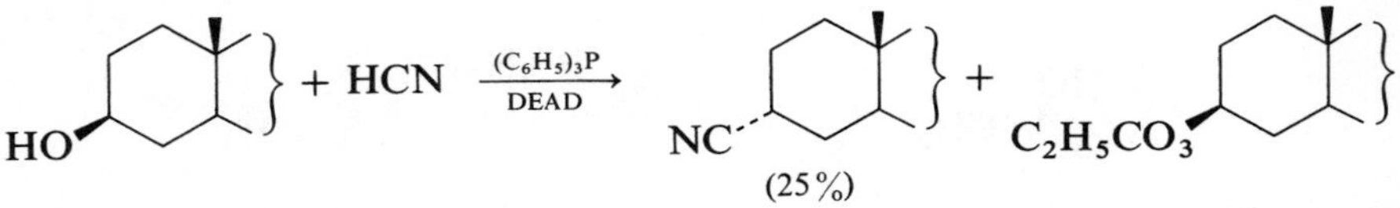

(Eq. 31)

Thiols cannot be used in place of alcohols with this reagent because they are oxidized to disulfides.[106]

Formation of Pentacoordinated Phosphoranes. Reaction of triphenylphosphine–DEAD with *cis*-1,2-diols and with 1,3-diols gives pentacoordinated

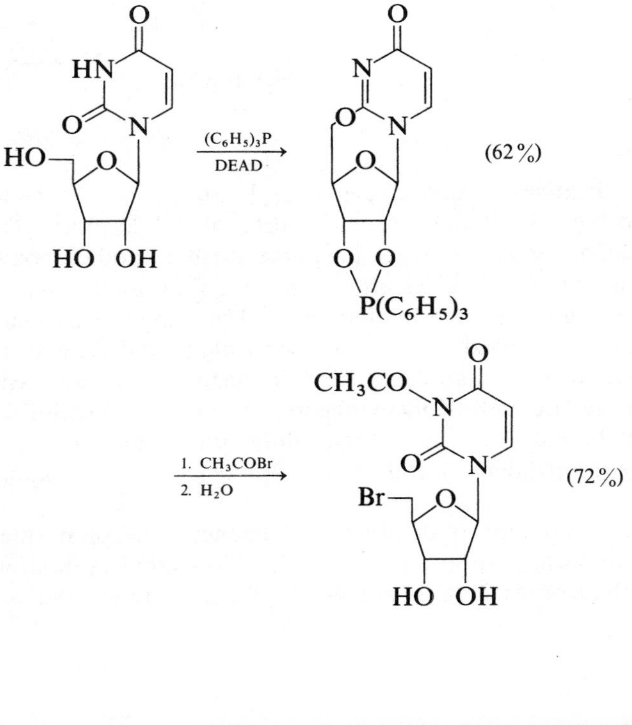

(62%)

(72%)

phosphoranes,[107,108] analogous to the reaction of diols with triphenylphos-phine–hexamethylphosphoramide.[35,109] Phosphorane formation occurs only with *cis*-1,2-diols; the *trans* diols form epoxides.[11,58,107,108] The phosphorane group can be used for temporary protection of *cis*-1,2-diols.[110–112]

Regiospecific splitting of such phosphoranes can be accomplished with hydrazoic acid or *p*-nitrobenzoic acid.[63]

$$A = -N_3, -O_2CC_6H_4NO_2\text{-}p$$
$$R = -Si(CH_3)_2C_4H_9\text{-}t$$

Gluco derivatives that bear free hydroxyl groups at positions 2 and 3 can be similarly substituted selectively at position 3.

The Triphenyl Phosphite–Methiodide Reagent

The adduct of triphenyl phosphite and methyl iodide was first described in 1898.[1] The oxyphosphonium salt **30** formed in this reaction is stable in the absence of moisture, in contrast to the intermediates **1** (Eq. 1) in the Michaelis–Arbuzov reaction, which undergo nucleophilic displacement with their own counterions unless very bulky substituents are involved.[113] The stability of the salt **30** results from the inability of the phenyl group to undergo nucleophilic substitution.

$$(C_6H_5O)_3P + CH_3I \longrightarrow \underset{\mathbf{30}}{(C_6H_5O)_3P^+CH_3, I^-}$$

In 1951 it was reported that Michaelis' salt **30** reacted with alcohols to form alkyl iodides with the liberation of phenol and methyl diphenylphosphonate.[114]

$$HOCH_2C(CH_3)_2CH_2OH \xrightarrow{(C_6H_5O)_3P^+CH_3,\ I^-} ICH_2C(CH_3)_2CH_2I$$

The reaction was used to prepare 1,3-diiodo-2,2-dimethylpropane, a key intermediate for the synthesis of caryophyllenic acid. The reaction appears to involve initial displacement of a phenoxy group by the hydroxyl group of the alcohol, followed by nucleophilic displacement to form the iodide.[115]

$$(C_6H_5O)_3P^+CH_3, I^- + ROH$$

$$\longrightarrow C_6H_5OH + (C_6H_5O)_2(RO)P^+CH_3, I^-$$

$$\longrightarrow RI + (C_6H_5O)_2P(O)CH_3$$

The ionic tetracovalent nature of phosphorus in salt **30** was confirmed by ^{31}P NMR studies.[116]

The triphenyl phosphite–methiodide reagent can be used to convert alcoholic hydroxyl groups into iodides in aliphatic compounds,[116] carbohydrates,[117,105,118,119] and nucleosides.[120–126] Some carbon skeleton rearrangement is observed with highly hindered compounds: t-amyl iodide (6%) is formed in the preparation of neopentyl iodide from neopentyl alcohol.[127]

It has been reported that a nonrearranged 3-deoxy-3-iodohexose (**31**) was obtained from the treatment of diacetone glucose with the triphenyl phosphite–methiodide reagent,[128] but the report was later questioned, and structure **32** was proposed for the product (Eq. 32).[105] However, neither of these references gives experimental conditions, and it now seems likely that structure **31** is correct in light of a recent report of the preparation of 1,2:5,6-di-O-isopropylidene-3-iodoallose.[103]

(Eq. 32)

A structural rearrangement also occurs to some extent with a 2,3-isopropylidene rhamnoside.[119,129]

Steroidal neopentylic alcohols undergo rearrangement followed by an elimination reaction.[130] Allylic and propargylic alcohols give products that result from S_N2' reactions, but products that arise from secondary rearrangements may also occur.[116]

$$HC{\equiv}CCHOHR \longrightarrow ICH{=}C{=}CHR$$

The 5'-hydroxyl group of nucleosides is substituted in high yield, and the reaction can exhibit primary/secondary selectivity.[120]

Substitution of 2'- or 3'-hydroxyl groups proceeds with retention of configuration because of participation by the heterocyclic nucleus.[120]

The participating intermediate (33) can be isolated in some reactions.

Reactions with vicinal glycols form phosphorane intermediates (**34, 35**) that yield a mixture of phosphonates (**36, 37**).[121]

Ur = Uridyl

Whereas primary alcohols undergo replacement by iodide with triphenyl phosphite–methiodide, secondary alcohols are selectively dehydrated by this reagent in hexamethylphosphortriamide at 75°.[131] Tertiary alcohols are unaffected under these conditions.

A modification of the triphenyl phosphite–methiodide reagent involves the use of methyl trifluoromethanesulfonate (triflate) in place of methyl iodide, thus producing a nonnucleophilic anion instead of iodide.[132]

$$(C_6H_5O)_3P + CF_3SO_3CH_3 \longrightarrow (C_6H_5O)_3P^+CH_3, CF_3SO_3^-$$

Reaction of this reagent with aliphatic alcohols gives symmetrical dialkyl ethers, whereas with sodium alcoholates, alkyl phenyl ethers are formed.

Reaction of benzyl chloride with triphenyl phosphite gives a reagent that can be used to form chlorides from alcohols.[115,133]

Reagents Formed from Halogens and Triphenylphosphine or Triphenyl Phosphite

Only a few studies report the use of the triphenyl phosphite–halogen reagent for the preparation of halides.[134–136] The reagent has two disadvantages:

it gives phenol as a by-product and usually produces mixtures of phosphorus species.[137,138]

Triphenylphosphine–halogen adducts have the advantage of containing only two replaceable groups,[139] and are useful and versatile reagents for converting alcohols into halides.[140] Their reaction with alcohols involves the rapid and irreversible formation of an alkoxyphosphonium salt that decomposes by a slow S_N2 reaction to form an alkyl halide.[141,142]

$$(C_6H_5)_3P^+X, X^- + ROH \xrightarrow{\text{Fast}} HX + (C_6H_5)_3P^+OR, X^-$$

$$\xrightarrow{\text{Slow}} (C_6H_5)_3PO + RX$$

Although acetonitrile or dimethylformamide is recommended as the solvent, the latter can react with the reagent to form halomethylene-N,N-dimethyliminium halides **38**.[124,143] These intermediates can react with alcohols to give

$$(C_6H_5)_3P^+X, X^- + OCHN(CH_3)_2$$

$$\longrightarrow (C_6H_5)_3P^+OCH=N^+(CH_3)_2, 2X^-$$

$$\longrightarrow (C_6H_5)_3PO + XCH=N^+(CH_3)_2, X^-$$
$$\textbf{38}$$

alkoxymethylene-N,N-dimethyliminium halides (**39**), which can then undergo either S_N2 replacement by halide ion (Eq. 33), especially with primary alcohols, or hydrolysis during workup to give a formate or to reform the starting material (Eq. 34).

$$\underset{\textbf{38}}{XCH=N^+(CH_3)_2, X^-} + ROH \longrightarrow HX + \underset{\textbf{39}}{ROCH=N^+(CH_3)_2, X^-}$$

$$\underset{\textbf{39}}{ROCH=N^+(CH_3)_2, X^-} \xrightarrow[\text{H}_2\text{O}]{\text{Heat}} \begin{array}{l} RX + OHCN(CH_3)_2 \qquad \text{(Eq. 33)} \\ ROCHO + H_2N^+(CH_3)_2, X^- \quad \text{(Eq. 34)} \end{array}$$

The replacement of alcoholic hydroxyl groups by halogen with the triphenylphosphine–halogen reagent generally involves inversion of configuration. No skeletal rearrangement is observed in the preparation of neopentyl halides from neopentyl alcohol[139] or in the S_N2 reactions of *endo*-norbornanol (Eq. 35) and 7-norbornanol.

$$\text{(Eq. 35)}$$

However, a similar reaction of *exo*-norbornanol appears to involve the non-classical Winstein intermediate (**40**).

Likewise, little rearrangement is observed in the halogenation of 3-methyl-2-butanol.[144]

$$(CH_3)_2CHCHOHCH_3$$

$$\xrightarrow{(C_6H_5)_3P^+Br,\ Br^-} (CH_3)_2CHCHBrCH_3 + (CH_3)_2CBrCH_2CH_3$$
$$\qquad\qquad\qquad\qquad\qquad (43.8\%)\qquad\qquad\quad (1.4\%)$$

Greater amounts of rearranged product are found in the same reaction with other reagents: $(C_6H_5)_3P^+CN$, Br^- (3%), $(C_6H_5O)_3P^+Br$, Br^- (55.6%), and $(C_6H_5)_3P^+CBr_3$, Br^- (6–10%).

The triphenylphosphine–halogen reagent has been used less than other phosphorus reagents for the halogenation of nucleosides.[124,145]

The triphenylphosphine–iodine reagent is less active than triphenyl phosphite–methiodide for the conversion of 5′-O-p-nitrobenzoylthymidine to 3′-desoxy-3′-iodo-5′-O-p-nitrobenzoylthymidine.[122]

Aryl halides are formed at temperatures up to 300°.[135,146]

β-Diketones react as β-hydroxy enones to give β-halo-α,β-enones.[147]

The selective halogenation of unsymmetrical diols can be achieved by conducting the reaction in dimethylformamide.[148] At temperatures below 0° the primary hydroxyl is converted to the halide, whereas the secondary hydroxyl is protected as the formate.

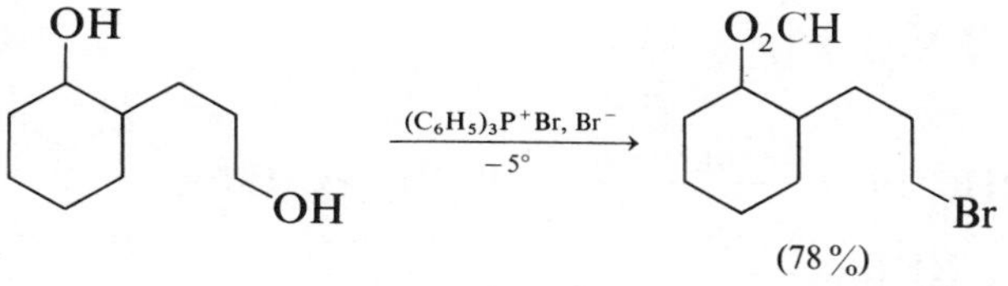

(78%)

Replacement of hydroxyl by bromine in a steroidal glycol is accompanied by Vilsmeier-type formylation of the dehydrated intermediate.[149]

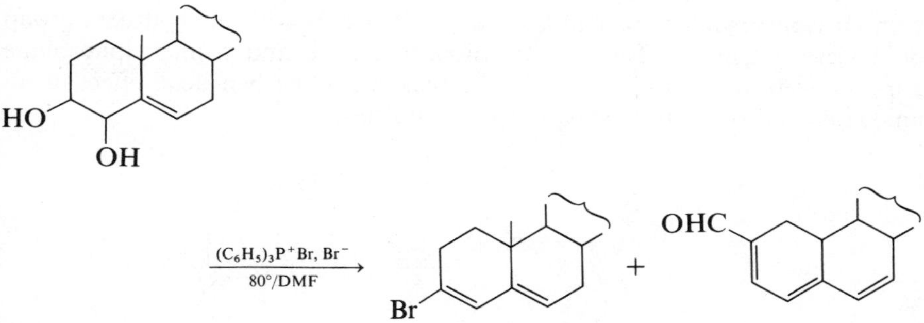

Dehydration of tertiary benzocyclobutanols proceeds with higher selectivity (Eq. 36) than the dehydration with p-toluenesulfonic acid (Eq. 37).[150]

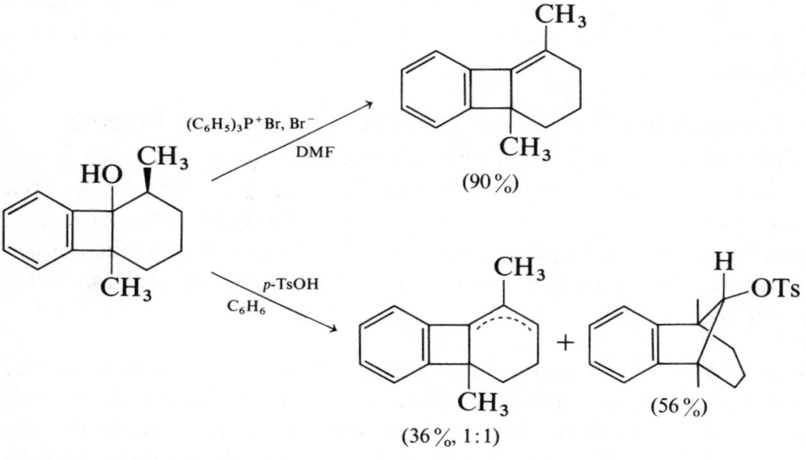

(Eq. 36)

(90%)

(Eq. 37)

(36%, 1:1)   (56%)

2-N-Alkylaminoalcohols are converted to N-alkylaziridines in good yield, with minor amounts of N,N-dialkylpiperazines as by-products.[151] Only piperazines are formed when R is hydrogen.

It has been shown recently that addition of imidazole to triphenylphosphine–iodine gives improved results in the conversion of alcohols to iodides, particularly with highly hindered secondary alcohols.[103] For example, substitution without isopropylidene rearrangement[120,121] occurs with the hydroxyl group of diacetone glucose. The use of triiodoimidazole and triphenylphosphine appears even more advantageous. The reason for the beneficial effect of an imidazole in this reaction has not been established.

An immobilized form of the triphenylphosphine–chlorine reagent has been prepared by treatment of cross-linked, chloromethylated or brominated polystyrene beads with lithium diphenyl phosphide, oxidation with peracetic acid, and reaction with phosgene.[152] This reagent converts benzyl alcohol into benzyl chloride in 88% yield, and can be regenerated after use by retreatment with phosgene.

The Triphenylphosphine–Carbon Tetrachloride and Related Reagents

The triphenylphosphine–carbon tetrachloride system has been used extensively for the transformation of alcohols into chlorides. Variations in the final stage of the reaction to substitute other nucleophiles have not been common, and in this respect the system appears less versatile than the triphenylphosphine–diethyl azodicarboxylate or the tris(dimethylamino)phosphine–carbon tetrachloride systems.

Synthesis of Halides. Halides have been synthesized from a wide variety of primary and secondary alcohols. With secondary or hindered alcohols inversion of configuration is often predominant or total.[153] However, the course of the reaction is dependent on the nature of both the substrate and the halide ion, and the reaction can produce partially racemized or rearranged products.

Carbon tetrachloride has been used most frequently as the solvent, but it is often not the most effective solvent for achieving a rapid, high-yield, selective reaction. Acetonitrile is an effective solvent, usually leading to rapid formation of the phosphonium salt intermediate.[27,154,155]

Use of pyridine as the solvent has provided high yields and selectivity in the preparation of halodeoxy sugars.[156] Twice the normal quantity of triphenyl-phosphine must be used to obtain high yields, indicating that the reaction proceeds by the ylid route (Eqs. 14 and 15). Primary hydroxyl groups are replaced selectively, and there is no undesirable formation of phosphoranes or epoxides from vicinal diols.

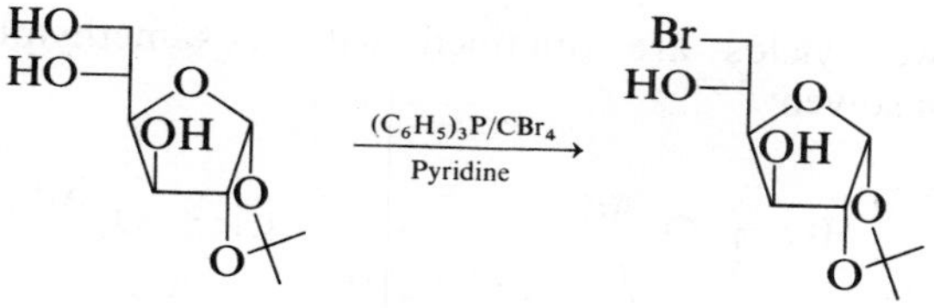

Dimethylformamide has been used as the solvent for most reactions of nucleosides.[124]

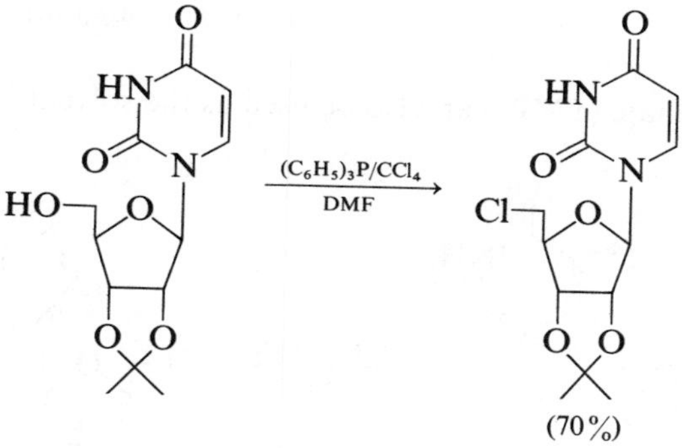

(70%)

It is necessary to use pyridine as the solvent when triphenylphosphine–carbon tetraiodide is the reagent.[143]

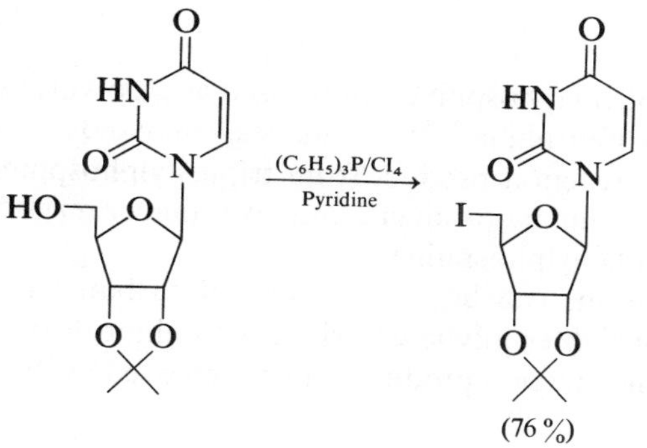

(76%)

Selective replacement of primary hydroxyl groups is observed in pyridine.[156]

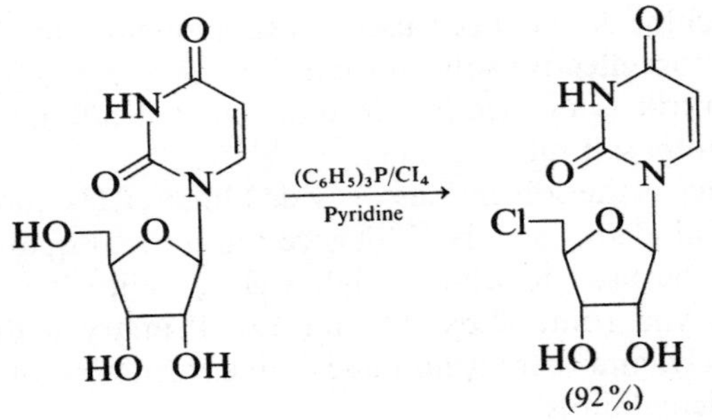

(92%)

Somewhat lower yields are obtained with hexamethylphosphortriamide (HMPA) as the solvent.[157]

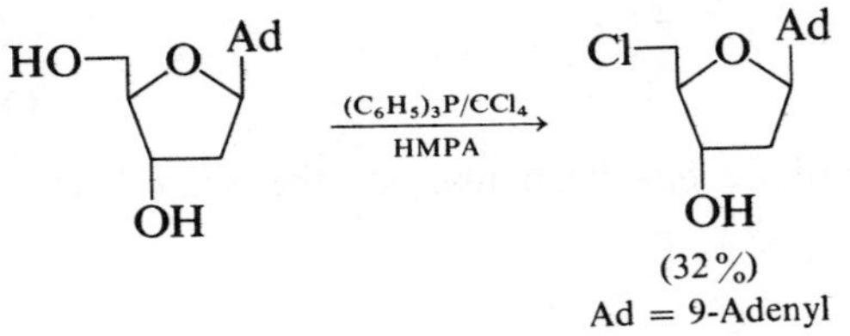

(32%)

Ad = 9-Adenyl

Trimethyl phosphate (TMP) can also be used as the solvent.[145,158]

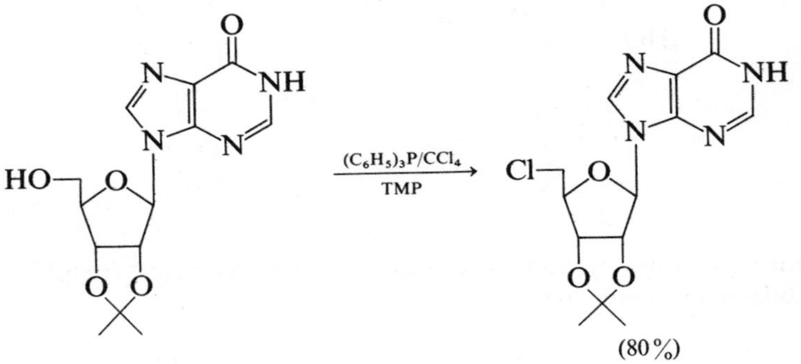

(80%)

　　　Polystyryl–diphenylphosphine can be used as a polymer-supported replacement for triphenylphosphine.[159,160] This reagent avoids the sometimes tedious separation of the reaction product from triphenylphosphine oxide and other phosphorus-containing by-products, and gives higher reaction rates than those obtained with triphenylphosphine.[161]

　　　Hexachloroacetone can be used instead of carbon tetrachloride to avoid contamination of lighter allylic chlorides with chloroform and carbon tetrachloride.[30,162] The reaction products can be removed by distillation as they are formed, thereby avoiding the isomerization that can occur if they remain in the reaction mixture.

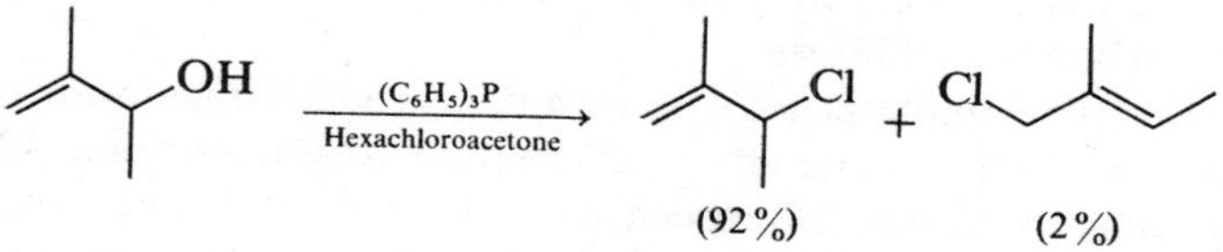

(92%)   (2%)

The reaction products are nevertheless contaminated with a minute quantity of chloroform, formed by an unknown mechanism.

Carbon tetraiodide can be replaced by 2,4,5-triiodoimidazole, with a slight improvement in yield.[103] Use of this reagent in substituting the 3-hydroxyl group of diacetone glucose gives inversion without skeletal rearrangement.

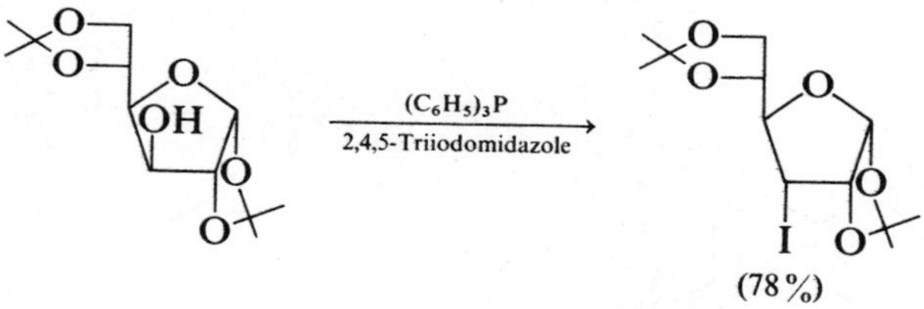

(78%)

Use of this reagent with vicinal diols gives eliminative deoxygenation,[163] in contrast with the substitution observed with triphenylphosphine–carbon tetrabromide.[156]

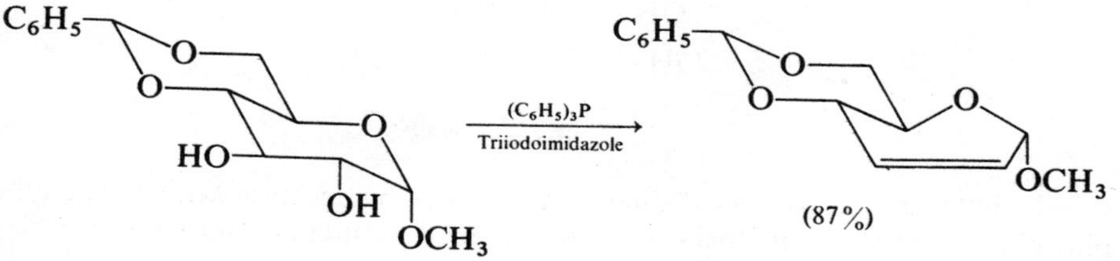

(87%)

Formation of Carbon–Nitrogen Bonds. A one-pot synthesis of alkyl azides can be carried out by performing the triphenylphosphine–carbon tetrabromide–alcohol reaction in the presence of lithium azide.[164–168]

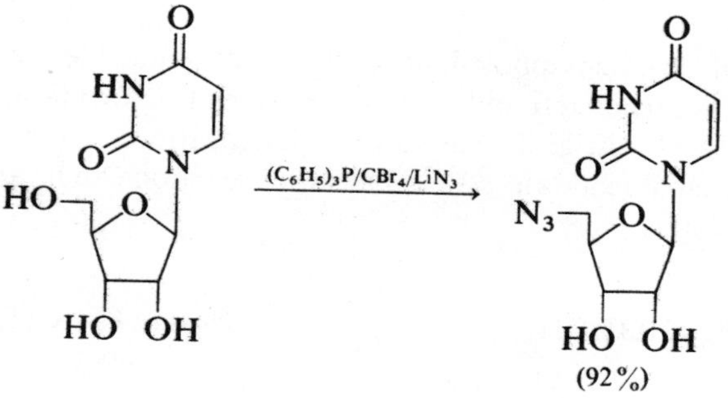

(92%)

Dimethylformamide must be used as the solvent, rather than pyridine or hexamethylphosphortriamide. Triphenyl phosphite–methiodide and triphenylphosphine–diethyl azodicarboxylate are much less effective than triphenylphosphine–carbon tetrachloride.

Reaction of the 2-aminoalcohol 41 with the carbon tetrachloride reagent in the presence of triethylamine effects cyclization to the aziridine 42,[169] and similar aziridine syntheses have been carried out.[170,171] The same reagent converts the 4-aminoalcohol 43 into octahydroindolizine (44).[172] Ring closure of 5-aminoalcohols to piperidines can be effected,[173] as can the analogous cyclization of 1,5-diols.[174]

$$C_6H_5CHOHCH_2NHC_4H_9\text{-}t \xrightarrow{\;(C_6H_5)_3P/CCl_4/NEt_3\;} \underset{\substack{\text{N}\\|\\C_4H_9\text{-}t}}{C_6H_5CH\!-\!CH_2} \quad (86\%)$$

41

42

$$\xrightarrow{\;(C_6H_5)_3P/CCl_4/NEt_3\;}$$

43   44   (42%)

$$\xrightarrow{\;(C_6H_5)_3P/CCl_4\;}$$

Alcohols can be converted directly to nitriles by reaction with triphenylphosphine–carbon tetrachloride in the presence of sodium or potassium cyanide and dimethyl sulfoxide.[153,175]

$$CH_3(CH_2)_3OH \xrightarrow[\;NaCN/Me_2SO\;]{(C_6H_5)_3P^+CCl_3,\ Cl^-} CH_3(CH_2)_3CN$$
$$(85\%)$$

Reaction of triphenylphosphine–carbon tetrachloride with secondary alcohols in excess carbon tetrachloride generally leads to formation of chlorides; however, with acetonitrile as the solvent dehydration of the alcohol is frequently the principal reaction.[176,177] The latter reaction can be a useful method for preparing olefins.

$$C_2H_5CHOHCH_3 \xrightarrow[\;CH_3CN\;]{(C_6H_5)_3P^+CCl_3,\ Cl^-} C_2H_5CH\!=\!CH_2$$
$$(74\%)$$

The Triphenylphosphine–N-Halosuccinimide Reagent

The triphenylphosphine–N-halosuccinimide reagent is used principally for replacement of primary hydroxyl groups, with good primary/secondary selectivity. The reaction is generally carried out in dimethylformamide, and the oxidant can be N-bromo-(NBS), N-chloro-, or N-iodosuccinimide. The reagent is useful in the preparation of halides from tetrahydrofurfuryl alcohol[178] and from carbohydrates and nucleosides.[179–184] An example of the reaction is shown in Eq. 38.[180] It is possible to achieve some selectivity in replacing only one of two primary hydroxyl groups by using a 2:1 ratio of reagent to substrate (Eq. 39).[183] A large excess is needed to achieve the disubstitution.

$$\text{(Eq. 38)}$$

(66%)

(37%)

(+18% 6,6'-Dibromohexaacetate)

$$\text{(Eq. 39)}$$

There is one report of the replacement of the 3-hydroxy group in a steroid with inversion of configuration (Eq. 40),[185] but generally mixtures of α and β isomers are obtained. Use of triphenyl phosphite and N-haloacetamides gives similar results.[185]

$$\text{(Eq. 40)}$$

(95%)

The one-step conversion of alcohols to azides does not occur with this reagent because the azide ion reacts with the phosphorus intermediate, attacking the phosphorus and deactivating the hydroxyl group. However, the conversion can be accomplished by carrying out the reaction in two steps without isolating the intermediate bromide.[184] This technique provides a route to aminoglycoside antibiotics; thus only the primary hydroxyl group of penta(benzyloxycarbonyl)neomycin can be replaced by an azido group in 73% yield.

Z = Benzyloxycarbonyl

The Tris(dimethylamino)phosphine–Carbon Tetrachloride Reagent

Tris(dimethylamino)phosphine reacts with carbon tetrachloride to give a trichloromethylphosphonium salt.[186] In the presence of an alcohol, alkyl chlorides are formed above 0°. When the reaction is carried out at −78°, an insoluble alkoxyphosphonium chloride is formed, which can be converted into a stable alkoxyphosphonium hexafluorophosphate by treatment with aqueous ammonium hexafluorophosphate.[187] The stability of the alkoxyphosphonium chloride is very solvent dependent, and is enhanced by polar solvents; the salts are completely stable in water.[188] The chloride ion can be replaced by stronger nucleophiles such as iodide, phenylthio, cyanide, thiocyanide, or azide, and even with a weaker nucleophile such as an amine, if the oxyphosphonium salt is isolated with a non-nucleophilic anion such as perchlorate or hexafluorophosphate.[189]

Alkoxy(trisdimethylamino)phosphonium salts are versatile reagents that can undergo substitution reactions with a variety of nucleophiles. Although a two-step reaction is required, it is characterized by high yields, good selectivity, and easy workup owing to the water solubility of the by-products, hexamethylphosphoramide and hexafluorophosphates.

The preparation of the salt is usually performed in the temperature range −20° to −45°. A dilute solution of the aminophosphine is prepared in a solvent such as tetrahydrofuran, dichloromethane, pyridine, acetonitrile, or dimethylformamide. This solution is added very slowly, e.g., using a motor-driven syringe, to a solution of the alcohol and carbon tetrachloride in the same solvent. A slight molar excess of carbon tetrachloride relative to alcohol is used.

The reaction is generally complete at the end of the addition; however, if alcohol can be detected by TLC, additional tris(dimethylamino)phosphine can be added. The reaction mixture is then poured into water, and solvents and unreacted alcohol are removed by washing with diethyl ether or ethyl acetate. Addition of potassium hexafluorophosphate usually precipitates the salt; if the salt is water-soluble it can be extracted with dichloromethane.

Some mono-oxyphosphonium salts of polyols are not extractable because of the remaining hydroxyl groups; these can be acetylated with acetic anhydride and pyridine, and they either precipitate or become extractable.[190] This technique is generally used with carbohydrates.[191]

Extraction with dichloromethane is feasible only with salts of large, hydrophobic anions such as perchlorate, hexafluorophosphate, iodide, azide, and thiocyanate; salts involving bromide and cyanide are borderline, and salts of chloride and fluoride are not extractable.[192] This characteristic permits the use of phase-transfer systems, in which the organic phase "catalyzes" the reaction. For example, a water solution of sodium azide and an alkoxy(trisdimethylamino)phosphonium chloride does not react at 60°; addition of chloroform extracts the alkoxyphosphonium azide into the organic phase, where nucleophilic substitution occurs.[188]

$$C_6H_5CH_2OP^+[N(CH_3)_2]_3, Cl^- + NaN_3 \xrightarrow[H_2O/30°]{CH_2Cl_2} C_6H_5CH_2N_3$$

Alternatively, the oxyphosphonium azide can first be extracted with dichloromethane, isolated, and suspended in boiling benzene; complete solution signals completion of the reaction.

More often, the isolated hexafluorophosphate is reacted with the nucleophile in dimethylformamide at 80–110°. This procedure permits the selective monoalkylation of primary amines (Eq. 41)[189] or the formation of azides from alcohols (Eq. 42).[40]

$$HOCH_2CH_2NH_2 + C_6H_5CH_2OP^+[N(CH_3)_2]_3, PF_6^-$$

$$\xrightarrow[80°]{DMF} C_6H_5CH_2NHCH_2CH_2OH \quad (Eq.\ 41)$$
$$(80\%)$$

(Eq. 42)

A similar technique can be used to prepare halides, using tetrabutylammonium halides as the source of halide ion.[193]

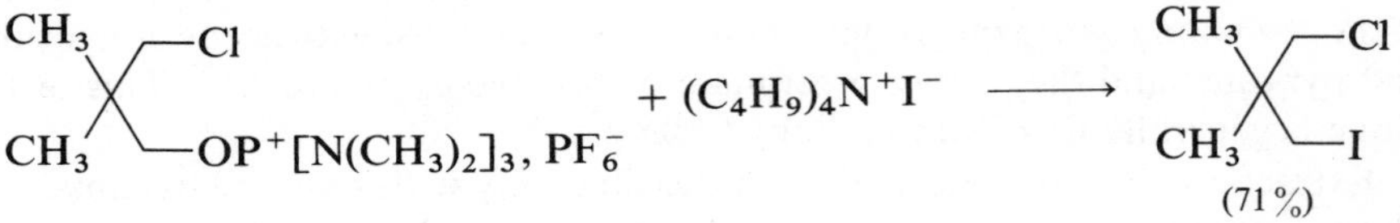

(71%)

Carbon–sulfur bonds can be formed by using triethylammonium thioacetate[190] or ammonium thiocyanate[188] as the nucleophile. Aryl thioethers are prepared by using triethylammonium[190] or potassium[194] aryl thiolates.

Nitriles are prepared by reaction of the oxyphosphonium salt with potassium cyanide. Reductive cleavage of the oxyphosphonium salt with lithium in liquid ammonia results in replacement of oxygen by hydrogen.[192]

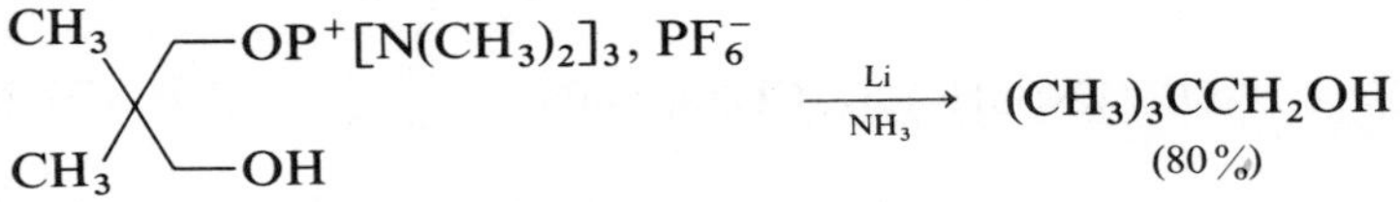

$$\xrightarrow[\text{NH}_3]{\text{Li}} (CH_3)_3CCH_2OH$$

(80%)

If an electronegative substituent is present in the β position to the salt group, reductive cleavage may be accompanied by elimination.[195]

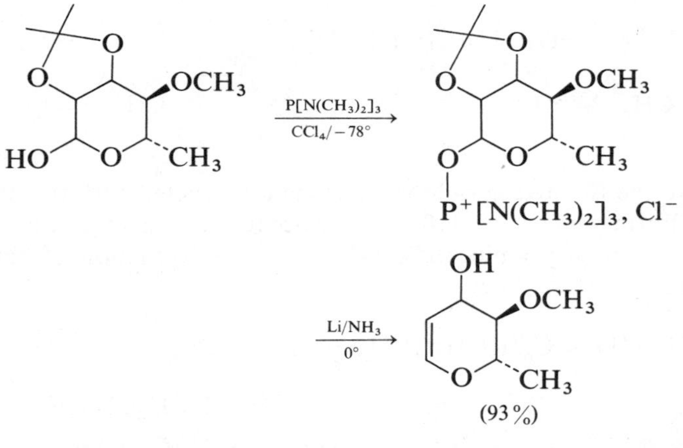

$$\xrightarrow[\text{CCl}_4/-78°]{\text{P[N(CH}_3)_2]_3}$$

$$\xrightarrow[0°]{\text{Li/NH}_3}$$

(93%)

This reaction probably involves the intermediate glycosyl chloride formed on warming to 0°. Reductive cleavage of oxyphosphonium salts can be effected in high yield with lithium triethyl borohydride.[196]

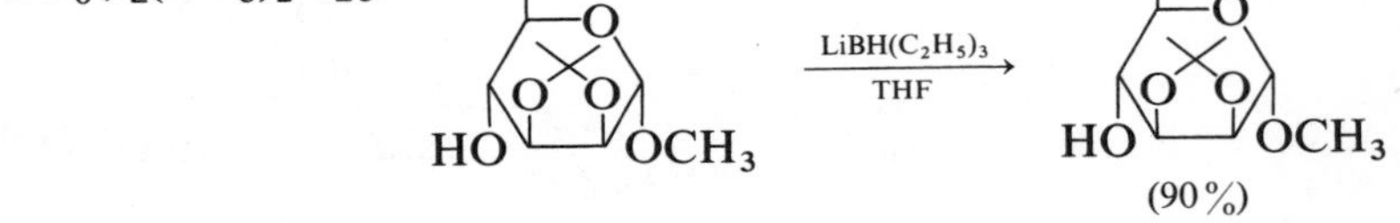

$$\xrightarrow[\text{THF}]{\text{LiBH(C}_2\text{H}_5)_3}$$

(90%)

Activation of the Anomeric Position of Carbohydrates. The behavior of oxyphosphonium salts of anomeric hydroxyl groups of carbohydrates depends largely on the structure of the carbohydrate. 2,3,4,6-Tetra-O-acetylglucose gives two anomeric salts, both stable at −40°. The α-*cis* isomer can be isolated as a hexafluorophosphate in 55% yield; the β-*trans* isomer is readily hydrolyzed to the starting materials. 2,3,4,6-Tetra-O-acetylmannose gives only one salt, with the α-*trans* configuration. This salt is stable only below −20°, and is converted into the chloride at higher temperatures. Similar behavior is observed with salts of carbohydrates with isopropylidene blocking groups: diacetone mannose, diacetone allose, and 5-O-trityl-2,3-O-isopropylideneribose. Whereas 2,3,4,6-tetra-O-benzylmannose gives a salt in the alpha configuration that is stable up to −20°, 2,3,4,6-tetra-O-benzylglucose gives the glycosyl chloride directly even at −100°; the existence of an oxyphosphonium salt could not be established by ^{31}P NMR.[197] The following oxyphosphonium salts of carbohydrates have been detected in solution below −20°:

$$Y = N(CH_3)_2 \qquad R = C_6H_5CH_2$$

In the presence of an aglycone the formation of glycosides requires precipitation of the chloride ion with a silver salt, such as silver tosylate[198] or silver hexafluorophosphate.[38] With the former reagent a reactive tosyl glycoside is formed. Glycosidation generally proceeds through an S_N1 process; although the stereoselectivity is poor with the lighter alcohols, it is better with hindered aglycones.[199]

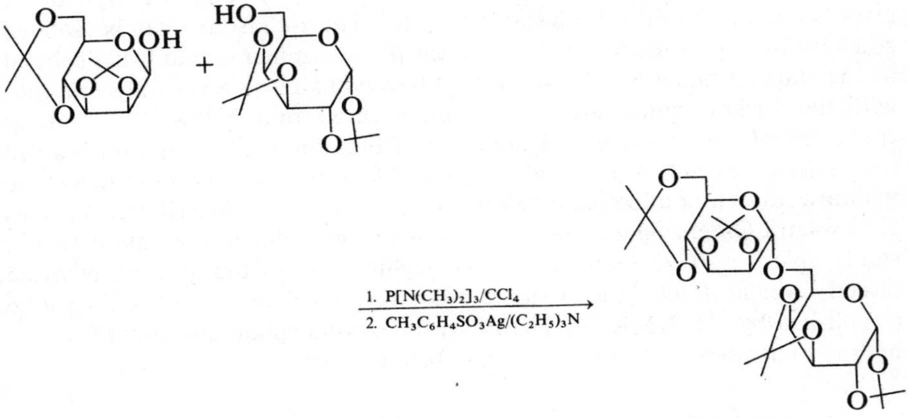

(45%)

Use of a tertiary base such as triethylamine can suppress the reaction; the following reaction does not proceed in the presence of triethylamine.[199]

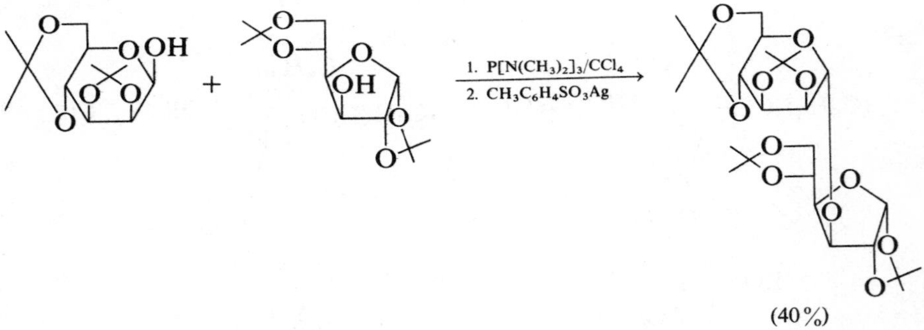

(40%)

Arylthioglycosylation can be effected in high yield directly from the oxyphosphonium chloride and an arylthiotriethylammonium salt. The reaction occurs with complete inversion of configuration, and can be used to determine the anomeric configuration of an oxyphosphonium salt.[199]

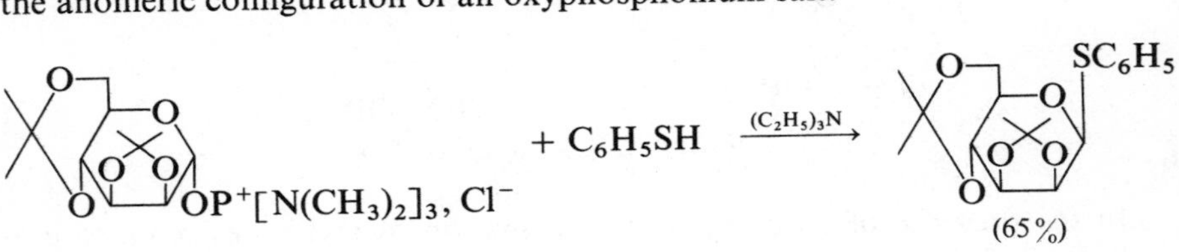

(65%)

Glycosyl azides are readily obtained, generally with inversion of configuration, by reaction of an azide anion with an oxyphosphonium chloride (Eq. 44). The azide ion is conveniently introduced as a soluble, crystalline, nonhygroscopic aryloxyphosphonium salt (45) (Eq. 43). This aryloxyphosphonium cation is completely unreactive, but provides a much more easily handled form of soluble azide than does tetrabutylammonium azide.[200] Glycosyl azides are easily converted into 1,2,3-triazolonucleosides.

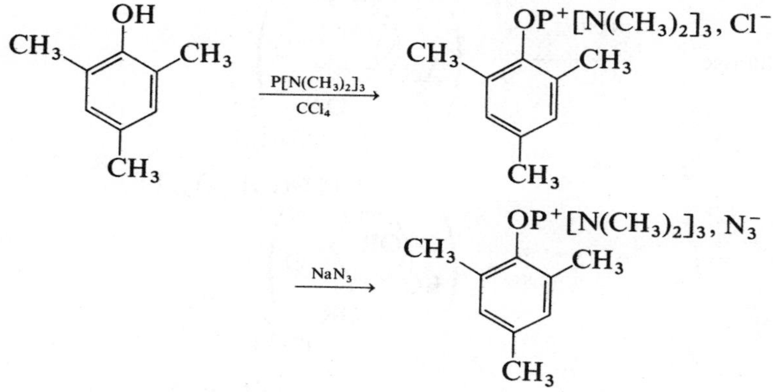

(Eq. 43)

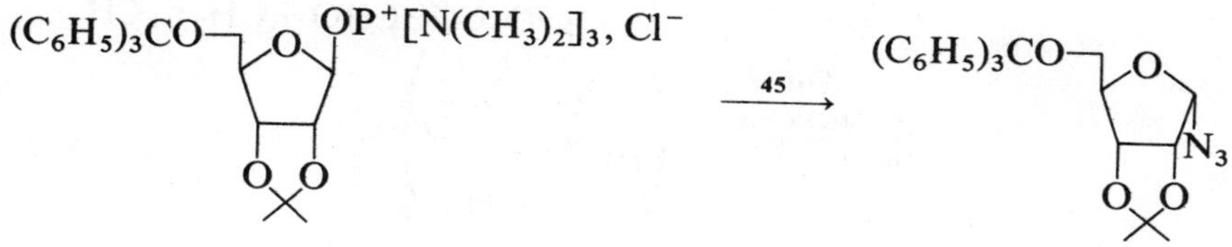

(Eq. 44)

Selectivity. The formation of alkoxy(trisdimethylamino)phosphonium salts is very selective for primary over secondary alcohols. This effect can be used in selective salt formation from carbohydrates.[201]

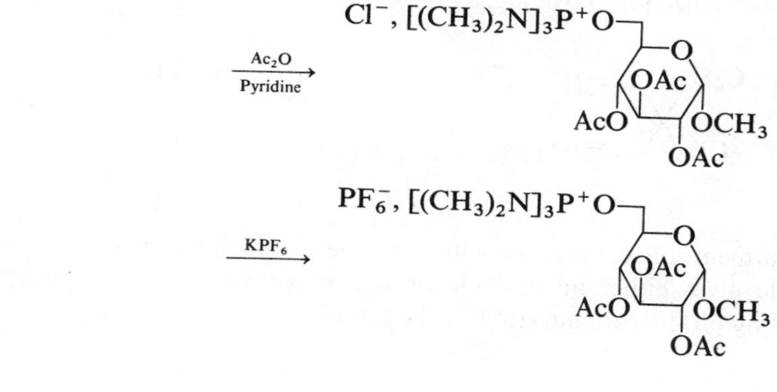

This technique has been used to synthesize Cord-Factor, an immunostimulant.[202]

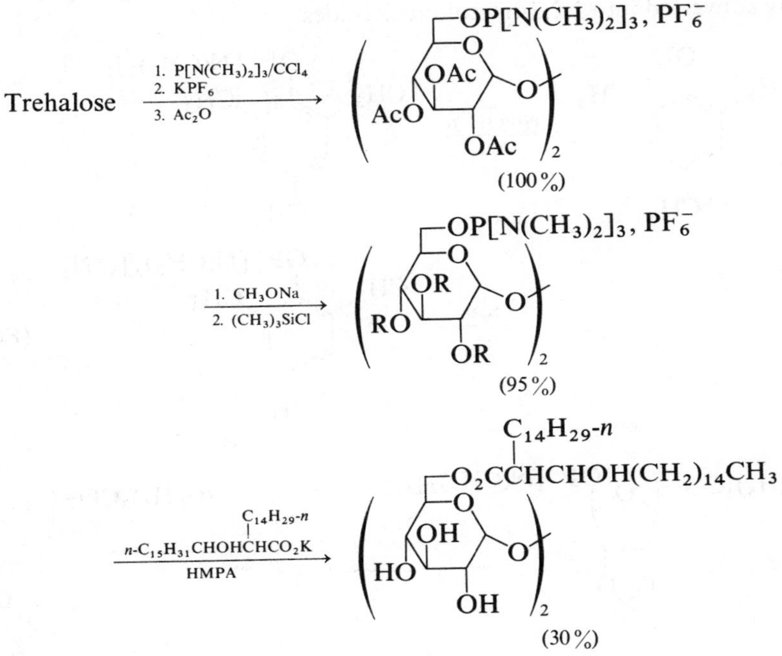

Trehalose $\xrightarrow[\text{3. Ac}_2\text{O}]{\begin{array}{l}\text{1. P[N(CH}_3)_2]_3/\text{CCl}_4 \\ \text{2. KPF}_6\end{array}}$

(100%)

$\xrightarrow[\text{2. (CH}_3)_3\text{SiCl}]{\text{1. CH}_3\text{ONa}}$

(95%)

$\xrightarrow[\text{HMPA}]{n\text{-C}_{15}\text{H}_{31}\text{CHOHCHCO}_2\text{K}}$

(30%)

$$R = -Si(CH_3)_3$$

Biprimary alcohols, such as 1,3-glycols, can be activated selectively at only

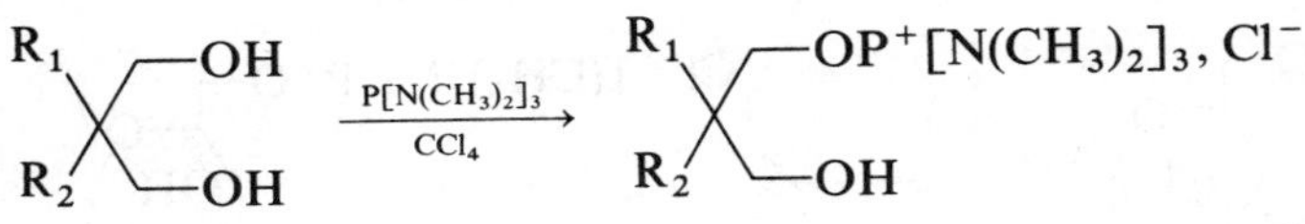

one of the two hydroxyl groups. These oxyphosphonium salts of 1,3-glycols are easily cyclized into oxetanes.[203]

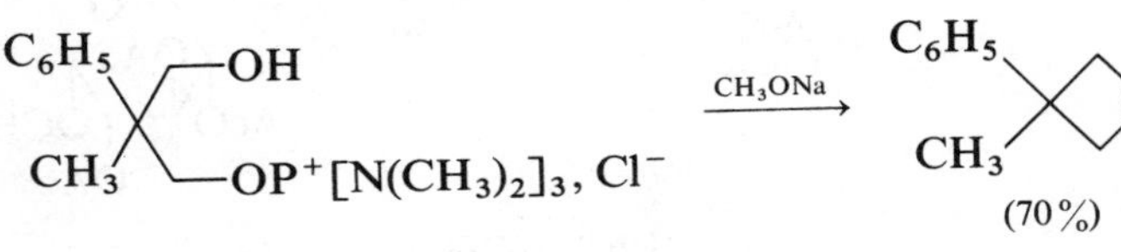

(70%)

Treatment of 1,2-glycols with hexamethylphosphoramide–carbon tetrachloride gives either an epoxide or a pentacoordinated spirophosphorane, depending on the configuration of the glycol.[204]

$$C_6H_5CHOHCHOHC_6H_5 \longrightarrow C_6H_5CH\text{—}CHC_6H_5$$

meso

trans (70%)

$$C_6H_5CHOHCHOHC_6H_5 \longrightarrow$$

threo-d, l

N(CH_3)_2

1,2-O-Isopropylideneglucofuranose is converted into an amidophosphite by nucleophilic substitution of the phosphorane intermediate.[205]

(60%)

Mono salts of ω-biprimary glycols can be obtained by using a solvent such as tetrahydrofuran that is a solvent for the glycol but not for the mono salt.[249]

$$HO(CH_2)_{10}OH \xrightarrow[\text{THF, }-20°]{P[N(CH_3)_2]_3,\ CCl_4} HO(CH_2)_{10}OP^+[N(CH_3)_2]_3,\ Cl^-$$

$$\xrightarrow{KPF_6} HO(CH_2)_{10}OP^+[N(CH_3)_2]_3,\ PF_6^- \xrightarrow{NaN_3} HO(CH_2)_{10}N_3$$

This synthesis provides an excellent route to ω-substituted alcohols.

EXPERIMENTAL PROCEDURES

1-(3′-Azido-3′-desoxy-2′,5′-di-O-trityl-β-D-xylofuranosyl)uracil (Use of Triphenylphosphine–Diethyl Azodicarboxylate–Hydrazoic Acid).[59] 2′,5′-Di-O-trityluridine (219 mg, 0.3 mmol) and triphenylphosphine (94 mg, 0.36 mmol) were dissolved in 4 mL of anhydrous benzene and mixed with a 0.36 millimoles of hydrazoic acid in benzene solution; diethyl azodicarboxylate (63 mg, 0.36 mmol) was then added, and the mixture was refluxed for 10 minutes. After evaporation of the solvent, the crude product was chromatographed on 30 g of Kieselgel and eluted with benzene–ethyl acetate (4:1). There was obtained 196 mg (87%) of 1-(3′-azido-3′-desoxy-2′,5′-di-O-trityl-β-D-xylofuranosyl)uracil. After solution in ether and precipitation with cyclohexane the compound had mp 136–139°; IR: 3370 (NH), 2115 (N_3), 1720, 1695 (CO) cm^{-1}; NMR (100 MHz) δ: 2.60 (d, 3′H), 3.12 (dd, 5′bH), 3.52 (dd, 5′aH), 4.01 (d, 2′H),

4.10 (ddd, $4'H$), 5.60 (dd, $5H$ of uracil), 6.46 (d, $1'H$), 7.30 (m, aromatic H), 9.32 (s, b, NH) ppm.

Methyl 3β-Formyloxy-12α-hydroxycholanate (Use of Triphenylphosphine–Diethyl Azodicarboxylate–Formic Acid).[67]

A solution of methyl deoxycholate (2.03 g, 5 mmol), triphenylphosphine (2.62 g, 10 mmol), and formic acid (0.46 g, 10 mmol) in dry tetrahydrofuran was stirred at room temperature. There was added dropwise a solution of 1.74 g (10 mmol) of diethyl azodicarboxylate in 10 mL of tetrahydrofuran. Removal of the tetrahydrofuran under reduced pressure afforded a syruplike product that was chromatographed on a Florisil column. The first five fractions (50 mL each) obtained with benzene–hexane (6:4) as the eluant gave 2.1 g (97%) of methyl 3β-formyloxy-12α-hydroxycholanate, mp 160–161°.

Triphenoxyphosphonium Methiodide.[120]

Triphenyl phosphite (52 mL, 0.2 mol) and methyl iodide (16 mL, 0.26 mol) were mixed in a 250-mL flask fitted with a very efficient 3-ft condenser and a thermometer well. The flask was placed in a 90° oil bath, and the temperature was slowly raised to 125° over 8 hours while the pot temperature rose slowly from 70° to 85° and then rapidly to 115°. This temperature was maintained for 12–14 hours and, on cooling and seeding, the mixture crystallized to a solid brown mass. Dry ether (100 mL) was added, the product was broken up with a spatula, and the resulting crystalline material was then washed repeatedly with fresh, dry ethyl acetate until the washings were only light-colored. The amber crystals were dried and stored *in vacuo*, giving 80 g (90%) of product suitable for direct use; NMR (rigorously dry $CDCl_3$) δ: 3.11 (d, $3H$, $J_{P,H} = 16.5$ Hz, PCH_3), 7.45 (m, $15H$, aromatic) ppm. If the sample is not prepared in a drybox, appreciable amounts of diphenyl methyl phosphonate are formed, as indicated by a doublet ($J_{P,H} = 18$ Hz) at δ 1.84 ppm. The reagent should always be weighed and handled in a drybox under a nitrogen atmosphere.

3α-Iodocholestane (Use of Triphenoxyphosphonium Methiodide).[120]

Cholestanol (0.78 g, 2.0 mmol) and triphenoxyphosphonium methiodide (1.81 g, 4.0 mmol) were dissolved in anhydrous dimethylformamide (10 mL), and the solution was stored for 2 hours at 25°. After addition of methanol (1 mL), the mixture was diluted with chloroform and extracted with dilute aqueous sodium thiosulfate followed by water. After drying (Na_2SO_4), the solvent was evaporated and the residue was chromatographed on a column of silicic acid with hexane. The product from the major peak was evaporated to give 640 mg of 3α-iodocholestane. Crystallization from acetone gave 562 mg (57%) of pure product, mp 112–113°; $[\alpha]_D^{23} + 36.9°$; NMR ($CDCl_3$) δ: 0.65 (s, $3H$, $C_{18}H_3$), 0.79 (s, $3H$, $C_{19}H_3$), 0.86 (d, $6H$, $J = 6$ Hz, $C_{26}H_3$ and $C_{27}H_3$), 0.90 (d, $3H$, $J = 6$ Hz, $C_{21}H_3$), 4.95 (m, $1H$, C_3H) ppm.

3-Bromo-2-cyclohexen-1-one (Use of Triphenylphosphine–Bromine).[147]

To an ice-cold, stirred solution of triphenylphosphine (576 mg, 2.2 mmol, freshly recrystallized from ethyl acetate–methanol) in 20 mL of dry benzene was added dropwise 2.2 mL of a 1.0 M solution of bromine in benzene. To the resulting suspension was added triethylamine (220 mg, 2.2 mmol, freshly distilled from lithium aluminum hydride) and 1,3-cyclohexanedione (224 mg, 2.0 mmol); stirring was continued at room temperature for 3 hours. The mixture was filtered through a short column of silica gel, and the column was eluted with ether. Concentration of the eluant, followed by distillation (air-bath temperature 95–105°) of the residual material under reduced pressure (11 mm), gave 342 mg (97%) of 3-bromo-2-cyclohexen-1-one as a clear, colorless oil; IR: 1682, 1605 cm^{-1}; NMR δ: 1.80–3.00 (m, 6H), 6.50 (m, 1H) ppm.

(S)-(−)-5-(Chloromethyl)-2-pyrrolidinone (Use of Triphenylphosphine–Carbon Tetrachloride).[206]

A solution of triphenylphosphine (6.75 g, 25.7 mmol) in 25 mL of dry carbon tetrachloride was added by syringe to a solution of 1.94 g (16.7 mmol) of (S)-(+)-5-(hydroxymethyl)-2-pyrrolidinone in 15 mL of dry chloroform. After 5 minutes the solution became cloudy. The mixture was then heated at 55° for 6 hours, cooled to room temperature, and the solvent was evaporated. The residue was extracted with benzene in a Soxhlet extractor for 15 hours. After cooling to room temperature, the benzene solution was filtered and concentrated. The resulting residue was triturated three times with 45 mL of distilled water. The combined aqueous extract was filtered through Celite, and the filtrate was concentrated to give 2.39 g of a semisolid. This yellow residue was vacuum-distilled to give 1.91 g (86%) of a colorless oil that crystallized to a white solid on cooling in ice, bp 106.5–107° (0.15 mm); mp 53–55°; $[\alpha]_D^{20} - 18°$ (c 2.5, ethanol); NMR (CDCl$_3$) δ: 2.2 (m, 4H), 3.5 (d, 2H, $J = 5$ Hz), 4.0 (m, 1H), 7.5 (br, 1H) ppm; IR (Nujol): 3200 (br), 1690 (s), 1282 (m), 765 (br), 645 (s) cm^{-1}. Anal. Calcd. for C$_5$H$_8$ClNO: C, 44.96; H, 6.04; N, 10.49. Found: C, 45.35; H, 6.24; N, 10.36.

Methyl 6-Chloro-6-desoxy-α-D-glucopyranoside (Use of Triphenylphosphine–Carbon Tetrachloride in Pyridine).[156]

Methyl α-D-glucopyranoside (1 g, 5 mmol) was dissolved in 10 mL of pyridine, and the solution was cooled to 0°. There was added 2.75 g (10.5 mmol) of triphenylphosphine followed by 0.8 g (5.2 mmol) of carbon tetrachloride. The mixture was protected from moisture, heated with stirring to 65° for 12 minutes, and cooled slightly; 10 mL of methanol was added to decompose any excess reagent. The solvent was evaporated, and the residue was chromatographed on silica gel with chloroform–methanol (20:1). Noncarbohydrate materials were removed by elution with chloroform, and further elution with chloroform–methanol (20:1) gave 1.07 g (98%) of methyl 6-chloro-6-desoxy-α-D-glucopyranoside. Crystallization from chloroform–hexane gave product, mp 111–112°; $[\alpha]_D^{25} + 153°$.

3-Chloro-2,2-dimethylpropanol [Use of Tris(dimethylamino)phosphine–Carbon Tetrachloride].[192,193]

To a well-stirred solution of 2,2-dimethylpropane-1,3-diol (5.20 g, 50 mmol) and 5 mL (8 g, 52 mmol) of carbon tetrachloride in

100 mL of tetrahydrofuran at $-30°$ there was added slowly a solution of 8.15 g (50 mmol) of tris(dimethylamino)phosphine in 15 mL of tetrahydrofuran. (Too rapid an addition caused the mixture to darken.) The mixture was allowed to come to room temperature and was stirred for 2 hours. Evaporation of the solvent gave crystalline (3-hydroxy-2,2-dimethyl)propoxytris(dimethylamino)-phosphonium chloride (14.7 g, 98%); after crystallization from methanol–ether, mp 61°d.

The salt was dissolved in 80 mL of dimethylformamide and heated at 90–100° for 6 hours. The reaction mixture was poured into 150 mL of water and extracted with ethyl acetate, the organic phase was washed with dilute hydrochloric acid, and the solvent was evaporated. Distillation gave 5.2 g (85%) of 3-chloro-2,2-dimethylpropanol, bp 67° (19 mm); mp 30°; n_D^{24} 1.448; IR (film): 720 (C–Cl), 3350 (OH) cm^{-1}; NMR (CCl$_4$) δ: 0.97 (s, CH$_3$), 3.38 (s, CH$_2$Cl), 3.4 (s, CH$_2$O) ppm.

N-(2,3:5,6-Di-O-isopropylidene-β-D-mannofuranosyl)phthalimide (Use of Triphenylphosphine–Diethyl Azodicarboxylate–Phthalimide).[51]

To a solution of 2,3:5,6-di-O-isopropylidene-D-mannofuranose (260 mg, 1 mmol), phthalimide (147 mg, 1 mmol), and triphenylphosphine (262 mg, 1 mmol) in 5 mL of tetrahydrofuran was added 190 mg (1.1 mmol) of diethyl azodicarboxylate. A slight exothermic reaction was observed. The reaction mixture was left at room temperature for 24 hours and then evaporated to dryness under reduced pressure. The residue was fractionated on a silica gel column, using benzene–diethyl ether (9:1). There was obtained 98 mg (25%) of N-(2,3:5,6-di-O-isopropylidene-β-D-mannofuranosyl)phthalimide, mp 133–134°; $[\alpha]_D$ +37°; IR (KBr): 1790, 1730, 1620, 725 cm^{-1}; ^{1}H NMR δ: 7.88 (phthalimide H), 5.82 (d, $J =$ 3.7 Hz, 1H) ppm. Anal. Calcd. for C$_{20}$H$_{23}$NO$_7$: C, 61.7; H, 6.0; N, 3.6. Found: C, 61.5; H, 6.0; N, 3.6.

There was also obtained 19 mg (5%) of the α isomer, mp 125°; $[\alpha]_D$ +19°; IR (KBr): 1780, 1720 (phthalimide C=O), 1610, 725 cm^{-1}; ^{1}H NMR δ: 7.90 (m, 4 aromatic H), 5.96 (s, 1H) ppm. Anal. Calcd. for C$_{20}$H$_{23}$NO$_7$: C, 61.7; H, 6.0; N, 3.6. Found: C, 61.8; H, 5.9; N, 3.5.

2-Phthalimidoacrylic Acid (Use of Triphenylphosphine–Diethyl Azodicarboxylate to Dehydrate an Alcohol).[100]

To a solution of 0.91 g (4.1 mmol) of (S)-3-hydroxy-2-phthalimidopropionic acid in 15 mL of tetrahydrofuran were added 1.07 g (4.1 mmol) of triphenylphosphine and 0.71 g (4.1 mmol) of diethyl azodicarboxylate, and the solution was stirred for 4 hours. The solvent was removed under reduced pressure, the oily residue was dissolved in benzene, and the precipitated N,N-diethoxycarbonylhydrazine was removed by filtration. The filtrate was chromatographed on a silica gel column with benzene, and the resulting 2-phthalimidoacrylic acid was recrystallized from diethyl ether–hexane (1:30). There was obtained 0.61 g (65%) of a solid, mp 111–112°; ^{1}H NMR (CDCl$_3$) δ: 3.7 (OCH$_3$), 5.9, 6.6 (vinyl H), 7.7 (phthalimido H) ppm.

3α-Cyanocholestane (Use of Triphenylphosphine–Diethyl Azodicarboxylate to Replace an Alcoholic Hydroxyl Group with Cyano).[11]   To a stirred solution of 3β-cholestanol (390 mg, 1 mmol) and triphenylphosphine (288 mg, 1.1 mmol) in 5 mL of dry benzene was added at room temperature a solution of hydrogen cyanide (30 mg, 1.1 mmol) in 2 mL of dry benzene and then a solution of diethyl azodicarboxylate (191 mg, 1.1 mmol) in 2 mL of dry benzene. The reaction mixture warmed slightly, and a precipitate appeared. After 1 hour the benzene was removed by distillation under reduced pressure, and the residue was chromatographed on a Kieselgel column with benzene–petroleum ether (1:1). The two principal fractions were 3α-cyanocholestane (25%) and ethyl 3β-cholestanyl carbonate (25%). These compounds were separated by preparative liquid chromatography using petroleum ether–ethyl acetate (9:1), and the 3α-cyanocholestane was recrystallized from diethyl ether–ethanol, mp 165–170°; $[\alpha]_D^{20}$ 20.8° (c 1%, CHCl₃); IR (CN): 2210 cm⁻¹. Anal. Calcd. for $C_{28}H_{47}N$: C, 84.56; H, 11.91; N, 3.52. Found: C, 84.66; H, 11.58; N, 3.4.

Methyl 6-Desoxy-2,3-O-benzyl-α-D-glucopyranoside [Use of Tris(dimethylamino)phosphine–Carbon Tetrachloride].[196]   A solution of tris(dimethylamino)phosphine (0.66 g, 4 mmol) in tetrahydrofuran was added dropwise to a cold (−45°), stirred solution of methyl 2,3-di-O-benzyl-α-D-glucopyranoside (1.125 g, 3 mmol) and carbon tetrachloride (0.912 g, 6 mmol) in 10 mL of tetrahydrofuran. The reaction was monitored by TLC, using silica gel and 1:4 methanol–ethyl acetate as eluant. When the reaction was complete, a 1 M solution (12 mL, 12 mmol) of lithium triethylborohydride in tetrahydrofuran was added, and the mixture refluxed under an inert atmosphere for 3 hours, then allowed to cool. Excess hydride was destroyed by the addition of 0.5 mL of water. The mixture was filtered, the solvent removed under reduced pressure, and the gummy product chromatographed on a silica gel column using diethyl ether–petroleum ether (25:75) as eluant. The product was dissolved in 50 mL of diethyl ether, and the solution was washed with 1 N hydrochloric acid (3 × 2 mL), 1 N aqueous sodium hydroxide (3 × 2 mL), and water to neutrality. The solution was dried over magnesium sulfate and filtered, and the solvent removed under reduced pressure to obtain methyl 6-desoxy-2,3-O-benzyl-α-D-glucopyranoside (0.968 g, 90%), n_D^{20} 1.5330; $[\alpha]_D^{25}$ +33.4° (c 5.3%, CCl₄).

2,3:5,6-Di-O-isopropylidene-β-D-mannofuranosyl Azide [Use of Tris(dimethylamino)phosphine–Carbon Tetrachloride].[200] A solution of 2,4,6-trimethylphenol (13.6 g, 0.1 mol) in dry methylene chloride (500 mL) containing carbon tetrachloride (20 g, 0.13 mol) was cooled to −40° under an inert atmosphere. The mixture was stirred while a solution of tris(dimethylamino)phosphine (19.56 g, 0.12 mol) in 20 mL of dry methylene chloride was added dropwise over 3 hours. The mixture was allowed to warm to room temperature, and a solution of 21 g (0.32 mol) of sodium azide in 20 mL of water was added with stirring. The organic layer was separated, dried over magnesium sulfate, and evaporated under reduced pressure to give a yellow syrup. The syrup was dissolved in 70 mL of water and washed with diethyl ether (3 × 100 mL) to remove excess tris(dimethylamino)phosphine. The aqueous layer was extracted with methylene chloride (3 × 200 mL), and the extracts were dried over magnesium sulfate and evaporated to give crude mesityloxy-tris(dimethylamino)-phosphonium azide. The crude product was dissolved in 20 mL of acetone and precipitated by addition of dry diethyl ether; yield, 23.1 g (68%); mp 110–112°; IR (film): 2030 (N_3^-) cm^{-1}; ^{1}H NMR (CDCl$_3$) δ: 2.29 (s, 9H, 3 CH$_3$), 2.87 (d, 18H, $J = 10.33$ Hz), 6.90 (s, 2H) ppm.

A solution of 2,3:5,6-di-O-isopropylidene-D-mannofuranose (260 mg, 1 mmol) in 15 mL of dry methylene chloride containing carbon tetrachloride (308 mg, 2 mmol) was cooled to −40° under an inert atmosphere. The mixture was stirred while a solution of tris(dimethylamino)phosphine (200 mg, 1.23 mmol) in dry methylene chloride was added dropwise over 1 hour. Mesityl-oxy-tris(dimethylamino)phosphonium azide (680 mg, 2 mmol) was added to the cold solution, and the mixture was stirred at −10° for 3 hours. The solvent was evaporated, and the glycosyl azide was dissolved in 200 mL of hexane, washed with water (3 × 20 mL), dried over magnesium sulfate, and evaporated. The resulting syrup was chromatographed on silica gel using diethyl ether–petroleum ether (1:4) as eluant. There was obtained 0.185 g (65%) of 2,3:5,6-di-O-isopropylidene-β-D-mannofuranosyl azide, $[\alpha]_D^{25}$ +2.7° (c 3.32%, CHCl$_3$); IR (film): 2150 cm^{-1} (C–N$_3$); ^{1}H NMR (CDCl$_3$) δ: 1.37, 1.44, 1.55 (3 s, 12H), 3.43–4.87 (m, 7H) ppm.

trans-2,3-Di-(*o*-methoxyphenyl)oxirane [Use of Tris(dimethylamino)phosphine–Carbon Tetrachloride].[204] To a solution of *meso*-1,2-di(*o*-methoxyphenyl)-1,2-ethanediol (2.74 g, 10 mmol) and carbon tetrachloride (3.24 g, 20 mmol) in methylene chloride cooled to −40° was added tris(dimethylamino)-phosphine (3.26 g, 20 mmol) dissolved in methylene chloride. After the addition, the solvent was removed under reduced pressure, the residue was hydrolyzed and extracted with pentane, and the solution was dried over magnesium sulfate. After evaporation of the solvent, the residue was crystallized from cyclohexane–ethyl acetate to obtain 1.5 g (66%) of *trans*-2,3-di-(*o*-methoxyphenyl)oxirane, mp 150–152°; ^{1}H NMR (CDCl$_3$) δ: 6.7–7.4 (m, aromatic H), 4.18 (s, CH), 3.78 (s, CH$_3$) ppm.

TABULAR SURVEY

The tables appear in the order in which the various reagent combinations are discussed in the text. Coverage of the literature extends to April 1981.

The Conditions column includes the temperature, solvent, and reaction time if provided in the reference; a dash indicates that the conditions were not given. Yields that were not given in the reference are marked (—) in the Products column.

Abbreviations used for a few large substituents in the structures of starting compounds and products are defined in footnotes to the tables. Abbreviations used in the Conditions column are as follows:

Ac	Acetyl
Bz	Benzoyl
Bzl	Benzyl
DMF	Dimethylformamide
DMAc	Dimethylacetamide
Et_2O	Diethyl ether
HMPA	Hexamethylphosphoroustriamide
Methyl carbitol	Diethylene glycol monomethyl ether
Py	Pyridine
Tetralin	1,2,3,4-Tetrahydronaphthalene
THF	Tetrahydrofuran
Triglyme	Triethylene glycol dimethyl ether
Ts	p-Toluenesulfonyl

TABLE I. REPLACEMENT OF HYDROXYL GROUPS BY NITROGEN NUCLEOPHILES USING TRIPHENYLPHOSPHINE–DIETHYL AZODICARBOXYLATE

	Alcohol	Nucleophile	Conditions	Products and Yields (%)	Ref.
C_2	C_2H_5OH	Phthalimide	25°, THF	N-Ethylphthalimide (91)	48
		$C_6H_5CONHCO_2CH_2C_6H_5$	"	$C_6H_5CON(C_2H_5)CO_2CH_2C_6H_5$ (66)	48
C_3	$CH_2{=}CHCH_2OH$	None	25°, Et_2O, overnight	$C_2H_5O_2CNCH_2CH{=}CH_2$ (73), with $C_2H_5O_2CNH$ attached (vertical bond)	8
	$n\text{-}C_3H_7OH$	Phthalimide	25°, THF	N-n-Propylphthalimide (93)	48
	$n\text{-}C_4H_9OH$	"	"	N-n-Butylphthalimide (81)	48
		Succinimide	"	N-n-Butylsuccinimide (76)	48
		$C_6H_5CONHCO_2CH_2C_6H_5$	"	$C_6H_5CON(C_4H_9\text{-}n)CO_2CH_2C_6H_5$ (66)	54
		Phthalimide	"	N-sec-Butylphthalimide (75)	48
C_5	$CH_3CHOHC_2H_5$	"	25°, THF, 12 hr	$C_2H_5O_2CCH(CH_3)N(CO)_2C_6H_4\text{-}o$ (58)	48
	$C_2H_5O_2CCHOHCH_3$	"	"	$(\pm)\text{-}CH_3CH(CO_2C_2H_5)N(CO)_2C_6H_4\text{-}o$ (58)	50
	$(\pm)\text{-}CH_3CHOHCO_2C_2H_5$	"	"	$(CH_3)_2C(CO_2CH_3)N(CO)_2C_6H_4\text{-}o$ (15)	50
	$(CH_3)_2COHCO_2CH_3$				
	[dihydropyranone–OH structure]	Phthalimide	25°, THF, 24 hr	[N-substituted phthalimide pyranone structure] (43)	51
C_6	Cyclohexanol	$(C_6H_5O)_2P(O)N_3$	"	$C_6H_{11}N_3$ (60)	65
	[cyclohexane with N_3 and OH structure]	HN_3	25°, C_6H_6, 1 hr	[cyclohexane with two N_3 structure] (81)	11
	[cyclohexane with Cl and OH structure]	"	"	[cyclohexane with Cl and N_3 structure] (—)	11
C_7	$C_6H_5CH_2OH$	None	Reflux, THF	$C_6H_5CH_2N(CO_2C_2H_5)NHCO_2C_2H_5$ (69) $+ C_6H_5CH_2N(CO_2C_2H_5)N(CO_2C_2H_5)_2$ (8)	44

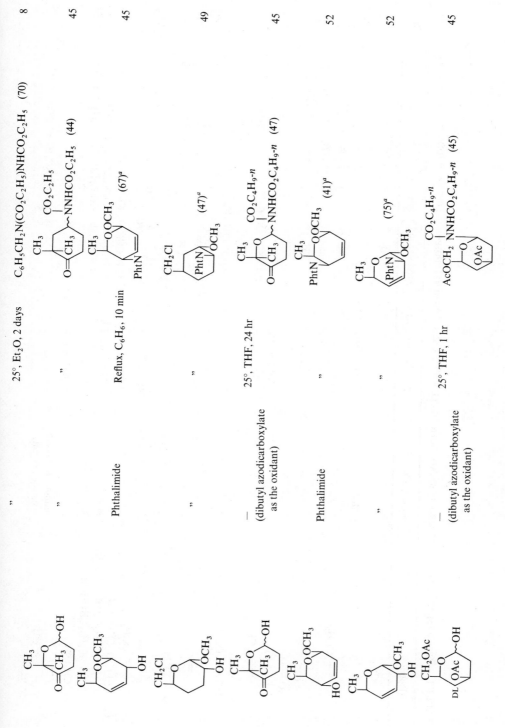

TABLE I. REPLACEMENT OF HYDROXYL GROUPS BY NITROGEN NUCLEOPHILES USING TRIPHENYLPHOSPHINE–DIETHYL AZODICARBOXYLATE (*Continued*)

Alcohol	Nucleophile	Conditions	Products and Yields (%)	Ref.
C$_7$ (*Contd.*) [pyranose structure with CH$_3$, HO, OCH$_3$]	Phthalimide	Reflux, C$_6$H$_6$, 10 min	[PhtN-pyranose structure with CH$_3$, OCH$_3$] (40)[a] + [PhtN-pyranose structure with CH$_3$, OCH$_3$] (17)[a]	52
[pyranose structure CH$_3$, HO, OCH$_3$]	"	Reflux, C$_6$H$_6$	[PhtN-pyranose structure CH$_3$, OCH$_3$] (44)[a]	52
[pyranose structure CH$_3$, HO, OOCH$_3$, OCH$_3$]	"	"	[PhtN-pyranose structure CH$_3$, OCH$_3$, OCH$_3$] (69)[a]	52
[pyranose structure CH$_3$, HO, OCH$_3$]	"	"	[PhtN-pyranose structure CH$_3$, OCH$_3$] (54)[a]	52
(±)-C$_6$H$_5$CHOHCH$_3$	HN$_3$	25°, THF	C$_6$H$_5$CH(CH$_3$)N(CO)$_2$C$_6$H$_4$-o (82)	48
C$_2$H$_5$O$_2$CCH$_2$CHOHCO$_2$C$_2$H$_5$	"	25°, C$_6$H$_6$, 1 hr	C$_2$H$_5$O$_2$CCH$_2$CH(N$_3$)CO$_2$C$_2$H$_5$ (75)	11
(S)-(+)-C$_6$H$_{13}$CHOHCH$_3$	Phthalimide	25°, THF	(R)-(−)-C$_6$H$_{13}$CH(CH$_3$)N(CO)$_2$C$_6$H$_4$-o (—)	48
C$_8$ [CH$_2$OAc-pyranose-OH structure]	(dibutyl azodicarboxylate as the oxidant)	Reflux, C$_6$H$_6$, 10 min	[CH$_2$OAc-pyranose-N(CO$_2$C$_4$H$_9$-n)NHCO$_2$C$_4$H$_9$-n structure] β (65) α (16)	45

	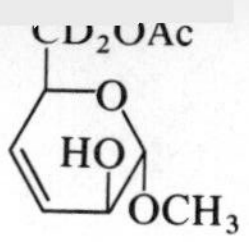	Phthalimide	"	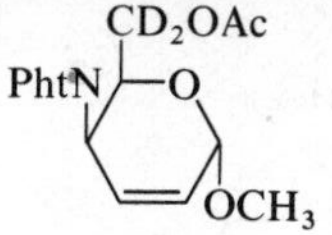 (62)[a] + 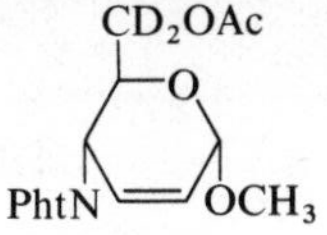 (12)[a]	52
				+ 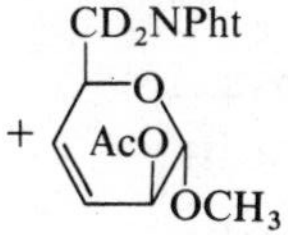 (4)	
	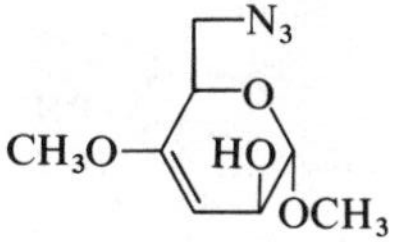	HN_3	25°, C_6H_6, 2 hr	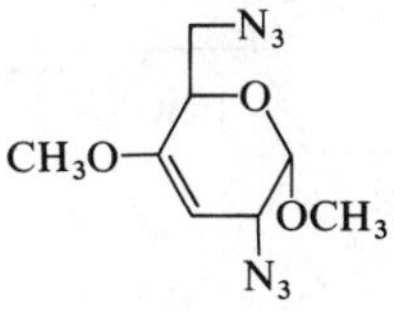 (61) + 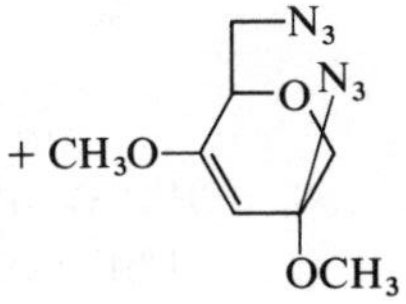 (8)	165
	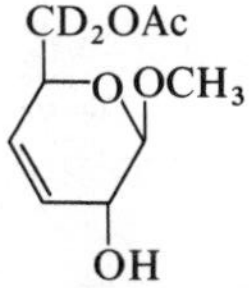	Phthalimide	Reflux, C_6H_6, 10 min	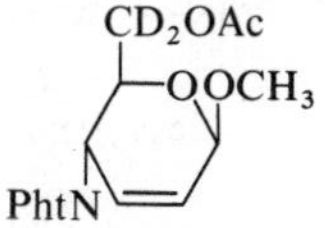 (63)[a]	52
C_9	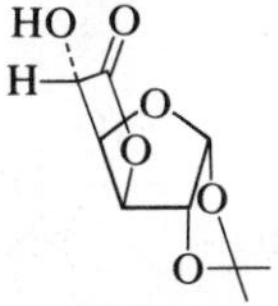	HN_3	25°, C_6H_6, 2 hr	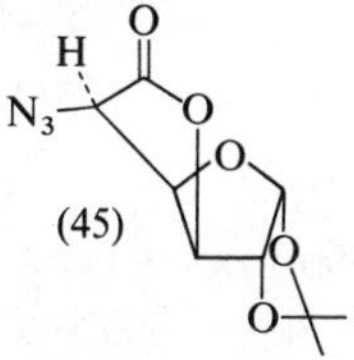 (45) + 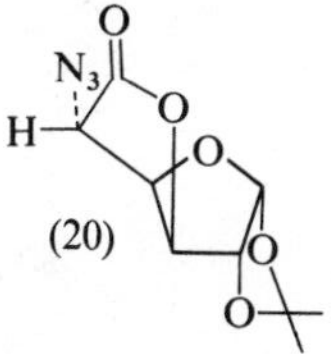 (20)	61
C_{10}	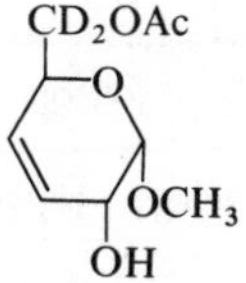	Phthalimide	Reflux, C_6H_6, 10 min	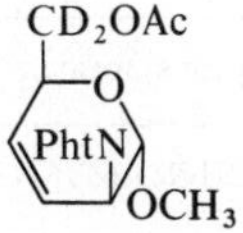 (100)[a]	52

TABLE I. REPLACEMENT OF HYDROXYL GROUPS BY NITROGEN NUCLEOPHILES USING TRIPHENYLPHOSPHINE–DIETHYL AZODICARBOXYLATE (*Continued*)

Alcohol	Nucleophile	Conditions	Products and Yields (%)	Ref.
C_{10} DL (*Contd.*) CH$_2$OAc ... OH ... OAc	— (dibutyl azodicarboxylate as the oxidant)	25°, THF, 1 hr	CH$_2$OAc, CO$_2$C$_4$H$_9$-n O—N—NHCO$_2$C$_4$H$_9$-n OAc (42)	52
	Phthalimide	25°, THF, 24 hr	—NPht β (43)[a] OAc	83
CH$_2$OAc ... O ... OAc ... OH	"	25°, THF, 24 hr	CH$_2$OAc O OAc ... NPht α (33)[a] β (8)[a]	51
CH$_2$OCH$_3$ O OCH$_3$ OCH$_3$ CH$_3$O ... OCH$_3$ ... OH	— (dibutyl azodicarboxylate as the oxidant)	25°, THF, 1 hr	CH$_2$OCH$_3$ CO$_2$C$_4$H$_9$-n O—N—NHCO$_2$C$_4$H$_9$-n OCH$_3$ CH$_3$O ... OCH$_3$ (55) + CH$_2$OCH$_3$ O OCO$_2$C$_4$H$_9$-n OCH$_3$ OCH$_3$ CH$_3$O ... OCH$_3$ (18)	45

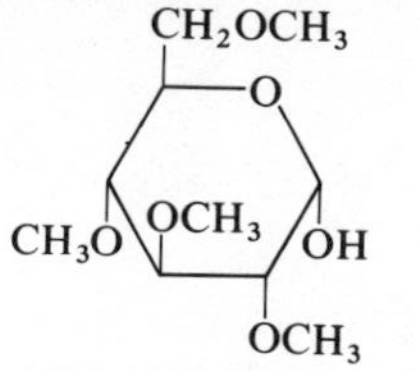

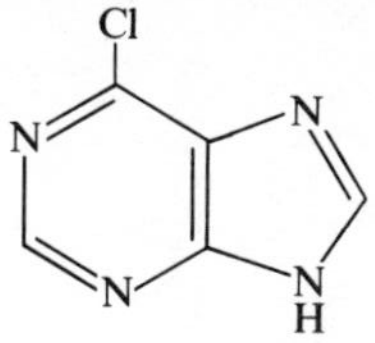

C_{11}

Substrate	Reagent	Conditions	Product	Ref.

Row 1:

Substrate: (structure) CH_2OCH_3 ... OH ... OCH_3 ... CH_3O ... OCH_3

Reagent: (purine structure) $=RH$, N, N_H, Cl, N, N

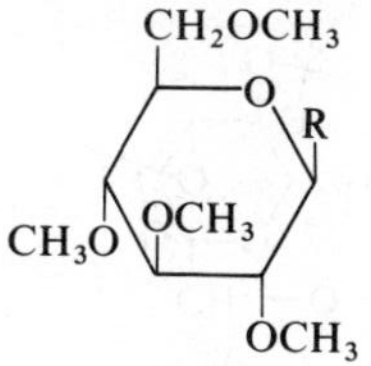

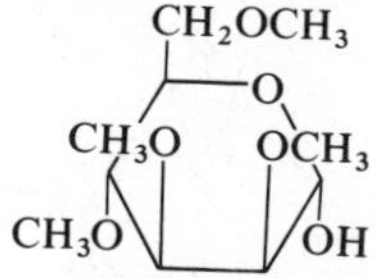

Conditions: 25°, THF, 12 hr, $CH_3P(C_6H_5)_2$ as the phosphine

Product: CH_2OCH_3 ... R ... OCH_3 ... CH_3O ... OCH_3 (66)

57

Row 2:

Substrate: CH_2OCH_3 ... OCH_3 ... OH ... CH_3O ... OCH_3

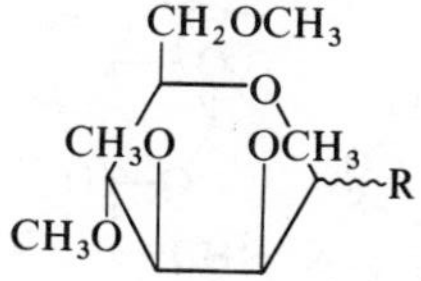

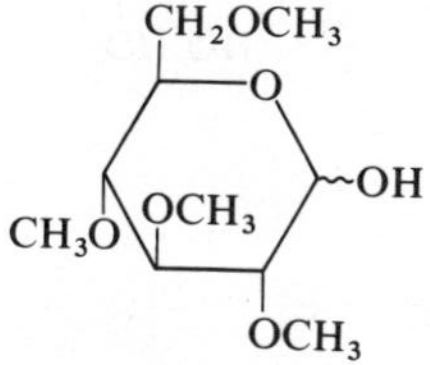

Reagent: "

Conditions: 25°, THF, 12 hr, $CH_3P(C_6H_5)_2$ as the phosphine

Product: CH_2OCH_3 ... OCH_3 ... CH_3O ... OCH_3 ... R α (34) β (37)

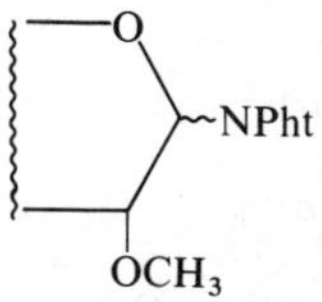

57

Row 3:

Substrate: CH_2OCH_3 ... OCH_3 ... CH_3O ... OCH_3 ... OH

Reagent: Phthalimide

Conditions: 25°, THF, 24 hr

Product: NPht ... OCH_3 α (4)[a] β (37)

51

Row 4:

Substrate: n-$C_{10}H_{23}OH$

Reagent: $(C_6H_5O)_2PON_3$

Conditions: "

Product: n-$C_{10}H_{23}N_3$ (68)

65

Row 5:

Substrate: 1-Menthol

Reagent: "

Conditions: "

Product: (+)-Menthyl azide (90)

65

Row 6:

Substrate: (S)-$(-)$-HO—C—$CH_2C_6H_5$ with H and $CO_2C_2H_5$

Reagent: Phthalimide

Conditions: 25°, THF, 24 hr

Product: (R)-$(+)$-$C_6H_5CH_2$—C—NPht with H and $CO_2C_2H_5$ (66)[a]

53

TABLE I. REPLACEMENT OF HYDROXYL GROUPS BY NITROGEN NUCLEOPHILES USING TRIPHENYLPHOSPHINE–DIETHYL AZODICARBOXYLATE (*Continued*)

Alcohol	Nucleophile	Conditions	Products and Yields (%)	Ref.
C_{12}	None	25°, THF, 1 hr	N—NHCO$_2$C$_2$H$_5$ / CO$_2$C$_2$H$_5$ (77) + OCO$_2$C$_2$H$_5$ (10)	45
CH$_2$OH	NH, O$_2$	25°, THF, 12 hr	CH$_2$— (33) + CH$_2$—O— (32)	56

54

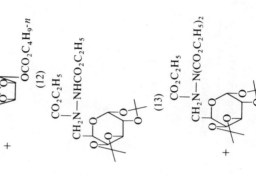

57

45

44

β (63)
α (16)

N—$NHCO_2C_4H_9\text{-}n$
$CO_2C_4H_9\text{-}n$
(79)

$OCO_2C_4H_9\text{-}n$
(12)

$CO_2C_2H_5$
CH_2N—$NHCO_2C_2H_5$
(13)

$CO_2C_2H_5$
CH_2N—$N(CO_2C_2H_5)_2$
(75)

25°, THF, 12 hr,
$CH_3P(C_6H_5)_2$
as the phosphine

25°, THF, 1 hr

Reflux, THF

$=RH$

None
(dibutyl azodicarboxylate
as the oxidant)

None

OH

OH

CH_2OH

TABLE I. REPLACEMENT OF HYDROXYL GROUPS BY NITROGEN NUCLEOPHILES USING TRIPHENYLPHOSPHINE–DIETHYL AZODICARBOXYLATE (*Continued*)

Alcohol	Nucleophile	Conditions	Products and Yields (%)	Ref.
C_{12} (*Cont.*)	None (dibutyl azodicarboxylate as the oxidant)	Reflux, THF	(35) + (58)	44
	Phthalimide	25°, THF, 24 hr	α (5)[a] β (25) (2)	51

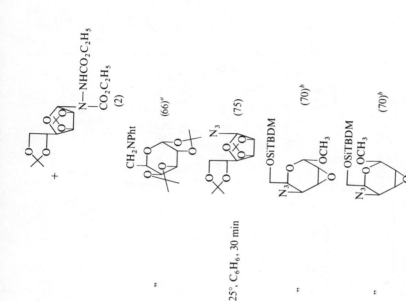

C₁₃

25°, C₆H₆, 30 min

HN₃

49

207

207

207

57

TABLE I. REPLACEMENT OF HYDROXYL GROUPS BY NITROGEN NUCLEOPHILES USING TRIPHENYLPHOSPHINE–DIETHYL AZODICARBOXYLATE (*Continued*)

Alcohol	Nucleophile	Conditions	Products and Yields (%)	Ref.
C$_{13}$ (*Contd.*)	C$_6$H$_5$CH$_2$O$_2$CNHCOC$_6$H$_5$	25°, C$_6$H$_6$, 30 min	(20)^c	53
C$_{14}$	HN$_3$	,,	(29) + (−)^c (57)^b	207
	,,	,,	(49) + (49)^b	207

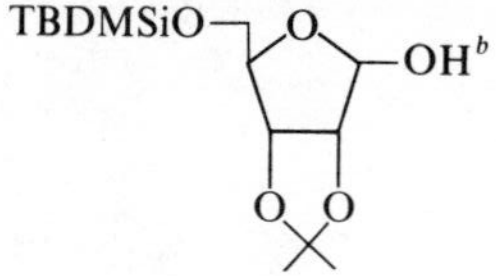

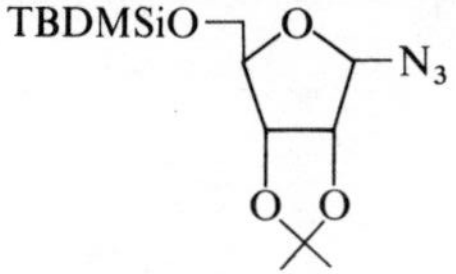

TBDMSiO —OH[b]

SiTBDM[b] —OH

"

"

"

"

TBDMSiO N$_3$ α (70)[b]
β (10)

207

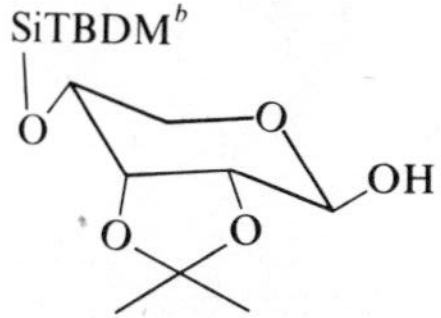

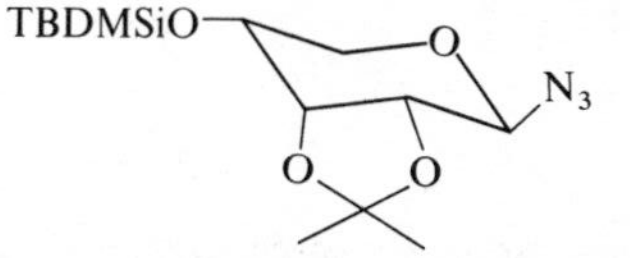

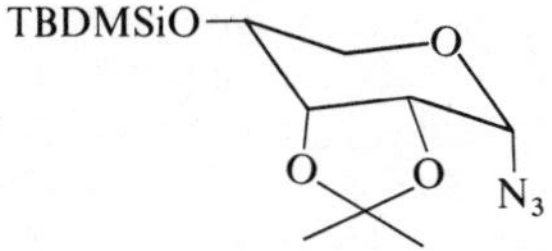

TBDMSiO N$_3$ (49)[b]

207

TBDMSiO N$_3$ (49)

+

R—O—O—CH$_2$OSiTBDM (48)[b]

57

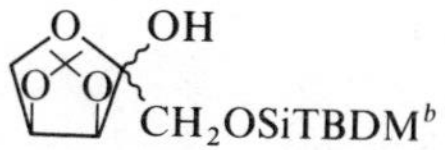

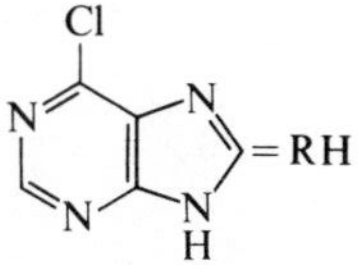

25°, THF, 12 hr,
CH$_3$P(C$_6$H$_5$)$_2$
as the phosphine

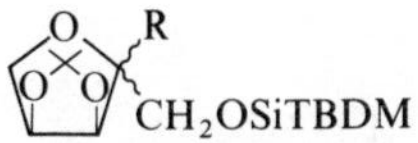

—RH

N N H

Cl N N

None

OH—CH$_2$OSiTBDM[b]

C$_{15}$ t-C$_4$H$_9$O$_2$CNH OH H

NHOCH$_2$C$_6$H$_5$

t-C$_4$H$_9$O$_2$CNH H OCH$_2$C$_6$H$_5$ (90)

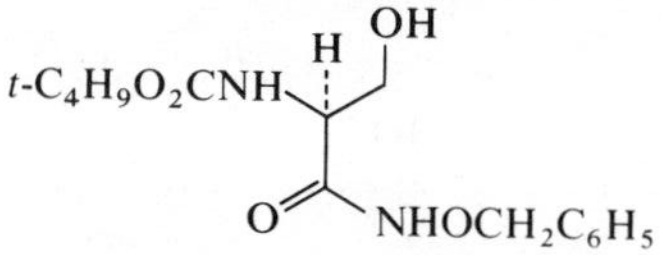

208

20°, THF, 20 hr

C$_{18}$ C$_6$H$_5$CH$_2$O$_2$CNH H OH NH OCH$_2$C$_6$H$_5$

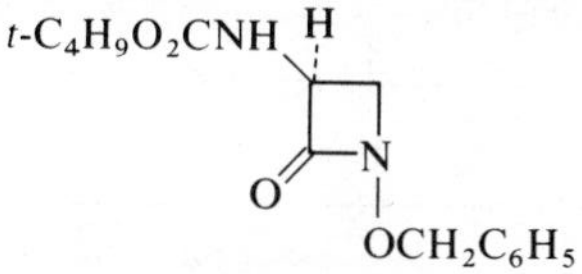

"

C$_6$H$_5$CH$_2$O$_2$CNH H OCH$_2$C$_6$H$_5$ (82)

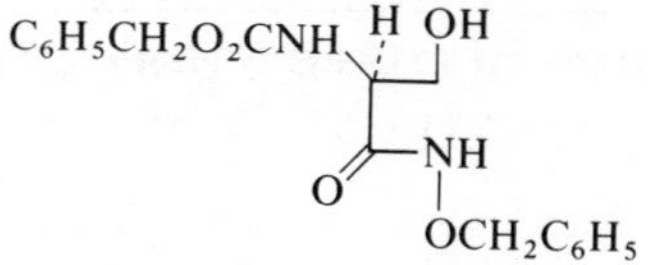

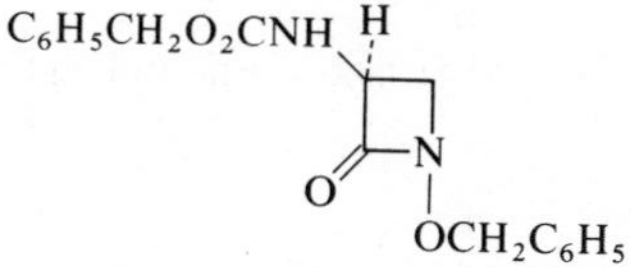

208

TABLE I. REPLACEMENT OF HYDROXYL GROUPS BY NITROGEN NUCLEOPHILES USING TRIPHENYLPHOSPHINE–DIETHYL AZODICARBOXYLATE (*Continued*)

Alcohol	Nucleophile	Conditions	Products and Yields (%)	Ref.
C_{18} (*Contd.*) $C_6H_5CH_2O_2CNH$ (structure with OH OH, CH_2OH, $CONHC_6H_4CH_3\text{-}p$)	None	25°, THF, 24 hr	$C_6H_5CH_2O_2CNH$ (structure) $NC_6H_4CH_3\text{-}p$ (53)	209
$C_6H_5CH_2O_2CNH$ (structure with OH OH, CH_3, $CONHC_6H_4CH_3\text{-}p$)	"	"	$O_2CCH_2C_6H_5$, CH_3––$CONHC_6H_4CH_3\text{-}p$ (aziridine, N) (57)	209
C_{19} 17β-Hydroxy-5α-androstan-3-one	HN_3	Reflux, C_6H_6, 10 min	17α-Azido-5α-androstan-3-one (73)	58
	$(C_6H_5O)_2P(O)N_3$	25°, THF, 24 hr	No reaction	65
3β-Hydroxy-5α-androstan-17-one	"	"	3α-Azido-5α-androstan-17-one (75)	65
3α-Hydroxy-5-androsten-17-one	"	"	3α-Azido-5-androsten-17-one (68)	65
Testosterone	HN_3	Reflux, C_6H_6, 10 min	17α-Azidoandrosten-3-one (86)	58
3β,17β-Diacetoxy-5α-androstan-6β-ol	"		3β,17β-Diacetoxy-6α-azido-5α-androstane (13) + 3β,17β-diacetoxyandrost-5-ene (63)	58
(nucleoside structure, $N\text{-}CH_2C_6H_5$, HO–)	Phthalimide	25°, THF, 18 hr	PhtN– (structure, B) (47)[a,c]	

60

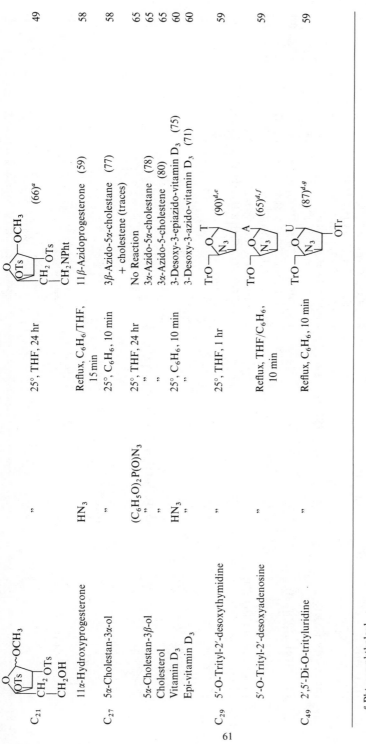

Compound	Reagent	Conditions	Product (yield %)	Ref.
C_{21} (structure: OTs, CH$_2$OTs, CH$_2$OH)	"	25°, THF, 24 hr	(structure: OCH$_3$, CH$_2$OTs) (66)[a]	49
11α-Hydroxyprogesterone	HN_3	Reflux, C_6H_6/THF, 15 min	11β-Azidoprogesterone (59)	58
C_{27} 5α-Cholestan-3α-ol	"	25°, C_6H_6, 10 min	3β-Azido-5α-cholestane (77) + cholestene (traces)	58
5α-Cholestan-3β-ol	$(C_6H_5O)_2P(O)N_3$	25°, THF, 24 hr	No Reaction	65
Cholesterol	"	"	3α-Azido-5α-cholestane (78)	65
Vitamin D_3	HN_3	25°, C_6H_6, 10 min	3α-Azido-5-cholestene (80)	65
Epi-vitamin D_3	"	"	3-Desoxy-3-epiazido-vitamin D_3 (75)	60
			3-Desoxy-3-azido-vitamin D_3 (71)	60
C_{29} 5'-O-Trityl-2'-desoxythymidine	"	25°, THF, 1 hr	(structure: TrO, O, T, N$_3$) (90)[d,e]	59
5'-O-Trityl-2'-desoxyadenosine	"	Reflux, THF/C_6H_6, 10 min	(structure: TrO, O, A, N$_3$) (65)[d,f]	59
C_{49} 2',5'-Di-O-trityluridine	"	Reflux, C_6H_6, 10 min	(structure: TrO, O, U, OTr) (87)[d,g]	59

[a] Pht = o-phthaloyl.
[b] TBDM = t-butyldimethyl.
[c] B = base group in the starting alcohol.
[d] Tr = trityl.
[e] T = 1-desoxythymidyl.
[f] A = 1-desoxyadenosyl.
[g] U = 1-uridyl.

TABLE II. FORMATION OF ESTERS USING TRIPHENYLPHOSPHINE–DIETHYL AZODICARBOXYLATE

	Alcohol	Acid	Conditions	Esters and Yields (%)	Ref.
C_2	C_2H_5OH	$n\text{-}C_4H_9CO_2H$	25°, Et_2O, 8 hr	$n\text{-}C_4H_9CO_2C_2H_5$ (34)	66
		$C_6H_5CO_2H$	"	$C_6H_5CO_2C_2H_5$ (85)	66
C_3	$i\text{-}C_3H_7OH$	$n\text{-}C_4H_9CO_2H$	"	$n\text{-}C_4H_9CO_2C_3H_7\text{-}i$ (43)	66
	$CH_2{=}CHCH_2OH$	"	"	$n\text{-}C_4H_9CO_2CH_2CH{=}CH_2$ (35)	66
		$C_6H_5CO_2H$	"	$C_6H_5CO_2CH_2CH{=}CH_2$ (85)	66
	$CH_3CHOHCH_2OH$	"	25°, THF, 14 hr	$CH_3CHOHCH_2O_2CC_6H_5$ $+ CH_3CH(O_2CC_6H_5)CH_2O_2CC_6H_5$ (5)	107
C_4	$CH_3CHOHCH_2CH_2OH$	"	"	$CH_3CHOHCH_2O_2CC_6H_5$ (70) $+ CH_3CH(O_2CC_6H_5)CH_2CH_2O_2CC_6H_5$ (7)	107
	$CH_3CHOHCHOHCH_3$	"	"	$CH_3CHOHCH(CH_3)O_2CC_6H_5$ (8)	107
C_5	$CH_3CHOHCH_2CHOHCH_3$	"	"	$CH_3CHOHCH_2CH(CH_3)O_2CC_6H_5$ (57) $+ CH_3CH(O_2CC_6H_5)CH_2CH(CH_3)O_2CC_6H_5$ (20)	107
C_6	[cyclohexane-1,2-diol]	"	"	[2-hydroxycyclohexyl benzoate, OH / $O_2CC_6H_5$] (22)	107
	[cyclohexane-1,2-diol]	$p\text{-}O_2NC_6H_4CO_2H$	Reflux, $(i\text{-}C_3H_7)_2O$, 14 hr	[OH / $O_2CC_6H_4NO_2\text{-}p$] (49)	11
	Cyclohexene oxide	$C_6H_5CO_2H$	25°, Et_2O, 18 hr	Cyclohexene oxide (68)	107
	[bicyclic structure, HO, OCH_3, OCH_3]	"	25°, THF, 1 hr	[BzO, OCH_3 structure] (94)	14
	[bicyclic structure, O–O, HO, OCH_3]	"	"	[BzO–O, OCH_3 structure] (90)	14

62

Substrate	Reagent	Conditions	Product(s) (Yield %)	Refs.
	"	"	(88)	14
Methyl-α-D-glucopyranoside	p-O$_2$NC$_6$H$_4$CO$_2$H	Reflux, dioxane/Py, 4 hr	(92)	210
	C$_6$H$_5$CO$_2$H	"	(71)	210
	C$_6$H$_5$CO$_2$H	25°, THF, 31 hr	(23) + (47)	13
Methyl-α-D-glucopyranoside	p-CH$_3$OC$_6$H$_4$CO$_2$H	Reflux, dioxane/Py, 4 hr	(12) + (75)	210

C$_7$

TABLE II. Formation of Esters Using Triphenylphosphine–Diethyl Azodicarboxylate (*Continued*)

	Alcohol	Acid	Conditions	Esters and Yields (%)	Ref.
C_7 (*Contd.*)	Methyl-α-D-glucopyranoside	$p\text{-ClC}_6\text{H}_4\text{CO}_2\text{H}$	Reflux, dioxane/Py, 4 hr	[pyranoside structure with $\text{O}_2\text{CC}_6\text{H}_4\text{Cl-}p$, OCH$_3$] (79)	210
C_8	$C_6H_5CHOHCH_2OH$	$C_6H_5CO_2H$	25°, Et$_2$O, 18 hr	$C_6H_5CHOHCH_2OBz$ (47) + $C_6H_5CH(OBz)CH_2OBz$ (5) + $C_6H_5CHOHCH_2OBz + C_6H_5CH—CH_2—O$ (27)	107
C_9	[indanol structure, *trans*, OH, OH] $[\alpha]_D^{20} = -14.1°$	"	"	[indanyl OBz structures] (29) (21)	107
C_9	[dithiane structure, HO...CH$_3$]	1. $C_6H_5CO_2H$ 2. CH_3ONa	—	[dithiane structure, HO...CH$_3$] (Quant.) $[\alpha]_D^{20} = +14.0°$	211
	Uridine	$C_6H_5CO_2H$	25°, Dioxane/HMPA, 18 hr	5'-O-Benzoyluridine (77)	68
		$p\text{-O}_2\text{NC}_6\text{H}_4\text{CO}_2\text{H}$	"	5'-O-p-Nitrobenzoyluridine (50)	68
C_{10}	Thymidine	$C_6H_5CO_2H$	25°, HMPA, 18 hr	5'-O-Benzoylthymidine (74)	80
		$p\text{-O}_2\text{NC}_6\text{H}_4\text{CO}_2\text{H}$	"	5'-O-p-Nitrobenzoylthymidine (84)	80

64

Reactant	Acid	Conditions	Product (Yield)	Refs.
	p-NCC₆H₄CO₂H	"	5'-O-p-Cyanobenzoylthymidine (79)	80
	p-CH₃OC₆H₄CO₂H	"	5'-O-p-Methoxybenzoylthymidine (65)	80
	p-CH₃C₆H₄CO₂H	"	5'-O-p-Methylbenzoylthymidine (65)	80
	CH₃CO₂H	25°, HMPA/dioxane, 18 hr	5'-O-Acetylthymidine (55)	80
Adenosine	p-O₂NC₆H₄CO₂H	Reflux, HMPA/dioxane, 18 hr	5'-O-p-Nitrobenzoyladenosine (43)	68
$ClCH_2CHOHCH_2O_2CC_6H_5$	$C_6H_5CO_2H$	25°, Et₂O, 18 hr	$ClCH_2CH(OBz)CH_2OBz$ (81)	107
[dithiane structure with CO₂H and OH]	None	−40°, +25°, Toluene, 52 hr	[macrocyclic structure] (60)	76
C₁₁ $CH_3CHOHCH_2CH_2OBz$	$C_6H_5CO_2H$	25°, THF, 31 hr	$CH_3CH(OBz)CH_2CH_2OBz$ (63)	107
[sugar structure, HO, OAc, OCH₃, OAc]	"	"	[sugar structure, OBz, OAc, OCH₃, OAc] (71)	13
[dithiane structure with CO₂H and OH]	None	−40°, −4°, Toluene, 84 hr	[macrocyclic structure] (84)	76

TABLE II. FORMATION OF ESTERS USING TRIPHENYLPHOSPHINE–DIETHYL AZODICARBOXYLATE (*Continued*)

Alcohol	Acid	Conditions	Esters and Yields (%)	Ref.
C₁₂	C₆H₅CO₂H	1. THF 2. CH₃ONa	(86)	212
	"	25°, THF, 48 hr	(18)	70
2′,3′-O-Isopropylideneuridine	"	25°, HMPA/dioxane, 18 hr	2′,3′-O-Isopropylidene-5′-O-benzoyluridine (43)	68
2′,3′-O-Isopropylideneuridine	p-O₂NC₆H₄CO₂H	"	2′,3′-O-Isopropylidene-5′-O-p-nitrobenzoyluridine (65)	68
2′,3′-O-Isopropylideneadenosine	"	"	2′,3′-O-Isopropylidene-5′-O-p-nitrobenzoyladenosine (63)	68
C₁₃	None	−28°; +12°, Toluene, 12 hr	(24)	76

Structure (OSiTBDM, OCH₃, HO, epoxide)	$p\text{-}O_2NC_6H_4CO_2H$	Reflux, THF, 1 hr	$COC_6H_4NO_2\text{-}p$, OSiTBDM, OCH₃ (56)[a]	64
Structure (OSiTBDM[a], OCH₃, HO, epoxide)	"	"	$COC_6H_4NO_2\text{-}p$, OSiTBDM, OCH₃ (46)[a]	64
$CH_3CH(OBz)CH_2CH_2CHOHCH_3$	$C_6H_5CO_2H$	"	$CH_3CH(OBz)CH_2CH_2CH(OBz)CH_3$ (75)	107
$CH_3CH(OBz)CH_2CH_2$ cyclohexane (OBz, OH)	"	"	(OBz, OBz cyclohexane) (60)	107
Structure ($OCH_2C_6H_5$, OCH₃, HO)	"	25°, THF, 1 hr	Structure ($OCH_2C_6H_5$, OCH₃, BzO) (87)	14
Cyclopentane ($CO_2C_2H_5$, CN, OH, pyridine)	CH_3CO_2H	25°, THF, 20 hr	Cyclopentane (AcO, $CO_2C_2H_5$, pyridine) (91)	213
C_{16} macrocyclic structure (O, OH, HO, CO)	None	−30°, Toluene, 15 hr	Macrocyclic lactone (14) + dimer (4)	214

C_{14}
C_{16}

TABLE II. FORMATION OF ESTERS USING TRIPHENYLPHOSPHINE–DIETHYL AZODICARBOXYLATE (*Continued*)

Alcohol	Acid	Conditions	Esters and Yields (%)	Ref.
C$_{16}$ (*Contd.*) [steroid/macrolactone structure with OH and CO$_2$H]	None	−20°, THF, 2 hr	[lactone structure] (37) + dimer (17)	214
C$_{19}$ 4-Androsten-17β-ol-3-one	C$_6$H$_5$CO$_2$H	25°, THF, 14 hr	No reaction	67
	"	Reflux, C$_6$H$_6$, 15 min	17α-Benzoyloxyandrost-4-en-3-one (64)	58
5-Androsten-3β, 16α-diol	"	25°, THF, 14 hr	5-Androsten-3α, 16β-diol dibenzoate (75)	67
5α-Androsten-3β-ol-17-one	HCO$_2$H	"	3α-Formyloxy-5α-androsten-17-one (85)	67
C$_{21}$ [sugar structure with C$_6$H$_5$, OCH$_3$, OCH$_2$C$_6$H$_5$, OCH$_2$, HO]	C$_6$H$_5$CO$_2$H	25°, THF	[sugar ester structure with O$_2$CC$_6$H$_5$, C$_6$H$_5$, OCH$_3$, OCH$_2$, OCH$_2$C$_6$H$_5$, HO] (69)	210
[sugar structure with C$_6$H$_5$, OCH$_3$, OCH$_2$C$_6$H$_5$, OCH$_2$, HO]	p-O$_2$NC$_6$H$_4$CO$_2$H	"	[sugar ester structure with O$_2$CC$_6$H$_4$NO$_2$-p, C$_6$H$_5$, OCH$_3$, OCH$_2$, OCH$_2$C$_6$H$_5$, HO] (83)	210
C$_{22}$ 16α-Methyl-5α-pregnan-3β-ol-20-one	HCO$_2$H	25°, THF, 14 hr	3α-Formyloxy-16α-methyl-5α-pregnan-20-one (75)	67

Substrate	Reagent	Conditions	Product (% yield)	Ref.
(benzyl-protected sugar: CH₂OCH₂C₆H₅, CH₂OCH₂C₆H₅, OCH₃, HO–)	C₆H₅CO₂H	25°, THF, 1 hr	(benzoyloxy sugar: CH₂OCH₂C₆H₅, CH₂OCH₂C₆H₅, OCH₃, BzO–) (82)	14
(benzyl-protected sugar: CH₂OCH₂C₆H₅, OCH₃, CH₂OCH₂C₆H₅, HO–)	"	"	(BzO–, OCH₃, CH₂OCH₂C₆H₅, CH₂OCH₂C₆H₅) (70)	14
(alkaloid structure: N–CH₃, H, HO, HO₂C, CH₃O, OCH₃, methylenedioxy)	None	20°, THF, 90 min	(lactone structure: N–CH₃, H, H, O, CH₃O, OCH₃, OCH₃, methylenedioxy)	215
C₂₄ Methyl 3α,12α-dihydroxycholanate	HCO₂H	25°, THF, 14 hr	Methyl 3β-formyloxy-12α-hydroxycholanate (97)	67
Methyl 3α,7α,12α-trihydroxycholanate	"	"	Methyl 3β-formyloxy-7α,12α-dihydroxycholanate (75)	67
C₂₇ 5α-Cholestan-3β,6β-diol	C₆H₅CO₂H	25°, THF, 15 hr	3α-Benzoyloxy-5α-cholestan-6β-ol (92)	69
5α-Cholestan-3β-ol	HCO₂H	"	3α-Formyloxy-5α-cholestane (97)	67
	C₆H₅CH₂CO₂H	"	5α-Cholestan-3α-ol phenylacetate (90)	67
	C₆H₅CO₂H	"	5α-Cholestan-3α-ol benzoate (100)	67
5α-Cholestan-3α-ol	"	"	No reaction	67
	"	25°, C₆H₆, 15 min	3β-Benzoyloxy-5α-cholestane (73) + 2-cholestene (25)	11
Δ⁴-Cholestan-3-ol	"	25°, THF, 10 min	Δ⁴-Cholesten-3α-ol benzoate (81)	14
	"	25°, THF, 24 hr	" (81)	15
	CH₃CO₂H	"	Δ⁴-Cholesten-3α-ol acetate (47)	15
	C₂H₅CO₂H	"	Δ⁴-Cholesten-3α-ol propionate (48)	15

TABLE II. FORMATION OF ESTERS USING TRIPHENYLPHOSPHINE–DIETHYL AZODICARBOXYLATE (*Continued*)

Alcohol	Acid	Conditions	Esters and Yields (%)	Ref.
C_{27} Δ^4-Cholesten-3-ol (*Contd.*)	$C_2H_5CO_2H$	25°, THF, 14 hr	Δ^4-Cholesten-3β-ol propionate (68)	15
	CH_3CO_2H	"	Δ^4-Cholesten-3β-ol acetate (85)	15
	$C_6H_5CO_2H$	"	Δ^4-Cholesten-3β-ol benzoate (72)	15
Cholesterol	"	0°, Et_2O, 3 hr	Cholest-5-en-3α-yl benzoate (11) + cholest-5-en-3β-yl benzoate (20) + 3,5-cyclocholest-6β-yl benzoate (23) + 3,5-cyclocholest-6α-yl benzoate (14) + cholesta-3,5-diene (10) + 3,5-cyclocholest-6-ene (10)	17
C_{28} (R)-HO—H, $CH_2OC_{18}H_{37}$-n (top), $CH_2O_2CC_6H_5$ (bottom)	CH_3CO_2H	25°, Et_2O	(S)-H—OAc, $CH_2OC_{18}H_{37}$-n (top), $CH_2O_2CC_2H_5$ (bottom)	16
"	$C_6H_5CO_2H$	"	(S)-H—OBz, $CH_2OC_{18}H_{37}$-n (top), $CH_2O_2CC_6H_5$ (bottom)	16
C_{37} $(\pm)$-H—OH, $CH_2O_2CC_{15}H_{31}$-n (top), $CH_2O_2CC_{17}H_{35}$-n (bottom)	"	"	$(\pm)$-H—OBz, $CH_2O_2CC_{15}H_{31}$-n (top), $CH_2O_2CC_{17}H_{35}$-n (bottom)	16
"	Oleic acid	"	$(\pm)$-H—$O_2CC_{17}H_{33}$, CH_2—$O_2CC_{17}H_{35}$-n (top), $CH_2O_2CC_{15}H_{31}$-n, $CH_2O_2CC_{17}H_{35}$-n	16

a TBDM = *t*-butyldimethyl.

TABLE III. FORMATION OF ETHERS AND GLYCOSIDES USING TRIPHENYLPHOSPHINE–DIETHYL AZODICARBOXYLATE

	Alcohol	Nucleophilic Hydroxyl Compound	Conditions	Products and Yields (%)	Ref.
C₁	CH₃OH		25°, THF, 12 hr	(90)	90
		Estrone	25°, THF, 1–4 days	(—)	93
C₂	C₂H₅OH		25°, THF, 2 hr	(76)	216
		"	"	(61)	216
		"	"	(55)	216
C₃	i-C₃H₇OH		25°, THF, 12 hr	(84)	90

71

TABLE III. FORMATION OF ETHERS AND GLYCOSIDES USING TRIPHENYLPHOSPHINE–DIETHYL AZODICARBOXYLATE (*Continued*)

Alcohol	Nucleophilic Hydroxyl Compound	Conditions	Products and Yields (%)	Ref.
C_3 i-C_3H_7OH (*Contd.*)	[2-t-C_4H_9-4-NO_2-6-t-C_4H_9-phenol]	25°, 24 hr	[quinone with ^-O—N^+=, OC_3H_7-i, t-C_4H_9] (79)	99
CH_2=$CHCH_2OH$	"	"	[OCH_2CH=CH_2, CO_2CH_3, OH aryl] (62)	216
C_4 n-C_4H_9OH	N-Hydroxyphthalimide	—	o-$C_6H_4(CO)_2NOCH_2CH$=CH_2 (80)	87
	N-Hydroxyphthalimide	25°, THF, 24 hr	o-$C_6H_4(CO)_2NOC_4H_9$-n (91)	87
	[chloro-benzotriazol-OH]	25°, THF, 12 hr	[benzotriazole-OC_6H_5-n, Cl] (80)	90
$CH_3CH_2CHOHCH_3$	N-Hydroxyphthalimide	25°, THF, 24 hr	o-$C_6H_4(CO)_2NOCH(CH_3)CH_2CH_3$ (76)	87
$CH_3CHOHCH_2CH_2OH$	[2-t-C_4H_9-4-NO_2-6-t-C_4H_9-phenol]	25°, 24 hr	[quinone with ^-O—N^+=, $OCH_2CH_2CHOHCH_3$, t-C_4H_9] (69)	99

C_5	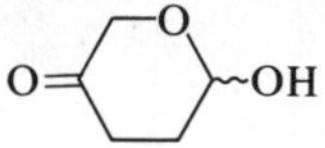	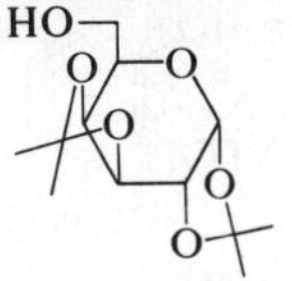	25°, THF, 12 hr, catalyst $HgBr_2$	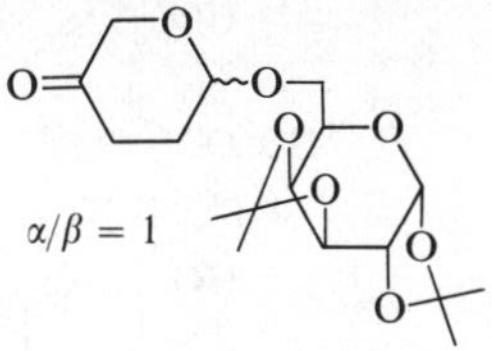 $\alpha/\beta = 1$ (36)	98
	$CH_3CHOHCH_2CHOHCH_3$	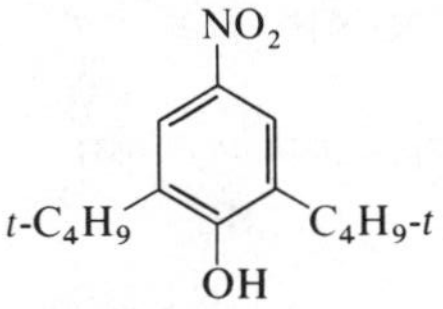	25°, 24 hr	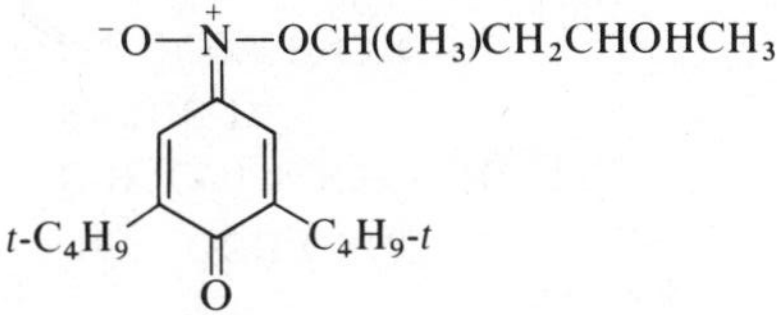 $^-O-\overset{+}{N}-OCH(CH_3)CH_2CHOHCH_3$ t-C_4H_9 C_4H_9-t O (68)	99
	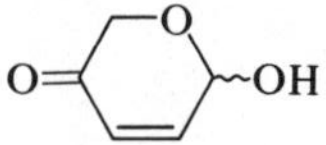	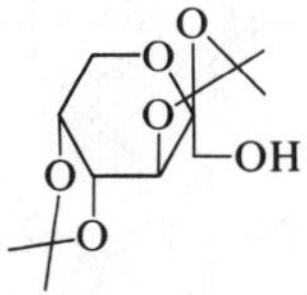	25°, THF, 12 hr, catalyst $HgBr_2$	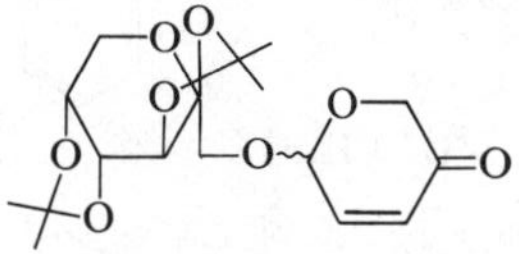 (17) $\alpha/\beta = 1$	98
		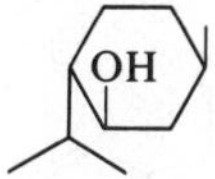	25°, THF, 12 hr,	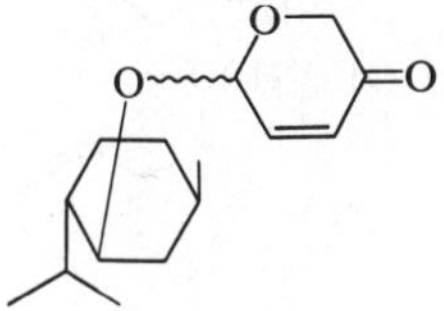 (48) $\alpha/\beta = 1$	98
		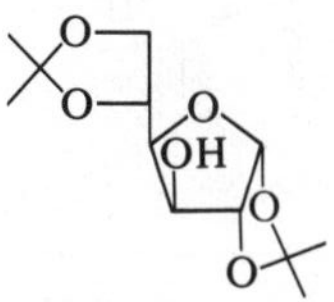	"	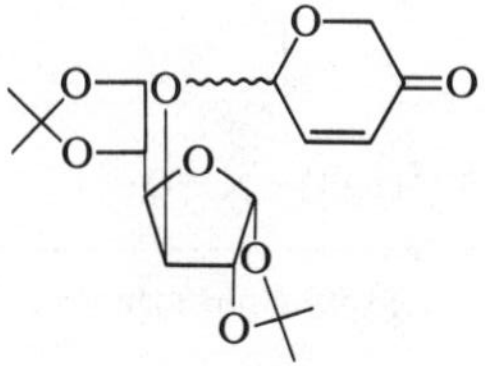 (32) $\alpha/\beta = 1$	98

TABLE III. FORMATION OF ETHERS AND GLYCOSIDES USING TRIPHENYLPHOSPHINE–DIETHYL AZODICARBOXYLATE (*Continued*)

Alcohol	Nucleophilic Hydroxyl Compound	Conditions	Products and Yields (%)	Ref.
C$_6$ HOCH$_2$(CH$_2$)$_4$CH$_2$OH	t-C$_4$H$_9$, C$_4$H$_9$-t, NO$_2$, OH (2,6-di-tert-butyl-4-nitrophenol)	25°, 24 hr	$^-$O—N$^+$—O(CH$_2$)$_5$CH$_2$OH quinone with t-C$_4$H$_9$ and C$_4$H$_9$-t (68)	99
C$_7$ C$_6$H$_5$CH$_2$OH	benzotriazinone N–OH (N=N–C–CN)	25°, THF, 12 hr	N–OCH$_2$C$_6$H$_5$ (68)	90
	Cl-substituted benzotriazole N–OH	,,	N–OCH$_2$C$_6$H$_5$ (85)	90
	benzotriazole N–OH	,,	N–OCH$_2$C$_6$H$_5$ (80)	90
	N-Hydroxyphthalimide	25°, THF, 24 hr	o-C$_6$H$_4$(CO)$_2$NOCH$_2$C$_6$H$_5$ (92)	87
	OH, CO$_2$CH$_2$COC$_6$H$_5$, OH	25°, THF, 2 hr	OCH$_2$C$_6$H$_5$, CO$_2$CH$_2$COC$_6$H$_5$, OH (70)	216
	OH, CO$_2$CH$_3$, OH	,,	OCH$_2$C$_6$H$_5$, CO$_2$CH$_3$, OH (73)	216

C$_8$ n-C$_6$H$_{13}$CHOHCH$_3$	(3,5-di-t-C$_4$H$_9$-4-OH-nitrophenol)	25°, 24 hr	$-$OCH(CH$_3$)C$_6$H$_{13}$-n (76)	99
C$_6$H$_5$CH$_2$CH$_2$OH	(3,5-di-t-C$_4$H$_9$-4-OH-nitrophenol)	"	$-$OCH$_2$CH$_2$C$_6$H$_5$ (83)	99
C$_6$H$_5$CHOHCH$_3$	(6-Cl-benzotriazol-1-ol)	25°, THF, 12 hr	OCH(CH$_3$)C$_6$H$_5$ (79) + α anomer (11)	90
C$_{12}$	N-Hydroxyphthalimide	25°, THF, 24 hr	ON(CO)$_2$C$_6$H$_4$-o (78)	88
	N-Hydroxyphthalimide	"	ON(CO)$_2$C$_6$H$_4$-o (54)	88
	N-Hydroxyphthalimide	"	ON(CO)$_2$C$_6$H$_4$-o (65)	88

TABLE III. FORMATION OF ETHERS AND GLYCOSIDES USING TRIPHENYLPHOSPHINE–DIETHYL AZODICARBOXYLATE (*Continued*)

Alcohol	Nucleophilic Hydroxyl Compound	Conditions	Products and Yields (%)	Ref.
C₁₂ (*Contd.*)	Cyclohexanol	Reflux, THF, 3 hr, catalyst HgBr₂	OC₆H₁₁ (72)	97
		25°, THF, 12 hr, catalyst HgBr₂	(21) α/β = 2.6	97
		25°, THF, 12 hr	+ α isomer (—) (66)	90
		"	+ α isomer (—) (90)	90

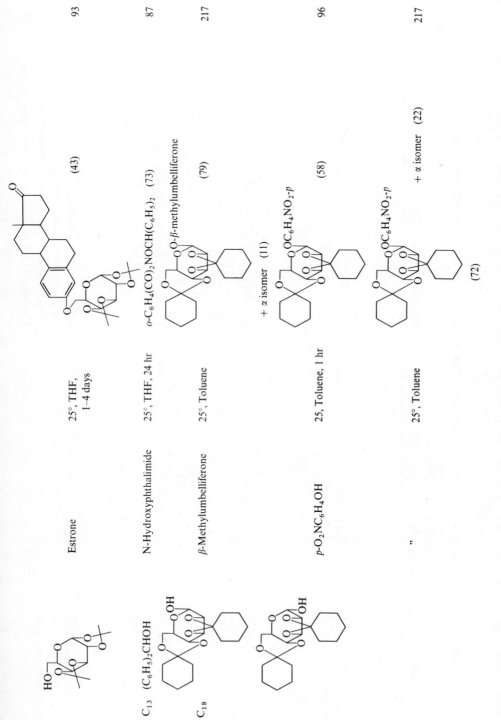

	Estrone	25°, THF, 1–4 days	(43)	93
C$_{13}$ (C$_6$H$_5$)$_2$CHOH	N-Hydroxyphthalimide	25°, THF, 24 hr	o-C$_6$H$_4$(CO)$_2$NOCH(C$_6$H$_5$)$_2$ (73)	87
C$_{18}$	β-Methylumbelliferone	25°, Toluene	O-β-methylumbelliferone (79) + α isomer (11)	217
	p-O$_2$NC$_6$H$_4$OH	25, Toluene, 1 hr	OC$_6$H$_4$NO$_2$-p (58)	96
	"	25°, Toluene	OC$_6$H$_4$NO$_2$-p + α isomer (22) (72)	217

77

TABLE III. FORMATION OF ETHERS AND GLYCOSIDES USING TRIPHENYLPHOSPHINE–DIETHYL AZODICARBOXYLATE (*Continued*)

Alcohol	Nucleophilic Hydroxyl Compound	Conditions	Products and Yields (%)	Ref.
C_{18} (*Contd.*)	m-$O_2NC_6H_4OH$	25°, Toluene	$OC_6H_4NO_2$-m (81) + α isomer (8)	217
	C_6H_5OH	"	OC_6H_5 (82) + α isomer (4)	217
	$2,4$-$(O_2N)_2C_6H_3OH$	25°, Toluene	$OC_6H_3(NO_2)_2$-$2,4$ (53) + α isomer (37)	217
C_{21} Hydrocortisone	N-Hydroxyphthalimide	25°, THF, 24 hr	21-Phthalimidoxyhydrocortisone (86)	89
Epi-Hydrocortisone	"	"	21-Phthalimidoxy-*epi*-hydrocortisone (80)	89
C_{24} Desoxycholic acid	"	"	3β-Phthalimidoxy-12α-hydroxy-24-O-cholanoyl-N-hydroxyphthalimide (75)	89

	Substrate	Reagent	Conditions	Product (Yield %)	Ref.
C$_{27}$	Cholestan-5α, 3β-ol	C$_6$H$_5$OH	25°, THF, 1–4 days	(80) —OC$_6$H$_5$	93
	Cholestan-5α,3β-ol	p-BrC$_6$H$_4$OH	"	(65) —OC$_6$H$_4$Br-p	93
	Cholesterol	C$_6$H$_5$OH	"	(65) C$_6$H$_5$O—	93
		p-BrC$_6$H$_4$OH	"	(60) p-BrC$_6$H$_4$O—	93
	Cholestan-3β-ol	N-Hydroxyphthalimide	25°, THF, 24 hr	3α-Phthalimidoxycholestane (96)	89
	Cholest-5-en-3β-ol	"	"	3α-Phthalimidoxycholest-5-ene (50)	89
C$_{34}$	(sugar) C$_7$H$_7$O, OC$_7$H$_7$, C$_7$H$_7$O, OH, OC$_7$H$_7$	Cyclohexanol	25°, THF, 12 hr, catalyst HgBr$_2$	(80)^a OC$_6$H$_{11}$, OC$_7$H$_7$, OC$_7$H$_7$, C$_7$H$_7$O, OC$_7$H$_7$	97

79

TABLE III. FORMATION OF ETHERS AND GLYCOSIDES USING TRIPHENYLPHOSPHINE–DIETHYL AZODICARBOXYLATE (*Continued*)

Alcohol	Nucleophilic Hydroxyl Compound	Conditions	Products and Yields (%)	Ref.
C$_{34}$ (*Contd.*)		25°, THF, 24 hr, catalyst HgBr$_2$	(80)^a	97
		"	(38)^a	97
	None	25°, THF, 48 hr, catalyst HgI$_2$, phosphine CH$_3$P(C$_6$H$_5$)$_2$	Octa-O-benzyl-α,α′-trehalose (40)	97

^{a} C$_7$H$_7$ = C$_6$H$_5$CH$_2$

TABLE IV. GLYCOL CYCLIZATIONS USING TRIPHENYLPHOSPHINE–DIETHYL DIAZODICARBOXYLATE

	Glycol	Conditions	Products and Yields (%)	Ref.
C_3	$CH_3CHOHCH_2OH$	0°, Toluene, 10 min	2-Methyloxirane (100)	94
	$HOCH_2CH_2CH_2OH$	20°, CDCl$_3$, 10 min	Oxetane (98)	94
C_4	$CH_3CHOHCHOHCH_3$	20°, THF, 10 min	2,3-Dimethyloxirane (100)	94
	$HOCH_2(CH_2)_2CH_2OH$	5°, Tetralin, 10 min	Tetrahydrofuran (100)	94
	$O(CH_2CH_2OH)_2$	25°, Tetralin, 10 min	1,4-Dioxane (96)	94
	$(CH_3)_2CH(NH_2)CH_2OH$	20°, Tetralin, 10 min	2,2-Dimethylaziridine (100)	94
C_5	$CH_2(CH_2CH_2OH)_2$	5°, CDCl$_3$, 10 min	Tetrahydropyran (95)	94
C_6	$HOCH_2(CH_2)_4CH_2OH$	''	Oxepane (94)	94
	trans-1,2-Cyclohexanediol	20°, THF, 10 min	Cyclohexene oxide (100)	94
		25°, Et$_2$O, 12 hr	(48)	107
		''	(68) (stereochemistry not specified)	107
		Reflux, C$_6$H$_6$	(75)	218
C_7	Methyl-α-D-galactopyranoside	Reflux, dioxane/Py	(80)	210

81

TABLE IV. Glycol Cyclizations Using Triphenylphosphine–Diethyl Diazodicarboxylate (*Continued*)

Glycol	Conditions	Products and Yields (%)	Ref.
C_7 Methyl-α-D-glucopyranoside (*Cont.*)	Reflux, dioxane/Py	(structure) HO, H, OCH$_3$, OH (85)	210
C_8 $C_6H_5CHOHCH_2OH$	21°, CDCl$_3$, 10 min; 25°, Et$_2$O, 12 hr	2-Phenyloxirane (100); " (27)	94; 107
C_{10} (structure) Adenyl-1, HO, HO, OH	70°, Dioxane, 40 min	(structure) Adenyl-1, HO (90)	108
C_{10} (structure) Adenyl-1, HO, OH, OH	70°, Dioxane, 6 hr	(structure) Adenyl-1, HO (60)	108
C_{12} (uracil structure) HN, O, HO		(structure) (80)	54

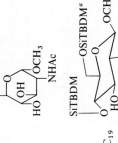

C_{13}

Reflux, C_6H_6, 30 min

(46) + (23)

64

"

(56) + (10)

64

C_{15}

0°, C_6H_6, 30 min

(80)

64

60°, C_6H_6, 30 min

(80)

64

C_{19}

Reflux, C_6H_6, 3 min

(71)

64

83

TABLE IV. GLYCOL CYCLIZATIONS USING TRIPHENYLPHOSPHINE–DIETHYL DIAZODICARBOXYLATE (*Continued*)

Glycol	Conditions	Products and Yields (%)	Ref.
C_{19} (*Contd.*) OSiTBDM[a] OCH$_3$ HO HO OSiTBDM	Reflux, C$_6$H$_6$, 30 min	OSiTBDM[a] OCH$_3$ OSiTBDM (54)	64
OSiTBDM[a] OCH$_3$ HO HO OSiTBDM	"	OSiTBDM OSiTBDM[a] OCH$_3$ OSiTBDM (68)	64
C_{27} 2β,3α-Dihydroxy-5α-cholestane	80°, C$_6$H$_6$, 10 min	2β,3β-Epoxy-5α-cholestane (—) + 2α,3β-Epoxy-5α-cholestane (—)	58

[a] TBDM = *t*-butyldimethyl.

84

TABLE V. PREPARATION OF ALKYL IODIDES USING TRIPHENYL PHOSPHITE–METHIODIDE

	Alcohol	Conditions	Products and Yields (%)	Ref.
C_3	$CH_3CHOHCH_3$	25°	CH_3CHICH_3 (75)	115
	$CH_2{=}CHCH_2OH$	0°, 30 min	$CH_2{=}CHCH_2I$ (84)	115
	$HC{\equiv}CCH_2OH$	0°, CH_2Cl_2, 3 hr	$HC{\equiv}CCH_2I + CH_2{=}C{=}CHI$ (65) (7)	116
		100°, DMF, 30 min	$HC{\equiv}CCH_2I + CH_2{=}C{=}CHI$ (11) (43)	116
	$HO(CH_2)_3OH$	—	$I(CH_2)_3I$ (95)	219, 115
C_4	$n\text{-}C_4H_9OH$	25°, 30 min	$n\text{-}C_4H_9I$ (72)	115
	$C_2H_5CHOHCH_3$	25°, 2 hr	$C_2H_5CHICH_3$ (60)	115
	$t\text{-}C_4H_9OH$	50°, 1 hr	$(CH_3)_3Cl$ (76)	115
	$CH_3CH{=}CHCH_2OH$	0°, 30 min	$CH_3CH{=}CHCH_2I$ (83)	115
	$HC{\equiv}CCHOHCH_3$	100°, DMF, 30 min	$ICH{=}C{=}CHCH_3$ (53)	116
	$(CH_3)_3CCH_2OH$	70°	$(CH_3)_3CCH_2I$ (74) $+ C_2H_5(CH_3)_2Cl$ (6)	115
			" (55)	127
C_5	$CH_3CHOHCO_2C_2H_5$	25°, 12 hr	$CH_3CHICO_2C_2H_5$ (92)	115
	$HC{\equiv}CCHOHCH{=}CH_2$	40°, 15 min	$ICH_2CH{=}CHC{\equiv}CH$ (46)	116
	$HC{\equiv}CCHOHC{\equiv}CH$	100°, DMF, 30 min	$ICH{=}C{=}CHC{\equiv}CH$ (42)	116
	$HOCH_2C(CH_3)_2CH_2OH$	—	$ICH_2C(CH_3)_2CH_2OH$ (—)	114a
C_6	Cyclohexanol	25°, 18 hr	Iodocyclohexane (75)	115
	$HC{\equiv}CCHOH(CH_2)_2CH_3$	80°, DMF, 2 hr	$ICH{=}C{=}CH(CH_2)_2CH_3$ (60)	116
	$HC{\equiv}CCHOHCH(CH_3)_2$	15°, DMF, 7 days	$ICH{=}CHCH{=}C(CH_3)_2 + ICH{=}C{=}CHCH(CH_3)_2$ (18) (25)	116
	bicyclic furofuran diol (HO–, –OH)	25°, C_6H_6, 24 hr	bicyclic furofuran diiodide (I–, –I) (100)	105

85

TABLE V. Preparation of Alkyl Iodides Using Triphenyl Phosphite–Methiodide (*Continued*)

	Alcohol	Conditions	Products and Yields (%)	Ref.
C$_7$	C$_6$H$_5$CH$_2$OH	25°, 1 hr	C$_6$H$_5$CH$_2$I (95)	115
	HC≡CCHOH(CH$_2$)$_3$CH$_3$	Not specified	ICH=C=CH(CH$_2$)$_3$CH$_3$ (20)	116
	HC≡CCHOHC(CH$_3$)$_3$	80°, DMF, 2 hr	ICH=C=CHC(CH$_3$)$_3$ (67)	116
C$_8$	C$_6$H$_{13}$CHOHCH$_3$	80°, 20 hr	C$_6$H$_{13}$CHICH$_3$ (80)	115
	C$_6$H$_5$CH$_2$OHCH$_3$	25°, 1 hr	C$_6$H$_5$CHICH$_3$ (95)	115
	C$_6$H$_5$CH$_2$CH$_2$OH	"	C$_6$H$_5$CH$_2$CH$_2$I (92)	115
	HC≡CCHOH(CH$_2$)$_4$CH$_3$	—	ICH=C=CH(CH$_2$)$_4$CH$_3$ (43)	116
C$_9$	HC≡CCHOHC≡CC(CH$_3$)$_3$	Distil.	ICH=C=CHC≡CC(CH$_3$)$_3$ (41)	116
	HC≡CCHOHC$_6$H$_5$	25°, DMF	ICH=C=CHC$_6$H$_5$ (21)	116
		25°, DMF, 12 hr	(84)	122,121
		25°, Py, 1 hr	(43)	120

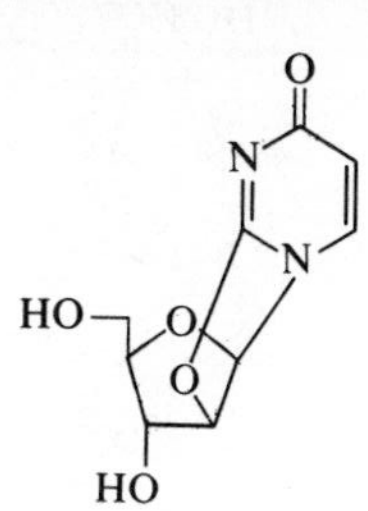

$(C_6H_5O)P(O)CH_3$

25°, DMF, 15 min
+ $(i\text{-}C_3H_7)_2NC_2H_5$

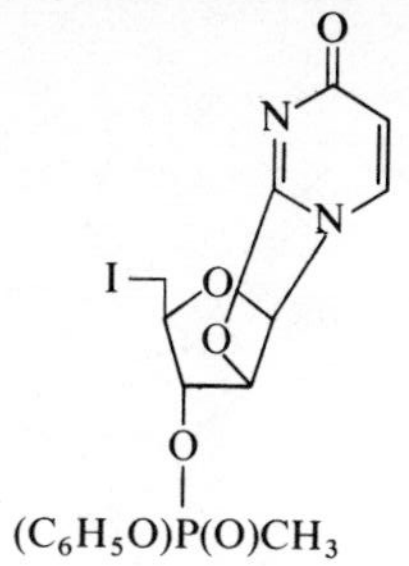

(42)

25°, DMF, 10 min

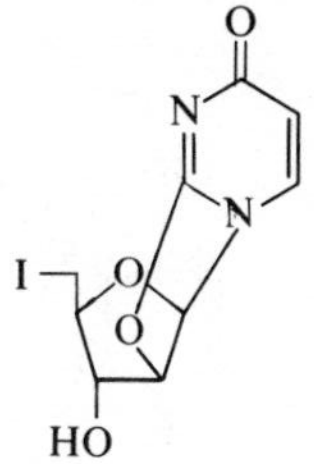

(31)

25°, DMF, 1 hr

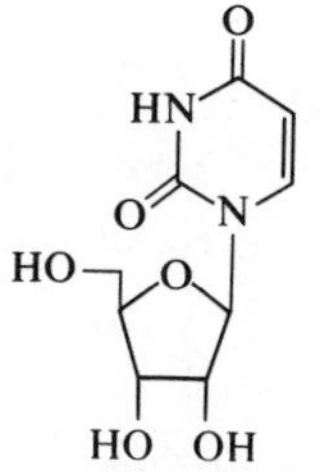

(65)

25°, DMF, 3 min

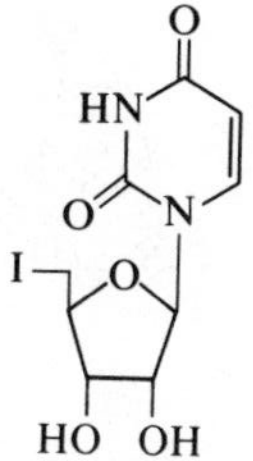

(82)

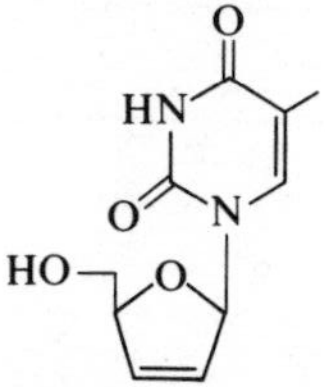

C_{10}

120

120

120

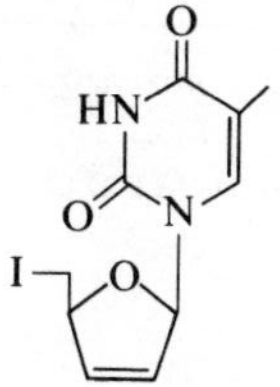

125

TABLE V. PREPARATION OF ALKYL IODIDES USING TRIPHENYL PHOSPHITE–METHIODIDE (*Continued*)

Alcohol	Conditions	Products and Yields (%)	Ref.
C$_{1}$° (*Contd.*)	25°, DMF, 10 min	(63)	122, 120
	25°, DMF, 12 hr	(76)	121
	25°, DMF, 5 min	(50)	120

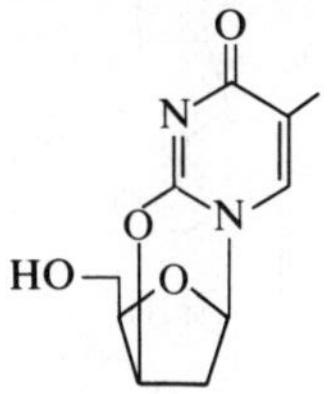

(58)

OCH$_3$

CH$_3$

(−)

HN

O

O

N

I

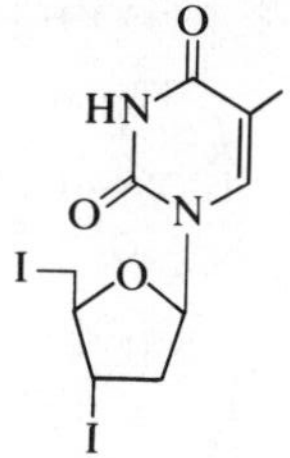

O

+

CH$_3$

I

H OCH$_3$

I

OCH$_3$

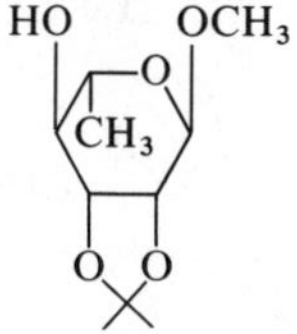

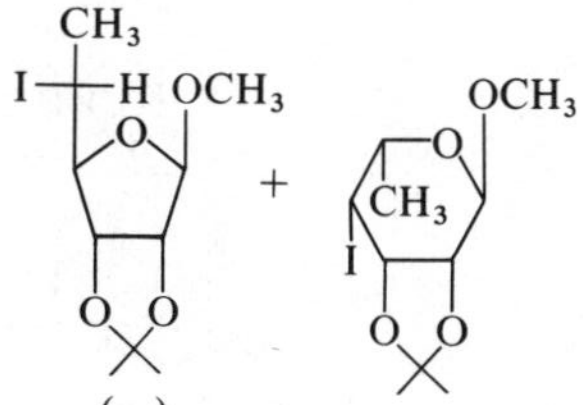

(−)

(60)

OCH$_3$

I

OH

OCH$_3$

OCH$_3$

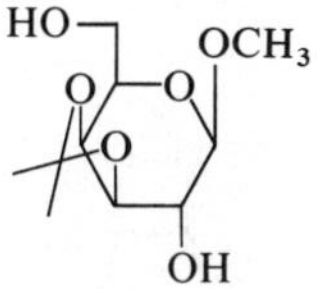

OCH$_3$

CH$_3$O

I

(31)

25°, DMF

—

25°, C$_6$H$_6$, 2 days

55°, C$_6$H$_6$, 12 hr

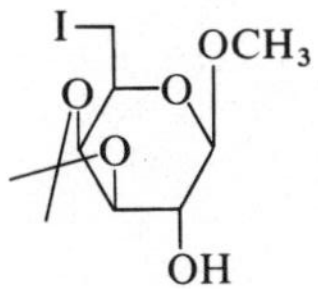

O

N

N

CH$_3$

O

HO

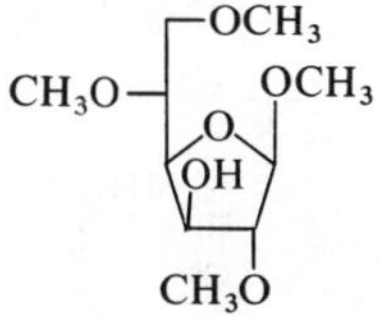

OCH$_3$

HO

CH$_3$

O

O

OCH$_3$

HO

OH

OCH$_3$

OCH$_3$

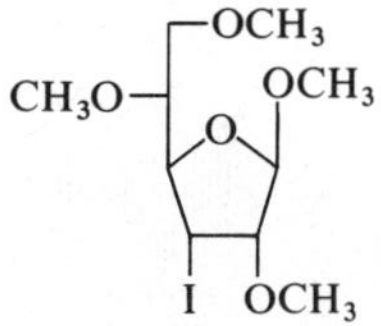

CH$_3$O

OH

OCH$_3$

CH$_3$O

TABLE V. PREPARATION OF ALKYL IODIDES USING TRIPHENYL PHOSPHITE–METHIODIDE (*Continued*)

Alcohol	Conditions	Products and Yields (%)	Ref.
C₁₁	25°, DMF, 30 min	(48)	120
	25°, DMF, 18 hr	(50)	122
	25°, DMF, 20 hr	(46)	121

120

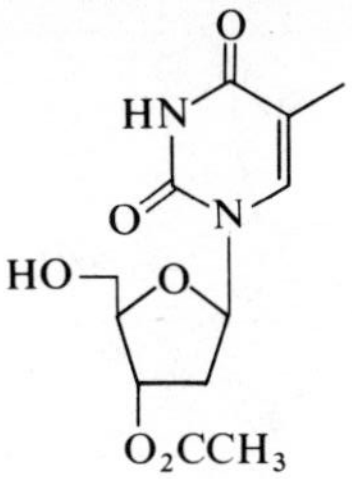

(88)

25°, DMF, 1 hr

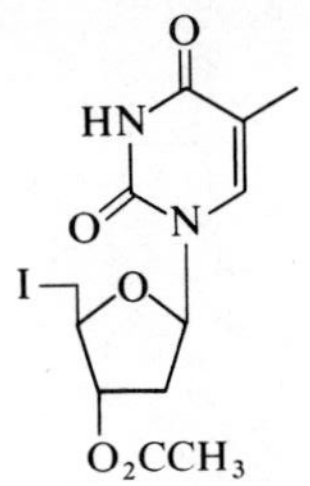

121

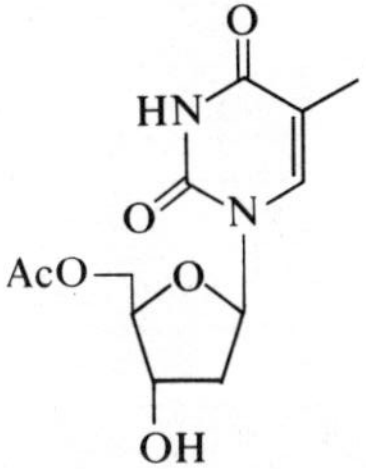

(50)

25°, DMF, 24 hr

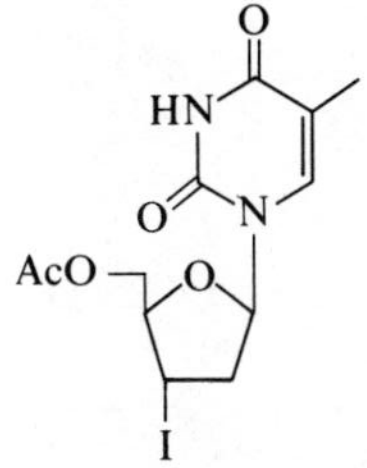

122, 120

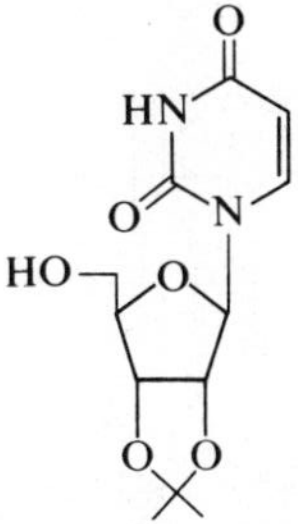

(96)

25°, DMF, 15 min

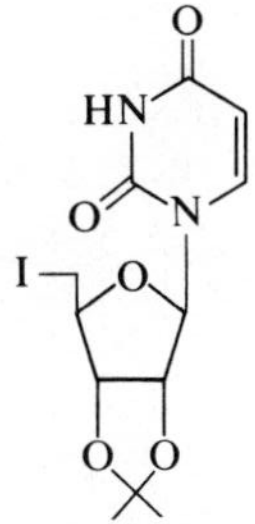

91

TABLE V. Preparation of Alkyl Iodides Using Triphenyl Phosphite–Methiodide (*Continued*)

Alcohol	Conditions	Products and Yields (%)	Ref.
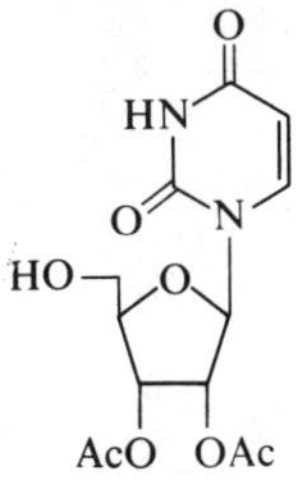	25°, DMF, 2 hr	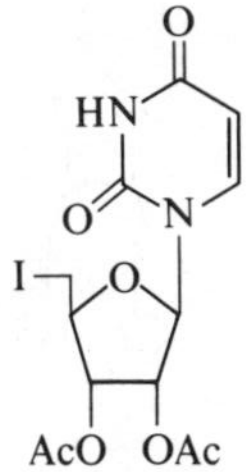 (84)	123, 120
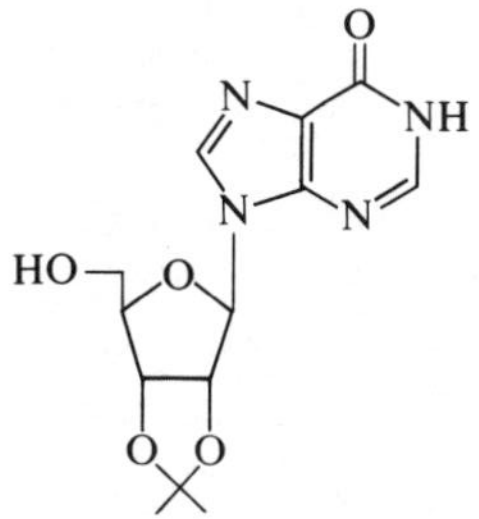	−78°, THF, 3 hr	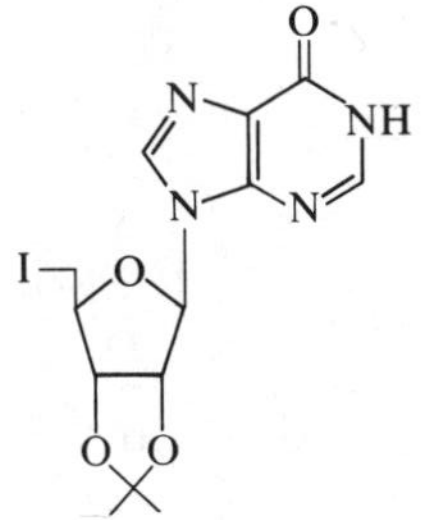 (87)	126
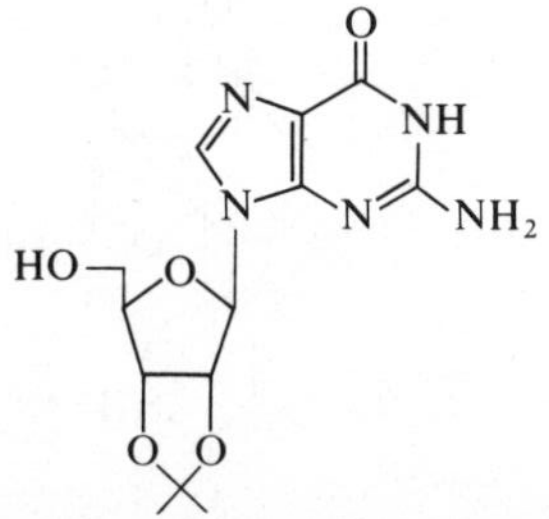	−78°, THF, 4 hr	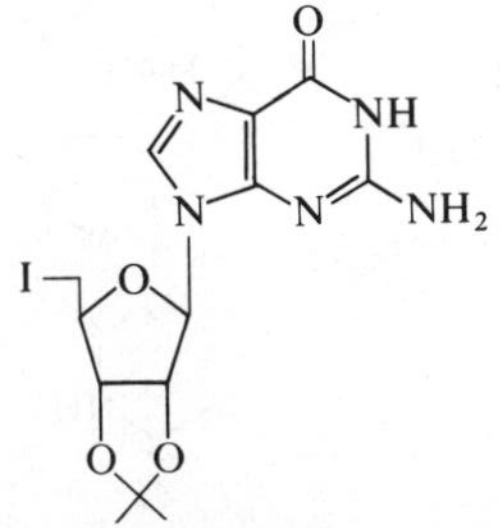 (58)	—

C₁₃

92

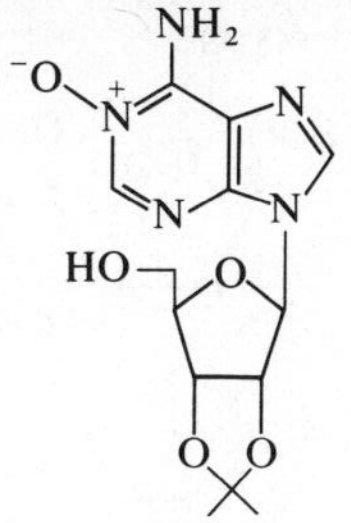

220

122, 120

122, 120

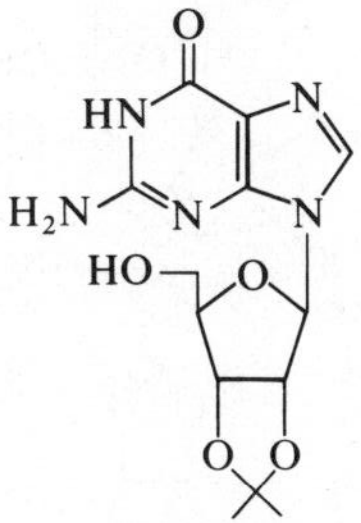

(74)

(87)

(96)

25°, DMF, 24 hr

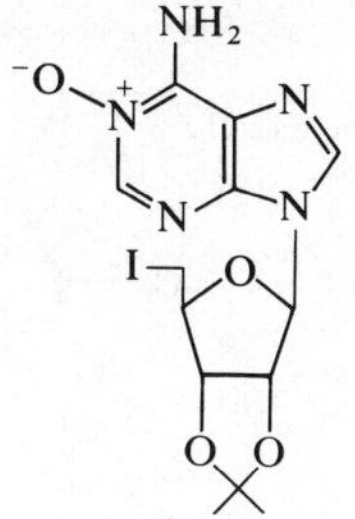

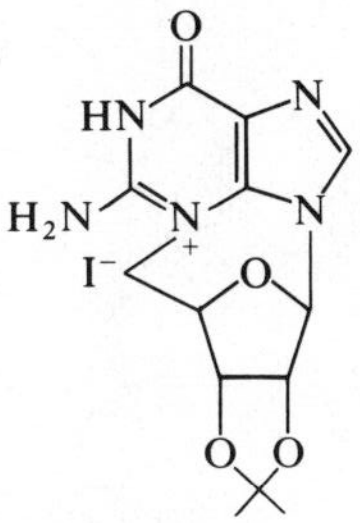

25°, DMF, 5 min

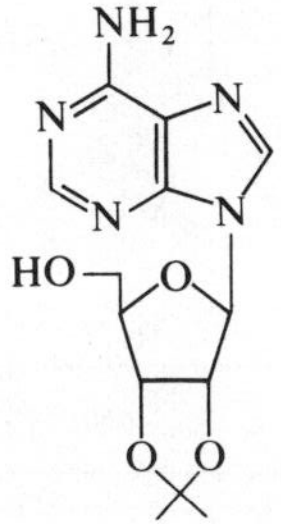

25°, DMF, 12 hr

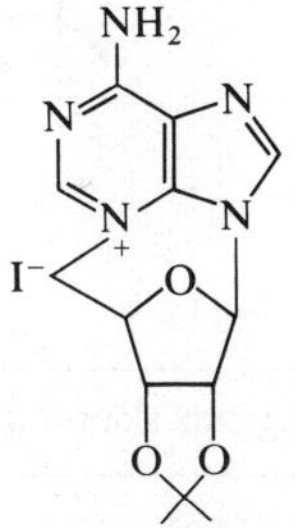

TABLE V. PREPARATION OF ALKYL IODIDES USING TRIPHENYL PHOSPHITE–METHIODIDE (*Continued*)

Alcohol	Products and Yields (%)	Conditions	Ref.
C₁₄	(91)	25°, DMF, 3 hr	125
C₁₅	(84)	−78°, THF, 1.5 hr	126
	(85)	50°, C₆H₆, 30 hr	118

C$_{16}$

118   (58)   50°, C$_6$H$_6$, 10 hr

105   (80)   55°, C$_6$H$_6$, 7 hr

125   (15)   25°, DMF, 10 min

120   (89)   25°, DMF, 12 hr

TABLE V. Preparation of Alkyl Iodides Using Triphenyl Phosphite–Methiodide (*Continued*)

Alcohol	Conditions	Products and Yields (%)	Ref.
C$_{17}$	25°, DMF, 24 hr	(85)	122, 121
	25°, DMF, 15 min	(53)	121
C$_{19}$	50°, C$_6$H$_6$, 24 hr	(—) + (—)	105

C$_{20}$ (OH, OH, C$_6$H$_5$ structure)	25°, C$_6$H$_6$, 80 hr	(I, I, C$_6$H$_5$ structure) (60)	105
C$_{23}$ (C$_6$H$_5$CO$_2$, HO, O$_2$CC$_6$H$_5$ structure)	50°, C$_6$H$_6$, 48 hr	(I, I, O$_2$CC$_6$H$_5$ structure) (90)	105
C$_{27}$ Cholesterol	50°, Neat, 1 hr	Cholesteryl iodide (30)	115
Cholestanol	25°, DMF, 2 hr	3-α-Iodocholestane (57)	120
C$_{29}$ (C$_6$H$_5$)$_3$CO (thymidine structure, OH)	25°, DMF, 24 hr	(I thymidine structure) (67)	121
C$_{35}$ (OH, C$_6$H$_5$, CH(SCH$_2$C$_6$H$_5$)$_2$ structure)	50°, C$_6$H$_6$, 30 hr	(I, C$_6$H$_5$, CH(SCH$_2$C$_6$H$_5$)$_2$ structure) (98)	105

TABLE VI. PREPARATION OF ALKYL HALIDES USING TRIPHENYL PHOSPHITE–HALOGEN

	Alcohol	Halogen	Conditions	Product and Yield (%)	Ref.
C_3	$CH_2=CHCH_2OH$	Br_2	$10°$	$CH_2=CHCH_2Br$ (75)	134
	$HO(CH_2)_3OH$	Cl_2	$0°$	$Cl(CH_2)_3Cl$ (63)	134
		Br_2	$0°$	$Br(CH_2)_3Br$ (92)	134
C_4	n-C_4H_9OH	I_2	$25°$	n-C_4H_9I (87)	134
	t-C_4H_9OH	"	$25°$, 30 min	t-C_4H_9I (83)	134
	n-C_4H_9OH	Br_2	$0°$	n-C_4H_9Br (76)	134
		Cl_2	$80°$	n-C_4H_9Cl (76)	134
C_5	$CH_3CHOHCO_2C_2H_5$	"	$0°$	$CH_3CHClCO_2C_2H_5$ (79)	134
		Br_2	$0°$	$CH_3CHBrCO_2C_2H_5$ (82)	134
C_6	Cyclohexanol	Cl_2	$25°$, 12 hr	Chlorocyclohexane (34)	134
C_8	$C_6H_5CH_2CH_2OH$	"	$0°$	$C_6H_5CH_2CH_2Cl$ (78)	134
		Br_2	$0°$	$C_6H_5CH_2CH_2Br$ (82)	134
C_{12}		"	1. $25°$ (C_6H_6), 48 hr 2. H_2SO_4	6-Bromo-6-desoxy-D-glucose (16)	105
		"	1. $25°$ (C_6H_6), 48 hr 2. H_2SO_4	6-Bromo-6-desoxy-D-galactose (20)	105
C_{27}	Cholesterol	"	$60°$, 15 min	Cholesteryl bromide (—)	134

98

TABLE VII. PREPARATION OF ALKYL HALIDES USING TRIPHENYLPHOSPHINE–HALOGEN

	Alcohol	Reagent	Conditions	Product and Yield (%)	Ref.
C$_2$	C$_2$H$_5$OH	(C$_6$H$_5$)$_3$PBr$_2$	80°, CCl$_4$, 20 min	C$_2$H$_5$Br (30)	140
C$_4$	n-C$_4$H$_9$OH	"	55°, DMF	n-C$_4$H$_9$Br (91)	139
C$_5$	(CH$_3$)$_2$CHCHOHCH$_3$	"	40°, C$_6$H$_6$, 1 hr	(CH$_3$)$_2$CHCHBrCH$_3$ (26)	144
	(CH$_3$)$_3$CCH$_2$OH	"	84°, DMF	(CH$_3$)$_3$CCH$_2$Br (91)	139
		(C$_6$H$_5$)$_3$PCl$_2$	Reflux, DMF	(CH$_3$)$_3$CCH$_2$Cl (92)	139
	CH$_3$CHOH(CH$_2$)$_3$OH	(C$_6$H$_5$)$_3$PBr$_2$	−5°, DMF, 36 hr	CH$_3$CH(O$_2$CH)(CH$_2$)$_3$Br (50)	148
C$_6$	n-C$_6$H$_{13}$OH	"	80°, CCl$_4$, 20 min	n-C$_6$H$_{13}$Br (55)	140
	p-Chlorophenol	"	200°, neat	p-Chlorobromobenzene (90)	139
	(2-methyl-3-hydroxycyclopent-2-enone)	"	25°, C$_6$H$_6$, 4 hr	(2-methyl-3-bromocyclopent-2-enone) (93)	147
	(2-methyl-3-hydroxycyclopent-2-enone)	(C$_6$H$_5$)$_3$PI$_2$	25°, C$_6$H$_6$, 24 hr	(2-methyl-3-iodocyclopent-2-enone) (71)	147
	(3-hydroxycyclohex-2-enone)	(C$_6$H$_5$)$_3$PCl$_2$	25°, C$_6$H$_6$, 1 hr	(3-chlorocyclohex-2-enone) (91)	147
		(C$_6$H$_5$)$_3$PBr$_2$	25°, C$_6$H$_6$, 4 hr	(3-bromocyclohex-2-enone) (97)	147

99

TABLE VII. PREPARATION OF ALKYL HALIDES USING TRIPHENYLPHOSPHINE–HALOGEN (*Continued*)

Alcohol	Reagent	Conditions	Product and Yield (%)	Ref.
C₆ (*Contd.*) 3-hydroxycyclohex-2-enone	(C₆H₅)₃PI₂	25°, CH₃CN, 4 days	3-iodocyclohex-2-enone (72)	147
2-methyl-3-hydroxycyclopent-2-enone	(C₆H₅)₃PCl₂	25°, C₆H₆, 4 hr	2-methyl-3-chlorocyclopent-2-enone (92)	147
C₇ C₆H₅CH₂OH	(C₆H₅)₃PBr₂ (C₆H₅)₃PBrCN	80°, CCl₄, 20 min 80°, C₆H₆, 20 min	C₆H₅CH₂Br (76) " (70)	140 140
methyl glycoside (sugar)	1. (C₆H₅)₃P/I₂/imidazole 2. Ac₂O	Toluene	iodo triacetyl methyl glycoside (80)	103
methyl glycoside (sugar)	1. (C₆H₅)₃P/I₂/imidazole 2. Ac₂O	Toluene	iodo triacetyl methyl glycoside (70)	103
2-methyl-3-hydroxycyclohex-2-enone	(C₆H₅)₃PCl₂	25°, C₆H₆, 4 hr	2-methyl-3-chlorocyclohex-2-enone (97)	147
	(C₆H₅)₃PBr₂	"	2-methyl-3-bromocyclohex-2-enone (96)	147

100

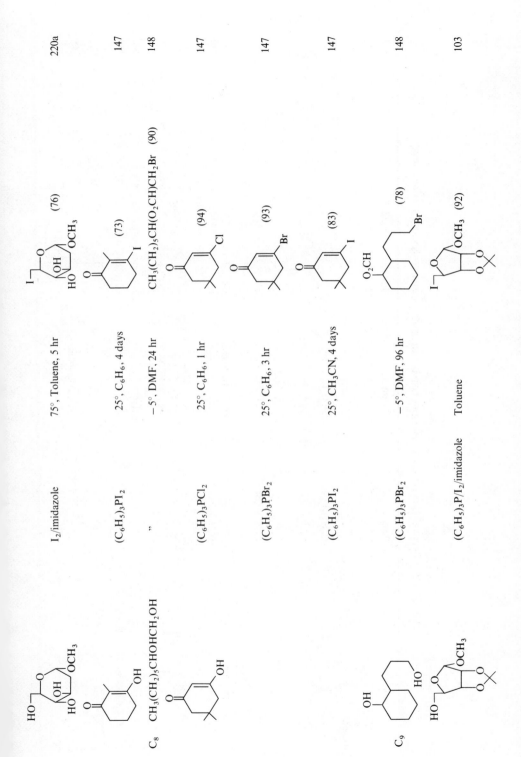

	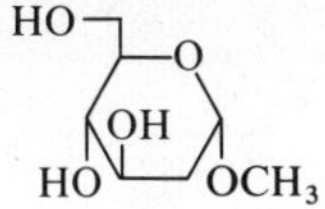	I_2/imidazole	75°, Toluene, 5 hr	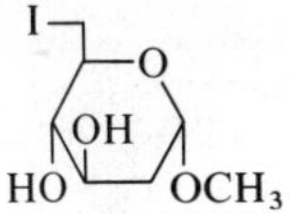 (76)	220a
	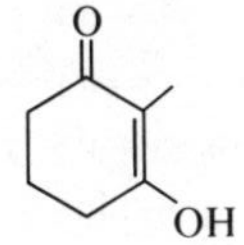	$(C_6H_5)_3PI_2$	25°, C_6H_6, 4 days	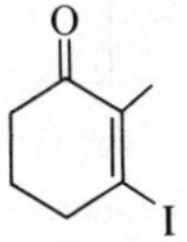 (73)	147
C_8	$CH_3(CH_2)_5CHOHCH_2OH$	"	−5°, DMF, 24 hr	$CH_3(CH_2)_5CH(O_2CH)CH_2Br$ (90)	148
	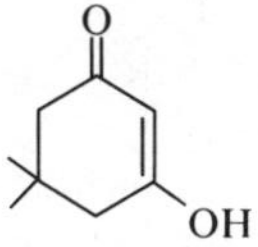	$(C_6H_5)_3PCl_2$	25°, C_6H_6, 1 hr	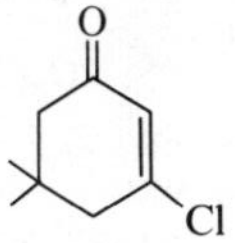 (94)	147
		$(C_6H_5)_3PBr_2$	25°, C_6H_6, 3 hr	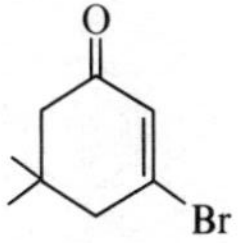 (93)	147
		$(C_6H_5)_3PI_2$	25°, CH_3CN, 4 days	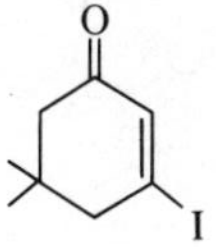 (83)	147
C_9	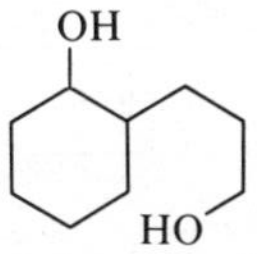	$(C_6H_5)_3PBr_2$	−5°, DMF, 96 hr	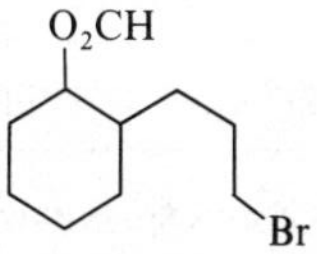 (78)	148
	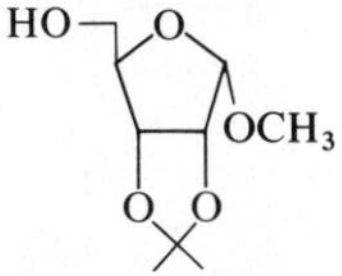	$(C_6H_5)_3P/I_2$/imidazole	Toluene	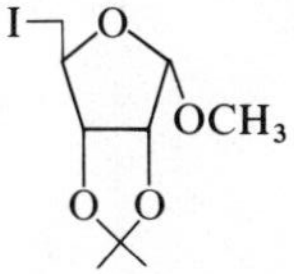 (92)	103

TABLE VII. Preparation of Alkyl Halides Using Triphenylphosphine–Halogen (*Continued*)

Alcohol	Reagent	Conditions	Product and Yield (%)	Ref.
C_9 (*Contd.*)	$(C_6H_5)_3PCl_2$	25°, C_6H_6, 7 hr	(90)	147
	$(C_6H_5)_3PBr_2$	25°, C_6H_6, 4 hr	(91)	147
C_{10} n-$C_{10}H_{21}OH$	"	80°, CCl_4, 20 min	n-$C_{10}H_{21}Br$ (73)	140
l-Menthol	$(C_6H_5)_3PCl_2$	—	*neo*-Menthyl chloride (93)	140
	$(C_6H_5)_3PBr_2$	—	*neo*-Menthyl bromide (45)	140
	"	−5°, DMF, 36 hr	(66)	148
β-Naphthol	"	CH_3CN	2-Bromonaphthalene (85)	146
C_{11} $CH_3CHOH(CH_2)_8CH_2OH$	"	−20°, DMF, 120 hr	$CH_3CH(O_2CH)(CH_2)_8CH_2Br$ (64)	148
C_{12}	"	25°, C_6H_6, 48 hr	(—)	105

C$_{13}$

Reactant	Reagent	Conditions	Product (yield)	Refs.
	$(C_6H_5)_3P/I_2$/imidazole	Toluene	(60)	103
	$(C_6H_5)_3P/I_2$/imidazole	Toluene	(94)	103
	$(C_6H_5)_3PBr_2$	80°, $OP(OC_2H_5)_3$, 15 min	(50)	145
	$(C_6H_5)_3PBrCN$	25°, $OP(OC_2H_5)_3$, 5 min	'' (60)	145
	$(C_6H_5)_3PI_2$	25°, $OP(OC_2H_5)_3$, 30 min	(60)	145

TABLE VII. Preparation of Alkyl Halides Using Triphenylphosphine–Halogen (*Continued*)

Alcohol	Reagent	Conditions	Product and Yield (%)	Ref.
C_{17} (thymidine deriv., p-$O_2NC_6H_4CO_2$, OH)	$(C_6H_5)_3PI_2$	23°, DMF, 14 days	(p-$O_2NC_6H_4CO_2$ deriv., I) (47)	124
C_{18} Lincomycin ($CH_3NCOC_6H_5$, OH, OH)	$(C_6H_5)_3PBr_2$	25°, CH_3CN, 3 hr	Clindamycin (73)	154
(adamantane deriv., $CH_3NCOC_6H_5$, OH)	"	1. 0°, DMF, 1 hr 2. H_2O	($CH_3NCOC_6H_5$, HCO_2, HCO_2) (33)	143
(adamantane deriv., $CH_3NCOC_6H_5$, OH)	"	0°, DMF, 15 min	($CH_3NCOC_6H_5$, $OCH=\overset{+}{N}(CH_3)_2$, Br^-) (80)	143

104

| C₂₇ | | | | 143 |

C_{27}

$CH_3NCOC_6H_5$

(—)

"

1. 0°, DMF, 15 min
2. 200°, neat, 15 min

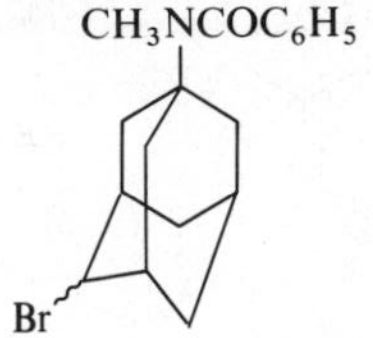

(—)   +   (—)

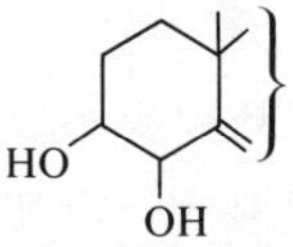

$(C_6H_5)_3PBr_2$ 80°, DMF, 16 hr   (—)

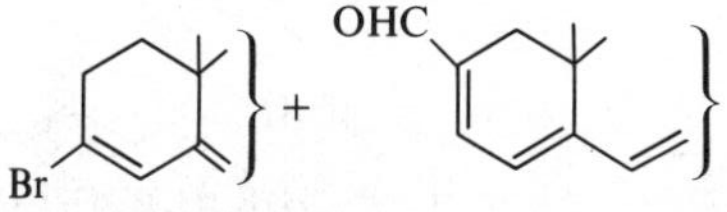

149

C_{28}

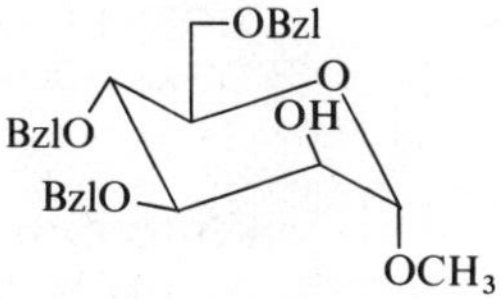

$(C_6H_5)_3PI_2$/imidazole   Toluene   (67)

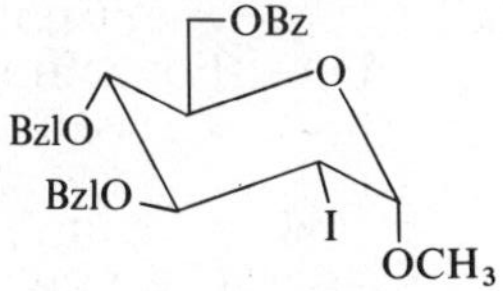

103

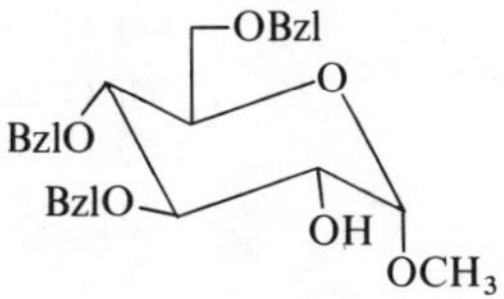

$(C_6H_5)_3PI_2$/imidazole   Toluene   (75)

103

OCH₃   OBz   OCH₃

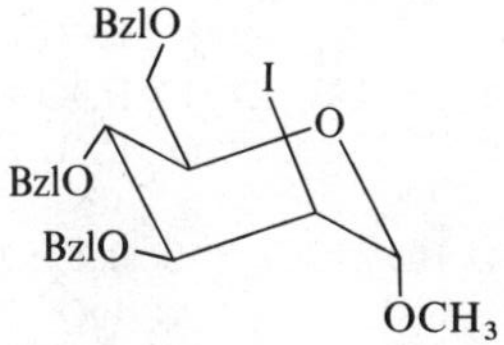

105

TABLE VIII. PREPARATION OF ALKYL HALIDES USING TRIPHENYLPHOSPHINE–CARBON TETRAHALIDES

	Alcohol	CX_4	Conditions	Products and Yields (%)	Ref.
C_1	CH_3OH	CCl_4	—	CH_3Cl (78)	23
C_2	C_2H_5OH	"	—	C_2H_5Cl (68)	23
	$HOCH_2CH_2OH$	"	80°, CCl_4, 12 hr	$ClCH_2CH_2OH + ClCH_2CH_2Cl$ (54) (20)	221
			—	$ClCH_2CH_2Cl + ClCH_2CH_2OH$ (30) (60)	
C_3	$n\text{-}C_3H_7OH$	"	—	$n\text{-}C_3H_7Cl$ (94)	23
	$CH_3OCH_2CH_2OH$	"	80°, CCl_4, 15 min	$CH_3OCH_2CH_2Cl$ (82)	221
C_4	$n\text{-}C_4H_9OH$	"	—	$n\text{-}C_4H_9Cl$ (89)	23
	$i\text{-}C_4H_9OH$	"	—	$i\text{-}C_4H_9Cl$ (—)	23
	$CH_3CHOHC_2H_5$	"	65°, CCl_4, 0.3 hr	$CH_3CHClC_2H_5$ (65)	176
		"	CCl_4, 5 min	" (60)	222
		$CCl_4/P(C_8H_{17})_3$	—	" (65)	23
	$HOCH_2CO_2C_2H_5$	"	80°, CCl_4, 15 min	$ClCH_2CO_2C_2H_5$ (—)	223
	$HO(CH_2)_3OH$	"	80°, CCl_4	$Cl(CH_2)_3Cl$ (—)	221
	$C_2H_5CHOHCH_2OH$	"	"	$C_2H_5CHClCH_2Cl$ (—)	221
	$CH_3CHOHCHOHCH_3$	"	"	$CH_3CHClCHClCH_3$ (—)	221
	(CH₂=CHCH₂CH=CHCH₂OH type)—OH	CCl_4	25°, CCl_4, 48 hr	—Cl (100)	224
	(butenol structure)	$(CCl_3)_2CO$	0–15°, $(CCl_3)_2CO$, 30 min	Cl structures (99.3) (0.7)	30
	(cis-butenol)—OH	"	"	Cl structures (98) (0.5)	30

106

Reactant	Reagent	Conditions	Product(s) (Yield %)	Refs.
CH₂=C(CH₃)CH₂OH	CCl₄	"	(93)	30
CH₃CH(OH)CH=CH₂	"	"	Cl–CH₂... (6) + ...Cl (94)	30
	CCl₄	25°, CCl₄, 48 hr	Cl–CH₂... (11) + ...Cl (89)	224
H...D, OH, CH₃, H (deuterated allylic alcohol)	"	"	CH₃ H Cl / H–D (—)	30
C₅ (CH₃)₃COH	CCl₄	—	(CH₃)₃CCl (76) (inversion ≥99%)	23
	"	—	" (—)	221
n-C₅H₁₁OH	CBr₄/P(C₈H₁₇)₃	80°, CCl₄, Et₂O, 5 min	n-C₅H₁₁Br (80)	222
	CCl₄	—	n-C₅H₁₁Cl (99)	23
(CH₃)₂CHCHOHCH₃	CBr₄	25°, THF, 1 hr	(CH₃)₂CHCHBrCH₃ + (CH₃)₂CBrCH₂CH₃ (10) (39)	144
HO(CH₂)₂CO₂C₂H₅	CCl₄	80°, CCl₄, 15 min	Cl(CH₂)₂CO₂C₂H₅ (—)	223
CH₃CHOHCO₂C₂H₅	"	80°, CCl₄, 15 min	CH₃CHClCO₂C₂H₅ (—)	223
lactone (CH₃), HO–	CBr₄	0°, CH₂Cl₂, 6 hr	lactone Br (68)	225
CH₃CH₂CH(OH)... OH	CCl₄	25°, CCl₄, 24 hr	...Cl (70) + ...Cl (14) + (16)	29

107

TABLE VIII. Preparation of Alkyl Halides Using Triphenylphosphine–Carbon Tetrahalides (*Continued*)

Alcohol	CX$_4$	Conditions	Products and Yields (%)	Ref.
C$_5$ (*Contd.*)				
(structure: CH$_3$COCH=C(OH)CH$_3$)	CCl$_4$	50°, CCl$_4$, 4 hr	(structure, Cl) (81)	22
(structure with CH$_3$, CH$_2$OH, H)	"	—	CH$_3$, CH$_2$Cl structure (Quant.)	226
(structure, OH)	(CCl$_3$)$_2$CO	0–15°, (CCl$_3$)$_2$CO, 30 min	Cl (98)	30
(structure, OH)	"	"	Cl (100)	30
(structure, OH)	"	"	Cl (91)	30
(structure, OH)	"	"	Cl structure (92) + Cl structure (2) + Cl structure (1)	30
(structure, OH)	"	"	Cl structure (80) + Cl structure (7) + structure (12)	30
(structure, OH)	"	"	Cl structure (21) + Cl structure (43) + structure (17)	30

108

Substrate	Reagent	Conditions	Product (% yield)	Refs.
![5-hydroxymethyl-2-pyrrolidinone]	CCl₄	55°, CCl₄/CHCl₃, 6 hr	![5-chloromethyl-2-pyrrolidinone] (86)	206
	CBr₄	20°, CH₃CN, 12 hr	![5-bromomethyl-2-pyrrolidinone] (74)	206
C₆ n-C₆H₁₃OH	CCl₄	—	n-C₆H₁₃Cl (83)	23
Cyclohexanol	CCl₄, polymer-supported phosphine	—	Chlorocyclohexane (40)	160
	"	80°, CCl₄, 2 hr	" (60)	159
C₂H₅CHOHCO₂C₂H₅	CCl₄	Reflux, CHCl₃, 4 hr	" (82)	227
	"	80°, CCl₄, 15 min	C₂H₅CHClCO₂C₂H₅ (—)	223
HO(CH₃)₃CO₂C₂H₅	"	"	Cl(CH₂)₃CO₂C₂H₅ (—)	223
n-C₄H₉OCH₂CH₂OH	CCl₄, polymer-supported phosphine	—	n-C₄H₉OCH₂CH₂Cl (94)	160
![dioxolane-methanol]	CCl₄	80°, CCl₄, 15 min	![dioxolane-chloromethyl] (76)	221
	CCl₄, polymer-supported phosphine	—	" (78)	160
(CH₃)₃SiC≡CCH₂OH	CCl₄	—	(CH₃)₃SiC≡CCH₂Cl	228
![3-hydroxycyclohexenone]	"	50°, CCl₄, 4 hr	![3-chlorocyclohexenone] (85)	22

TABLE VIII. PREPARATION OF ALKYL HALIDES USING TRIPHENYLPHOSPHINE–CARBON TETRAHALIDES (Continued)

Alcohol	CX_4	Conditions	Products and Yields (%)	Ref.
C_6 (Contd.) 2-methyl-3-hydroxy-cyclopent-2-enone	CCl_4	50°, CCl_4, 3 hr	2-methyl-3-chlorocyclopent-2-enone (73)	22
	CBr_4	50°, $CHCl_3$, 3 hr	2-methyl-3-bromocyclopent-2-enone (75)	22
hex-1-en-3-ol	CCl_4	25°, CCl_4, 24 hr	(71) + (14) + (14)	29
2-methylhex-5-en-3-ol	"	"	(23) + (18) + (58)	29
(chiral alcohol)	"	80°, CCl_4, 2.5 hr	H Cl (40) (Racemization ≥24%)	229
oct-2-en-4-ol	$(CCl_3)_2CO$	0–15°, $(CCl_3)_2CO$, 30 min	(53) + (26)	30

Substrate	Conditions	Product(s) and Yield(s) (%)	Refs.
(C$_7$ structure, OH)	"	(27) + (15) + (61)	30
(OH)	"	(15) + Cl	30
(OH)	"	(7) + ClCH$_2$~ (15) (Z) + (E) (50/50)	30
		(9) (37); (Z) + (E)	30
C$_7$ n-C$_7$H$_{15}$OH	CCl$_4$	n-C$_7$H$_{15}$Cl (—)	23
C$_6$H$_5$CH$_2$OH	"/P(C$_8$H$_{17}$)$_3$ (CCl$_4$), 5 min	C$_6$H$_5$CH$_2$Cl (100)	222
	CCl$_4$	" (83)	23
	CCl$_4$, polymer-supported phosphine	" (88)	160
Cycloheptanol	" 80°, CCl$_4$, 2 hr	" (99)	159
	CCl$_4$, polymer-supported phosphine	Chlorocycloheptane (92)	159
n-C$_3$H$_7$CHOHCO$_2$C$_2$H$_5$	CCl$_4$ 80°, CCl$_4$, 15 min	n-C$_3$H$_7$CHClCO$_2$C$_2$H$_5$ (—)	223
(with CH$_3$, OH)	" 50°, CCl$_4$, 4 hr	(with CH$_3$, Cl) (81)	22

(—) ratio = 18:8

TABLE VIII. PREPARATION OF ALKYL HALIDES USING TRIPHENYLPHOSPHINE-CARBON TETRAHALIDES (*Continued*)

Alcohol	CX₄	Conditions	Products and Yields (%)	Ref.

C₇
(*Contd.*)

	CCl₄	1. 25°, CCl₄, 24 hr 2. 175°	(53)	24
	(CCl₃)₂CO	0–15°, (CCl₃)₂CO, 30 min	(—) + (—) ratio = 62:38 (—) + (—) ratio = 89:11	30
	CCl₄	25°, CCl₄, 24 hr	n-C₄H₉ (70) + n-C₄H₉ (18)	29
	''	''	n-C₃H₇ (11) + (50) + (26) + (23)	29

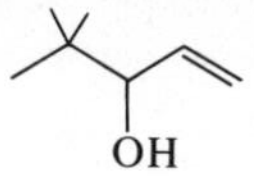	”	”	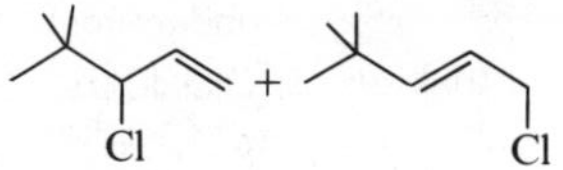 (21) (79)	29
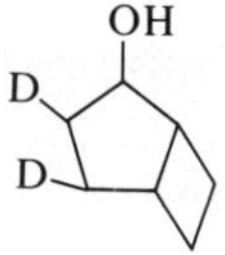	”	25°, CCl_4, 65 hr	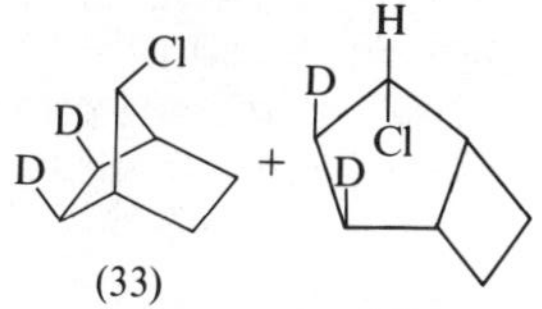 (33) (7)	24
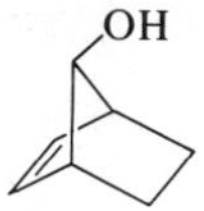	”	60°, CCl_4, 2.5 hr	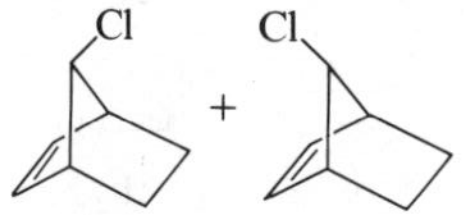 (42) (6)	230
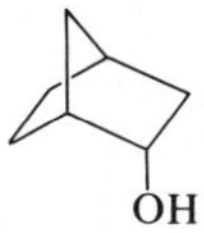	”	80°, CCl_4, 3.5 hr	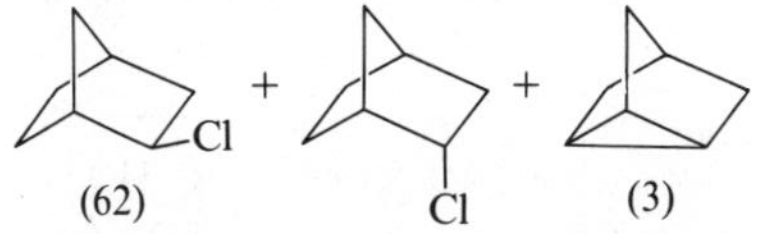 (62) (3) (3)	230
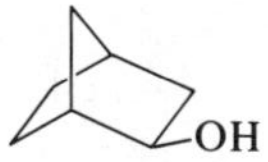	”	80°, CCl_4, 1.75 hr	 (8) (28) (27)	230
$(CH_3)_3SiC(CH_3)=CHCH_2OH$	”	—	$(CH_3)_3SiC(CH_3)=CHCH_2Cl$ (60)	228

TABLE VIII. Preparation of Alkyl Halides Using Triphenylphosphine–Carbon Tetrahalides (*Continued*)

Alcohol	CX_4	Conditions	Products and Yields (%)	Ref.
C_7 (*Contd.*)	CCl_4	65°, Py, 0.2 hr	(98)	156
	CBr_4	"	(93)	156
	Cl_4	25°, Py, 18 hr	(97)	156
C_8 $n\text{-}C_8H_{17}OH$	$CCl_4/P(C_8H_{17})_3$	CCl_4, 5 min	$n\text{-}C_8H_{17}Cl$ (88)	222
	CCl_4, polymer-supported phosphine	—	" (98)	160
$CH_3CHOHC_6H_{13}\text{-}n$	CBr_4	80°, CCl_4, 2 hr	" (90)	159
	"	0°, CH_2Cl_2	$n\text{-}C_8H_{17}Br$ (91)	231
	"	"	$CH_3CHBrC_6H_{13}\text{-}n$ (90)	231
D-$CH_3CHOHC_6H_{13}\text{-}n$	CCl_4	80°, CCl_4, 6 hr	L-$CH_3CHClC_6H_{13}\text{-}n$ (45)	133
	CBr_4	—	L-$CH_3CHBrC_6H_{13}\text{-}n$ (38)	133
	CCl_4	60°, CCl_4, 4 hr	L-$CH_3CHClC_6H_{13}\text{-}n$ (50) $(\alpha)_D^{20} = -34.2°$	230
	$CCl_4/P(C_8H_{17})_3$	CCl_4, 5 min	L-$CH_3CHClC_6H_{13}\text{-}n$ (80) (optical purity 90%)	222

114

Substrate	Reagent	Conditions	Product (Yield)	Refs.
$n\text{-}C_4H_9CH(C_2H_5)CH_2OH$	CCl_4	—	$n\text{-}C_4H_9CH(C_2H_5)CH_2Cl$ (82)	23
$(-)\text{-}C_2H_5O_2CCH_2CHOHCO_2C_2H_5$	"	80°, CCl_4	$D(+)\text{-}C_2H_5O_2CCH_2CHClCO_2C_2H_5$ (−)	223
$threo\text{-}C_6H_5CHDCHDOH$	"	60°, CCl_4, 6 hr	$erythro\text{-}C_6H_5CHDCHDCl$ (75)	230
$C_6H_5CH_2CH_2OH$	$CCl_4/P(C_8H_{17})_3$	CCl_4, 5 min	$C_6H_5CH_2CH_2Cl$ (66)	222
[spiro[3.4] alkenol, HO]	CCl_4	80°, CCl_4, 1 hr	[spiro[3.4] alkenyl chloride] (−)	232
[3-hydroxy-5,5-dimethylcyclohex-2-enone]	"	50°, CCl_4, 4 hr	[3-chloro-5,5-dimethylcyclohex-2-enone] (82)	22
	CBr_4	50°, $CHCl_3$, 3 hr	[3-bromo-5,5-dimethylcyclohex-2-enone] (85)	22
C_9 $C_6H_5(CH_2)_3OH$	CCl_4	—	$C_6H_5(CH_2)_3Cl$ (80)	23
$endo$-Norborneol	"	80°, CCl_4, 6 hr	exo-Norbornyl chloride (57)	133
	CBr_4	—	exo-Norbornyl bromide (64)	133
$HO(CH_2)_6CO_2C_2H_5$	CCl_4	80°, CCl_4, 15 min	$Cl(CH_2)_6CO_2C_2H_5$ (−)	223
$(Z)\text{-}n\text{-}C_6H_{13}CH{=}CHCH_2OH$	CBr_4	0°, CH_2Cl_2	$(Z)\text{-}n\text{-}C_6H_{13}CH{=}CHCH_2Cl$ (89)	231
[2-(2-cyanoethyl)-3-hydroxycyclohex-2-enone]	CCl_4	50°, CCl_4, 4 hr	[2-(2-cyanoethyl)-3-chlorocyclohex-2-enone] (79)	22

115

TABLE VIII. PREPARATION OF ALKYL HALIDES USING TRIPHENYLPHOSPHINE–CARBON TETRAHALIDES (*Continued*)

Alcohol	CX$_4$	Conditions	Products and Yields (%)	Ref.
C$_9$ (*Contd.*)	CCl$_4$	5°, Py, 20 hr	(96)	156
	CBr$_4$	60°, Py, 0.15 hr	(95)	156
	″	23°, DMF, 24 hr	(55)	124
	CCl$_4$	25°, Py, 18 hr	(92)	156

116

	Reagent	Conditions	Product(s) and Yield(s) (%)	Refs.
	"	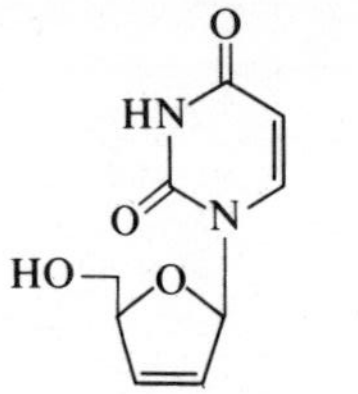 25°, DMF, 12 hr	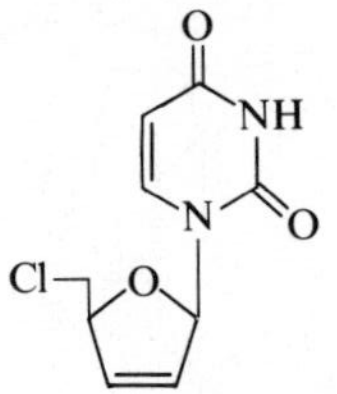 (57)	233
$n\text{-}C_{10}H_{21}OH$	"	23°, DMF, 4 hr	(27)	124
	CCl$_4$, polymer-supported phosphine	80°, CCl$_4$, 2 hr	$n\text{-}C_{10}H_{21}Cl$ (89)	159
C$_6$H$_5$CHOHCO$_2$C$_2$H$_5$	CCl$_4$	80°, CCl$_4$, 15 min	C$_6$H$_5$CHClCO$_2$C$_2$H$_5$ (—)	223
Geraniol	"	80°, CCl$_4$, 2 hr	Geranyl chloride (54)	234
	CBr$_4$	0°, CH$_2$Cl$_2$	Geranyl bromide (82)	231
C$_{10}$	"	"	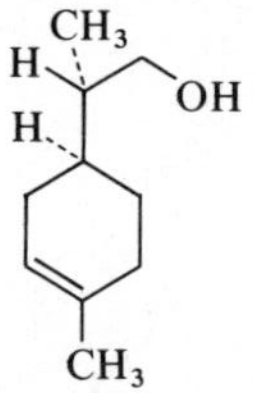 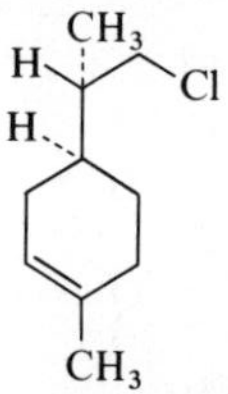 (93)	235
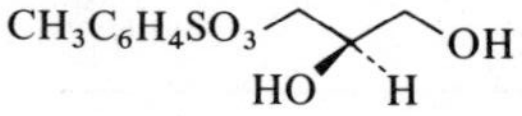	CCl$_4$	50°, DMF, 3 hr	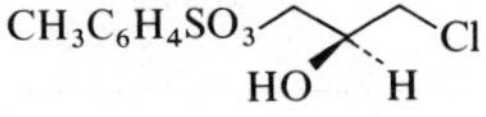	236

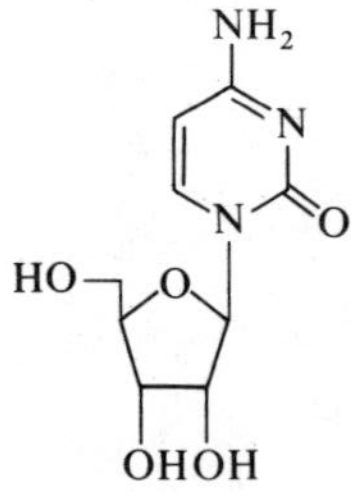

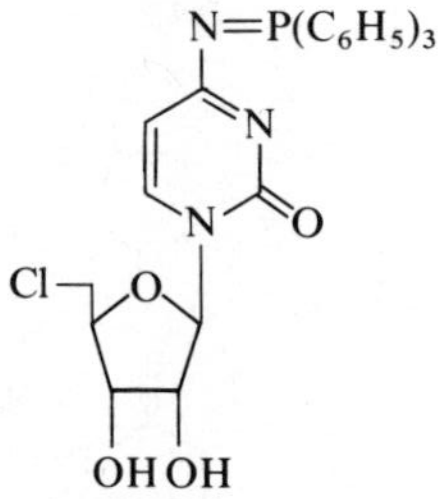

CH$_3$C$_6$H$_4$SO$_3$

$\text{N}=\text{P}(\text{C}_6\text{H}_5)_3$

117

TABLE VIII. Preparation of Alkyl Halides Using Triphenylphosphine–Carbon Tetrahalides (*Continued*)

Alcohol	CX₄	Conditions	Products and Yields (%)	Ref.
C₁₀ (*Contd.*)	CBr₄	80°, C₆H₆, 1 hr		237
	CCl₄	80°, CCl₄		237
	CCl₄	23°, CCl₄, 24 hr		124
	"	23°, DMF, 10 hr		124

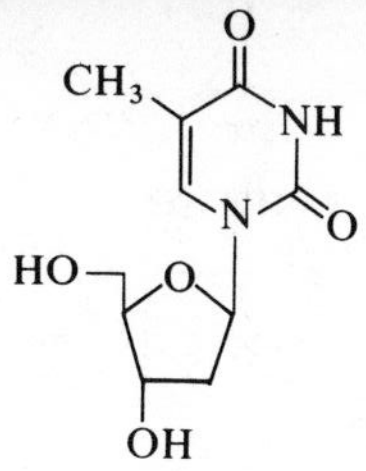

124

(4)

+

(12)

+ (60)

CBr₄ — 23°, DMF, 7 hr

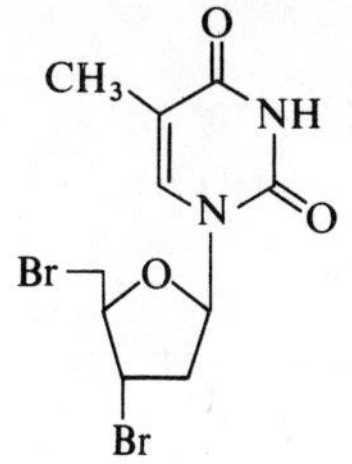

CH₃

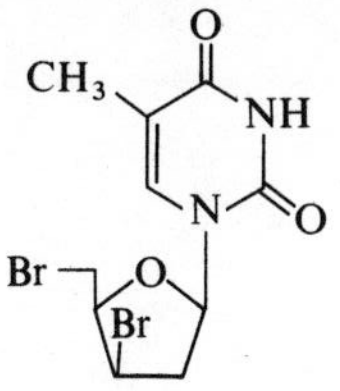

CH₃

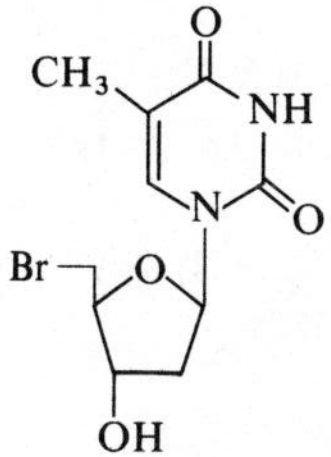

CH₃

156

(94)

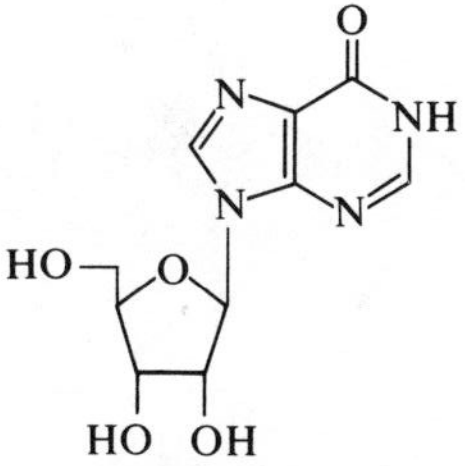

CCl₄ — 65°, Py, 0.2 hr

238

(47)

" — 20°, CH₃CN, 12 hr

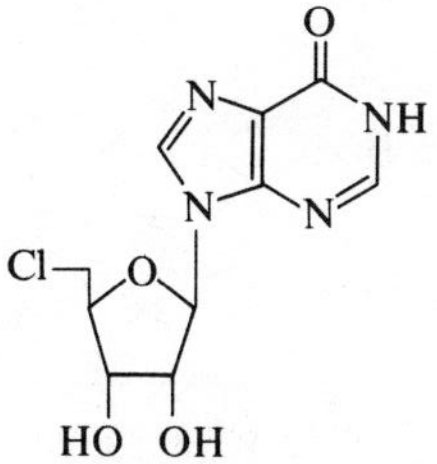

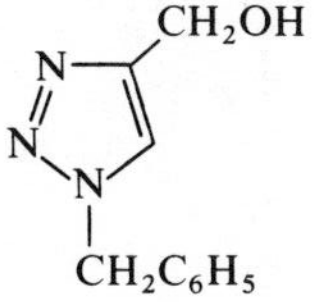

CH₂Cl

CH₂C₆H₅

CH₂OH

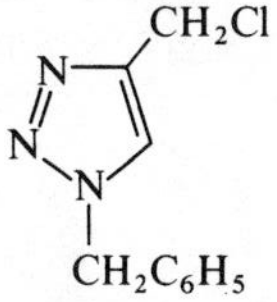

CH₂C₆H₅

TABLE VIII. PREPARATION OF ALKYL HALIDES USING TRIPHENYLPHOSPHINE–CARBON TETRAHALIDES (*Continued*)

Alcohol	CX$_4$	Conditions	Products and Yields (%)	Ref.
C$_{10}$ (*Contd.*)	CCl$_4$	80°, CCl$_4$, 3 hr	(93)	195
	"	"	(89)	195
C$_{11}$ n-C$_{11}$H$_{23}$OH	CCl$_4$, polymer-supported phosphine	80°, CCl$_4$, 2 hr	n-C$_{11}$H$_{23}$Cl (99)	159
C$_6$H$_{13}$CH=C=CH(CH$_2$)$_2$OH	CBr$_4$	0°, CH$_2$Cl$_2$	C$_6$H$_{13}$CH=C=CH(CH$_2$)$_2$Br (88)	231
	CCl$_4$	20°, DMF, 48 hr	(58)	124
C$_{12}$ n-C$_{12}$H$_{25}$OH	CCl$_4$, polymer-supported phosphine	80°, CCl$_4$, 2 hr	n-C$_{12}$H$_{25}$Cl (71)	159

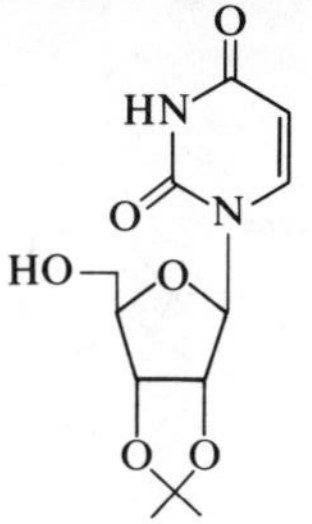	CCl_4	23°, DMF, 18 hr	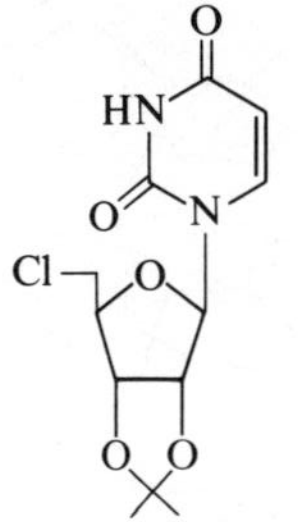 (70)	124
	CBr_4	”	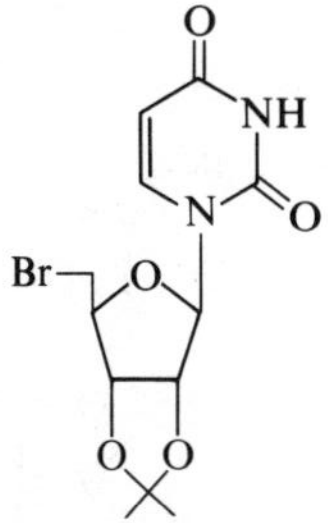 (55)	124
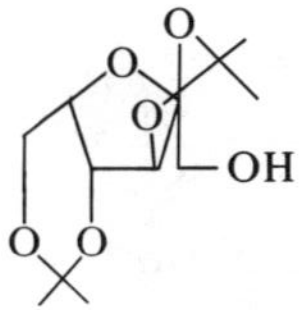	CCl_4	80°, CCl_4, 48 hr	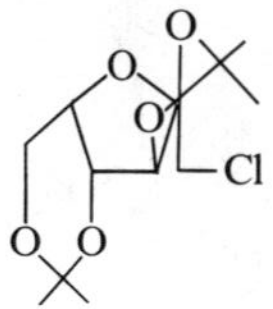 (56)	104
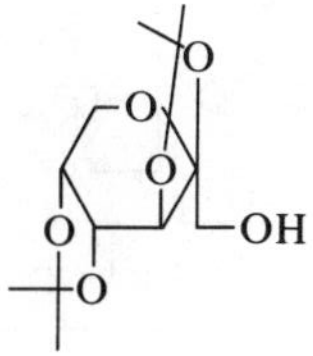	”	”	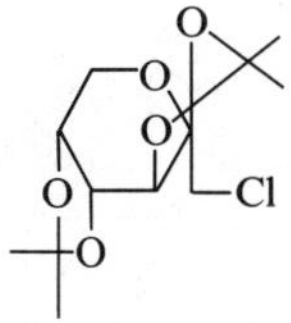 (57)	104

TABLE VIII. PREPARATION OF ALKYL HALIDES USING TRIPHENYLPHOSPHINE–CARBON TETRAHALIDES (*Continued*)

Alcohol	CX_4	Conditions	Products and Yields (%)	Ref.
C_{12} (*Contd.*) 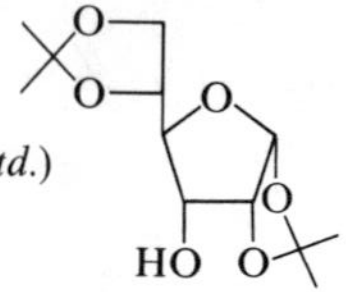	CCl_4	80°, CCl_4, 48 hr	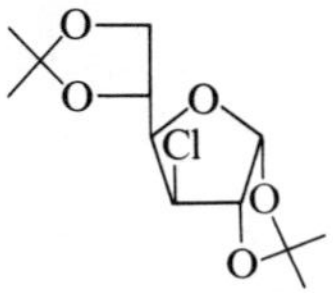 (85)	104
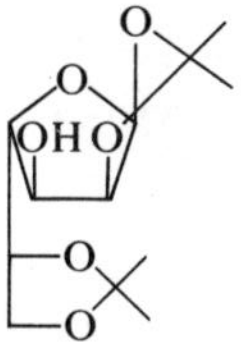	"	"	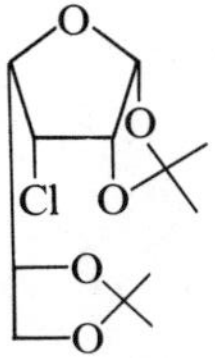 (57)	104
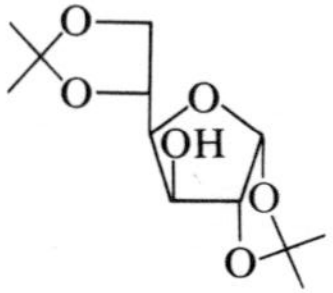	"	"	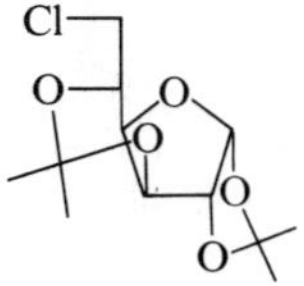 (79)	104
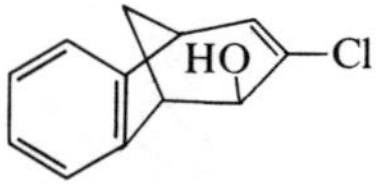	CCl_4	25°, CCl_4, 24 hr	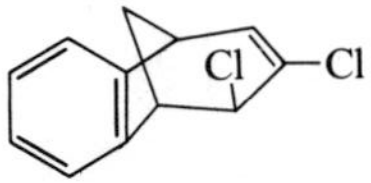 (70)	27
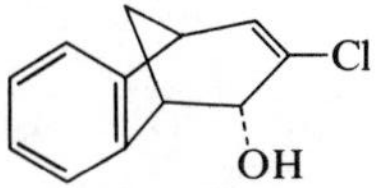	"	25°, CCl_4, 5 days	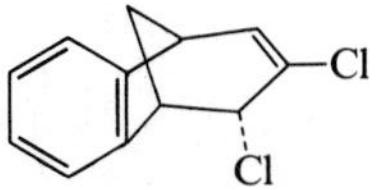 (20)	27
	"	70°, CCl_4, 25 hr	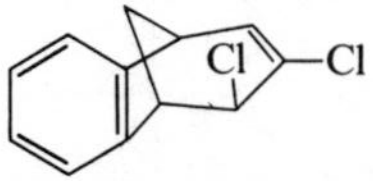 (68)	27

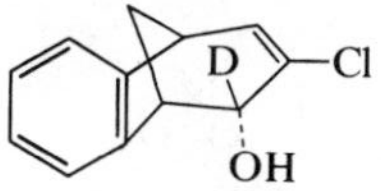	”	”	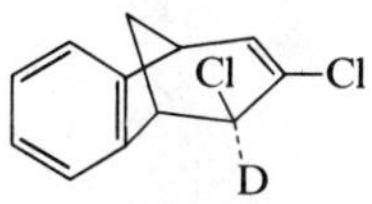 (64) + 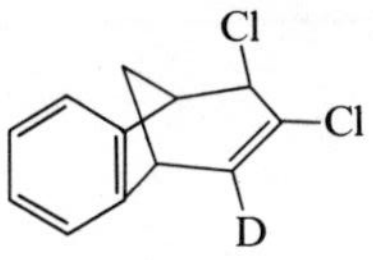 (16)	27
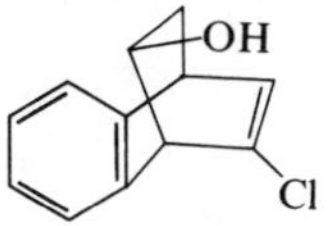	”	25°, CH_3CN, 3 hr	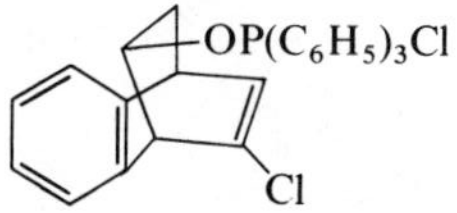 (Quant.)	27
	”	70°, CH_3CN, 1 hr	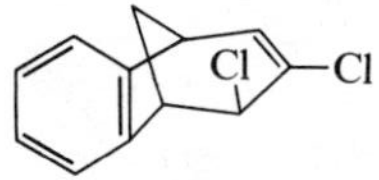 (Quant.)	27
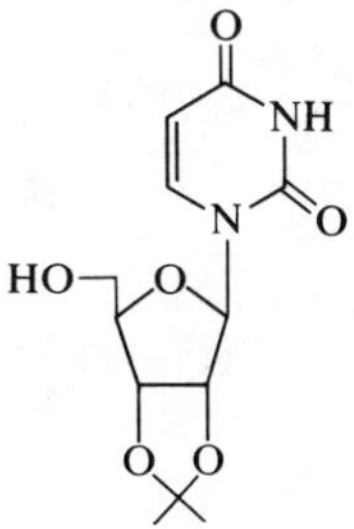	CI_4	23°, Py, 16 hr	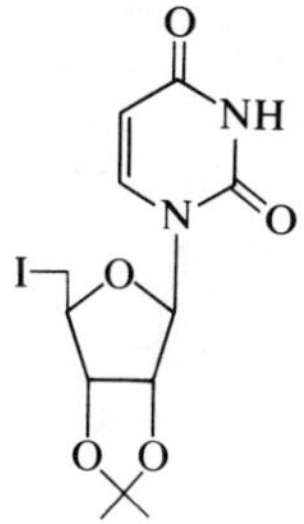 (76)	124
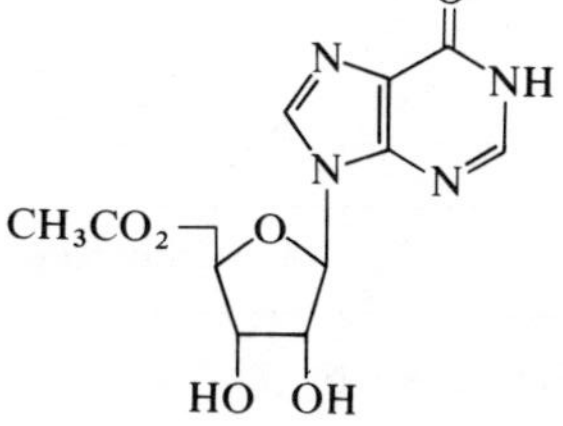	CCl_4	11°, $OP(OC_2H_5)_3$, 6 hr	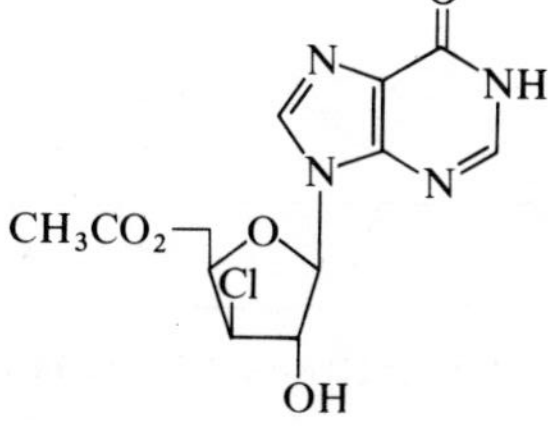 (19)	143

TABLE VIII. Preparation of Alkyl Halides Using Triphenylphosphine–Carbon Tetrahalides (*Continued*)

Alcohol	CX$_4$	Conditions	Products and Yields (%)	Ref.
C$_{12}$ (*Contd.*) [structure]	CCl$_4$	80°, CCl$_4$, 3 hr	[structure, Cl]	221
[structure]	"	80°, CCl$_4$, 45 min	[structure] (50)	31
[structure]	"	70°, Py, 2 hr	[structure] (92)	156
C$_{13}$ n-C$_8$H$_{17}$CH=C=CHCH$_2$CH$_2$OH	CBr$_4$	0°, CH$_2$Cl$_2$	n-C$_8$H$_{17}$CH=C=CHCH$_2$CH$_2$Br (76)	231
[structure]	CCl$_4$	23°, DMF, 12 hr	[structure] (80)	124
	"	100°, OP(OC$_2$H$_5$)$_3$, 5 min	" (86)	145

124

C$_{15}$ *trans, trans*-Farnesol	CCl$_4$	20°, CCl$_4$, 18 hr	*trans,trans*-Farnesyl bromide (90)	155

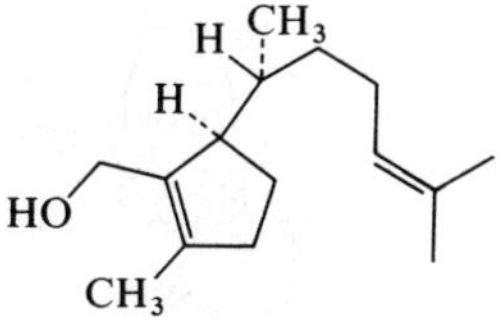

CBr$_4$ 23°, DMAc, 16 hr   (49)   124

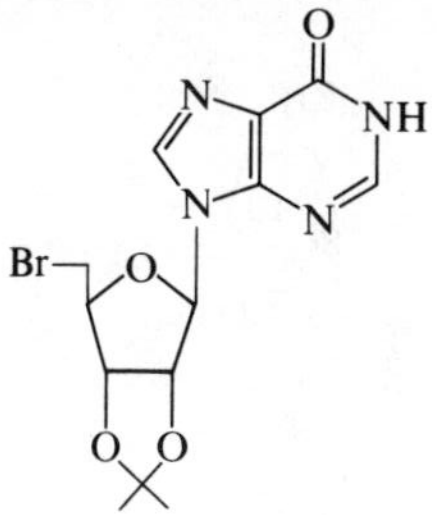

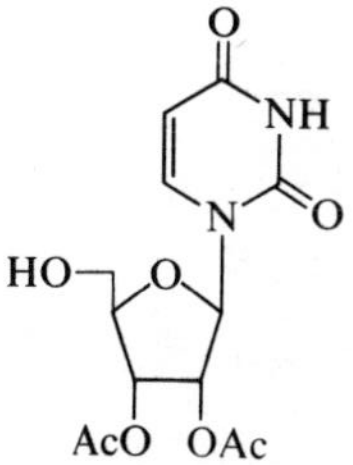

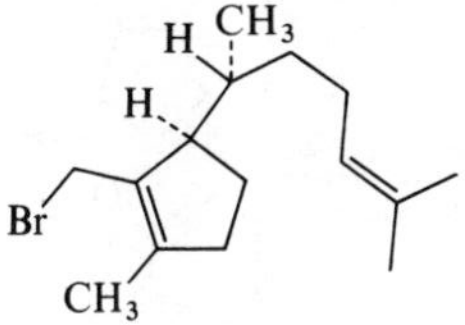

CCl$_4$ 20°, CCl$_4$, 18 hr   (50)   124

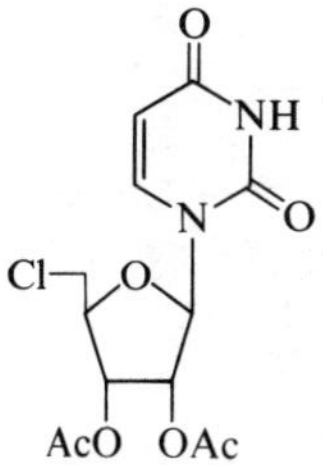

"   25°, CH$_3$CN, 3 hr

CBr$_4$ "   (−)   235

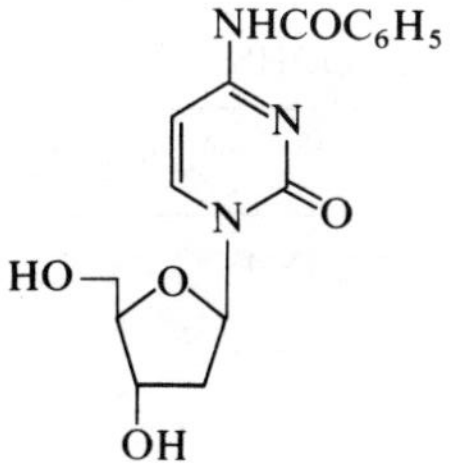

CCl$_4$ 23°, DMF, 16 hr   (10)   124

C$_{16}$

NHCOCH$_3$

NHCOC$_6$H$_5$

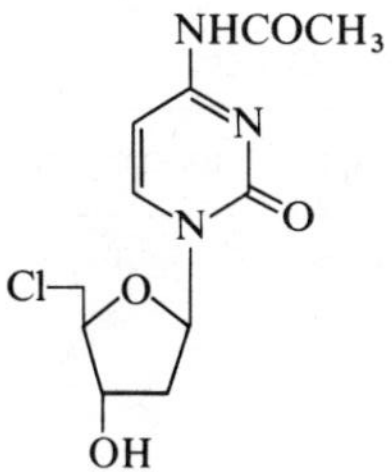

TABLE VIII. PREPARATION OF ALKYL HALIDES USING TRIPHENYLPHOSPHINE–CARBON TETRAHALIDES (*Continued*)

Alcohol	CX_4	Conditions	Products and Yields (%)	Ref.
C_{16} (*Contd.*)	CBr_4	25°, DMAc, 4 hr	(57)	125
	CCl_4	80°, $(C_2H_5O)_3PO$, 4 hr	(64)	158
	"	80°, $(C_2H_5O)_3PO$, 5 hr	(—)	158

Substrate	Reagent/Solvent	Conditions	Product(s) (%)	Refs.
	CCl$_4$	80°, CCl$_4$, 5 hr	(72)	239
Hexadecyl alcohol	CCl$_4$, polymer-supported phosphine	—	Hexadecyl chloride (98)	160
C$_{17}$	CCl$_4$	50°, CCl$_4$, 19 hr	(75) + (20)	27
	"	70°, CCl$_4$, 36 hr	(5) + (70) + (17)	27
	"	80°, CCl$_4$, 6 hr	(47)	240

127

TABLE VIII. Preparation of Alkyl Halides Using Triphenylphosphine–Carbon Tetrahalides (*Continued*)

Alcohol	CX_4	Conditions	Products and Yields (%)	Ref.
C_{18} (triazolyl peracetylated sugar, OH OH)	CCl_4	20°, CH_3CN, 12 hr	(triazolyl peracetylated sugar, CH_2Cl, CH_2Cl) (80)	238
Lincomycin	"	25°, CH_3CN, 18 hr	Clindamycin (67)	154
C_{19} Ethyl D-12-hydroxystearate	"	80°, CCl_4, 6 hr	Ethyl L-12-chlorostearate (92)	133
C_{21} (sugar structure, HO, HO, $OCH_2C_6H_5$, C_6H_5, OCH_3)	"	80°, CCl_4, 36 hr	(sugar structure, Cl, $OCH_2C_6H_5$, C_6H_5, OCH_3) (84)	241
C_{24} (purinone nucleoside, HO, BzO, OBz)	"	23°, DMF, 24 hr	(purinone nucleoside, Cl, BzO, OBz) (86)	124

128

C_{27} Cholesterol	80°, CCl_4, 8 hr	3-α-Chlorocholest-5-ene (8) + 3-β-chlorocholest-5-ene (28) + cholesta-3,5-diene (13) + 3,5-cyclocholest-6-ene (34)	26
Isocholesterol	"	3-β-Chlorocholest-5-ene (30) + 3,5-cyclocholest-6-ene (48)	26
5-β-Cholan-24-ol	CCl_4, polymer-supported phosphine	24-Chloro-5-β-cholane (78)	160
C_{29} (C_6H_5)_3CO— structure with dioxolane ring, —OH	CCl_4	(C_6H_5)_3CO— structure with dioxolane ring, —Cl (55)	242
Thymidine structure CH_3, (C_6H_5)_3CO—, —OH	25°, DMF, 1 hr	Thymidine structure CH_3, (C_6H_5)_3CO—, —Cl (15)	
	23°, DMF, 24 hr	structure (35) + (C_6H_5)_3CO— structure —Cl	124
C_{39} $CH_2OCOC_{17}H_{35}$-n / $CHOH$ / $CH_2OCOC_{17}H_{35}$-n	80°, CCl_4	$CH_2OCOC_{17}H_{35}$-n / $ClCH$ / $CH_2OCOC_{17}H_{35}$-n (95)	243

129

TABLE IX. PREPARATION OF ALKYL HALIDES AND AZIDES USING TRIPHENYLPHOSPHINE–N-HALOSUCCINIMIDE

	Substrate	N-Halosuccinimide	Conditions	Products and Yields (%)	Ref.
C_2	C_2H_5OH	Br	—	C_2H_5Br (—)	244
C_7	Cycloheptanol	"	25°, THF, 2 hr	Bromocycloheptane (78)	185
	(sugar structure, OCH₃, OH, HO)	"	1. 50°, DMF, 2 hr 2. Ac₂O	(sugar structure, Br, OAc, OCH₃, OAc, AcO) (66)	180
		"	1. 55°, DMF, 2 hr 2. NaN₃, 80°, CH₃OH 3. Ac₂O	(sugar structure, N₃, OAc, OCH₃, OAc, AcO) (75)	184
C_8	(S)-(−)-D— (CH₂OH)—C—H, C₆H₅	"	—	(S)-(−)-D— (CH₂Br)—C—H, C₆H₅ (—)	245
C_9	(sugar structure, OCH₃, NHAc, OH, HO)	"	1. 50°, DMF, 2 hr 2. Ac₂O	(sugar structure, Br, OAc, OCH₃, NHAc, AcO) (55)	180
	(sugar structure, COCF₃, NH, OCH₃, HO, HO)	"	DMF	(sugar structure, COCF₃, NH, OCH₃, Br, HO) (—)	246

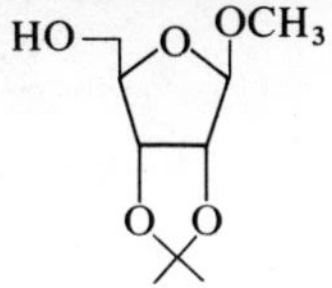

25°, CH₂Cl₂ or
50°, DMF, 15 min — (95)

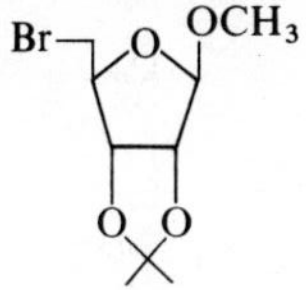

"

50°, DMF, 20 min — I — (71)

" — Cl — (95)

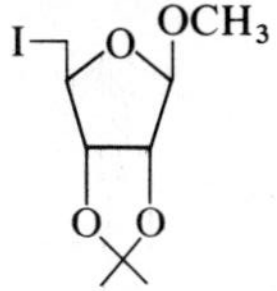

1. 50°, DMF, 3 hr
2. (CH₃)₂CO — Br — (70)

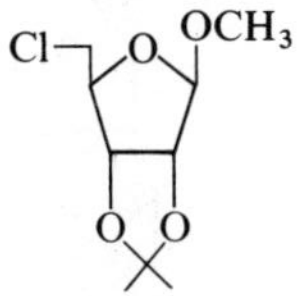

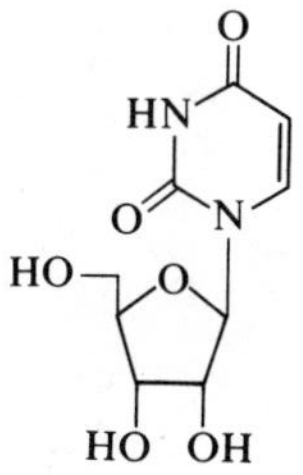

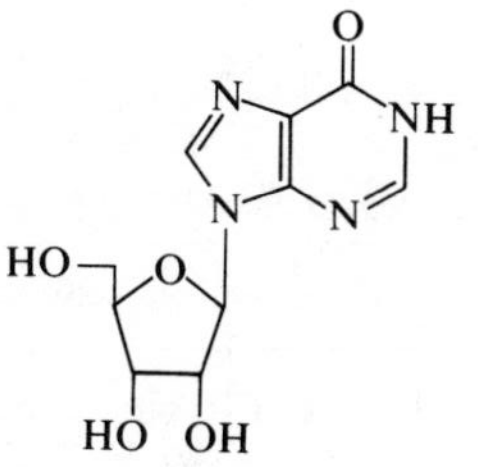

1. 50°, DMF, 2.5 hr
2. Ac₂O — " — (59)

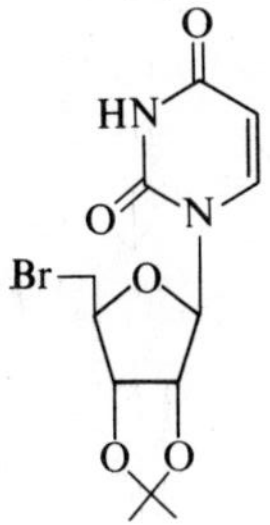

180

180

180

180

180

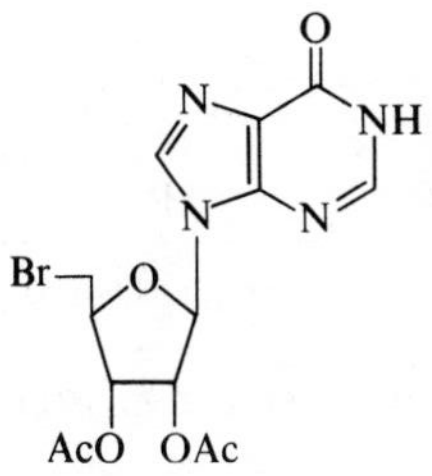

C₁₀

TABLE IX. PREPARATION OF ALKYL HALIDES AND AZIDES USING TRIPHENYLPHOSPHINE–N-HALOSUCCINIMIDE (*Continued*)

Substrate	N-Halosuccinimide	Conditions	Products and Yields (%)	Ref.
C$_{12}$	Br	50°, DMF, 3.5 hr	(79)	180
	I	50°, DMF, 2.5 hr	(67)	180
	Cl	"	(89)	180

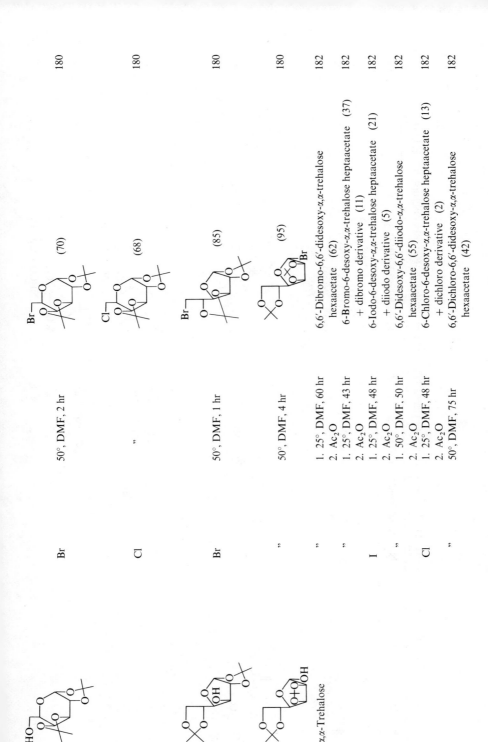

Starting material	Halide	Reaction conditions	Product (yield %)	Ref.
	Br	50°, DMF, 2 hr	(70)	180
	Cl	"	(68)	180
	Br	50°, DMF, 1 hr	(85)	180
	"	50°, DMF, 4 hr	(95)	180
	"	1. 25°, DMF, 60 hr / 2. Ac₂O	6,6'-Dibromo-6,6'-didesoxy-α,α-trehalose hexaacetate (62)	182
	"	1. 25°, DMF, 43 hr / 2. Ac₂O	6-Bromo-6-desoxy-α,α-trehalose heptaacetate (37) + dibromo derivative (11)	182
	I	1. 25°, DMF, 48 hr / 2. Ac₂O	6-Iodo-6-desoxy-α,α-trehalose heptaacetate (21) + diiodo derivative (5)	182
	"	1. 50°, DMF, 50 hr / 2. Ac₂O	6,6'-Didesoxy-6,6'-diiodo-α,α-trehalose hexaacetate (55)	182
	Cl	1. 25°, DMF, 48 hr / 2. Ac₂O	6-Chloro-6-desoxy-α,α-trehalose heptaacetate (13) + dichloro derivative (2)	182
	"	50°, DMF, 75 hr	6,6'-Dichloro-6,6'-didesoxy-α,α-trehalose hexaacetate (42)	182

α,α-Trehalose

TABLE IX. PREPARATION OF ALKYL HALIDES AND AZIDES USING TRIPHENYLPHOSPHINE–N-HALOSUCCINIMIDE (*Continued*)

Substrate	N-Halosuccinimide	Conditions	Products and Yields (%)	Ref.
C$_{13}$	Br	50°, DMF, 1 hr	(77)	180
	"	50°, DMF, 1.5 hr	(50)	180
C$_{15}$	"	1. 25°, DMF, 2 hr 2. NaN$_3$, CH$_3$OH, 80° 3. Ac$_2$O	(84)	184
	"	1. 25°, DMF, 2 hr 2. Ac$_2$O	(84)	184

C$_{16}$ [structure: HN, O, N, HO, O, C$_6$H$_5$]	"	50°, DMF, 4 hr	[structure with Br] (64)	180
C$_{18}$ 4-Estren-17β-ol-3-one	Br	25°, THF, 1 hr	17α-Bromo-4-estren-3-one (75)	185
C$_{19}$ 5-Androsten-3β-ol-17-one	"	"	3-Bromo-5-androsten-17-one (75) (3α/3β = 4:1)	185
C$_{27}$ 3β-Cholestanol	Br	"	3α-Cholestanyl bromide (95)	185
	I	"	3α-Cholestanyl iodide (85)	185
Cholesterol	Br	"	Cholesteryl bromide (3α/3β = 3:1) (65)	185
[steroid structure, R = isooctyl, HO, AcO]	I	25°, THF, 90 min	[steroid structure, R, I, AcO] (19)	247
C$_{62}$ Penta-N-benzyloxycarbonylneomycin	Br	1. 55°, HMPA, 4 hr 2. NaN$_3$, 55°, HMPA, 4 hr	5'-Azido-5'-desoxypenta-N-benzyloxycarbonylneomycin (73)	181
	"	55°, HMPA, 4 hr	5'-Bromo-5'-desoxypenta-N-benzyloxycarbonylneomycin (91)	181

135

TABLE X. PREPARATION OF ALKYL HALIDES USING ALKOXY(TRISDIMETHYLAMINO)PHOSPHONIUM SALTS

	Alcohol	Anion for Salt Isolation and Yield (%)	Nucleophile	Conditions	Products and Yields from Alcohol (%)	Ref.
C$_4$	n-C$_4$H$_9$OH	Cl⁻	None	25°, Et$_2$O	n-C$_4$H$_9$Cl (81)	187
C$_5$	n-C$_5$H$_{11}$OH	"	"	"	n-C$_5$H$_{11}$Cl (61)	187
	(structure, OH)	F⁻ (benzoylated)	"	1. 135°, DMF, 15 hr 2. C$_6$H$_5$COCl	(structure, F / O$_2$CC$_6$H$_5$) (20)	192, 193
	(structure, Cl/OH)	Cl⁻ (98)	"	90°, DMF, 6 hr	(structure, Cl/OH) (85)	192, 193
		ClO$_4^-$ (95)	n-C$_4$H$_9$N(C$_2$H$_5$)$_3$Br	100°, DMF, 6 hr	(structure, Br/OH) (60)	188, 189
		ClO$_4^-$ (95)	KI	100°, DMF, 10 hr	(structure, I/OH) (87)	192, 193
		I⁻ (extraction)	None	80°, C$_6$H$_6$, 10 hr	(structure, Cl/I) (81)	192, 193
	(structure, N$_3$/OH)	Cl⁻	"	"	(structure, N$_3$/Cl) (75)	192, 193
	(structure, OH/OH)	I⁻ (extraction)	"	110°, toluene, 20 hr	(structure, N$_3$/I) (65)	192, 193
	(structure, OH/OH)	Cl⁻ (93)	"	100°, DMF, 8 hr	(structure, Cl/OH) (80)	192, 193

136

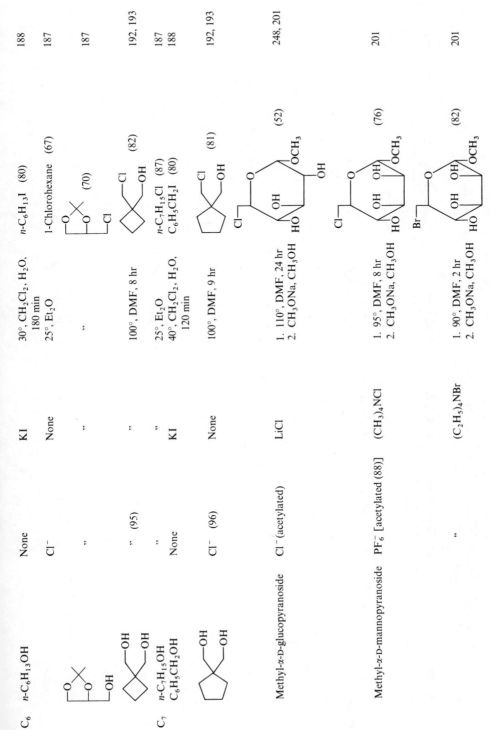

	Reactant	Anion source	Conditions	Product (yield)	Refs.
C₆	n-C₆H₁₃OH	None	30°, CH₂Cl₂, H₂O, 180 min	n-C₆H₁₃I (80)	188
	(2-substituted dioxolane, OH)	Cl⁻	25°, Et₂O	1-Chlorohexane (67)	187
	(dioxolane, OH) (95)	"	"	(dioxolane, Cl) (70)	187
	(cyclobutane diol)	"	100°, DMF, 8 hr	(cyclobutane, Cl, OH) (82)	192, 193
C₇	n-C₇H₁₅OH	"	25°, Et₂O	n-C₇H₁₅Cl (87)	187
	C₆H₅CH₂OH	None	40°, CH₂Cl₂, H₂O, 120 min	C₆H₅CH₂I (80)	188
	(cyclopentane diol)	Cl⁻ (96)	None	100°, DMF, 9 hr	(cyclopentane, Cl, OH) (81)	192, 193
	Methyl-α-D-glucopyranoside	Cl⁻ (acetylated)	LiCl	1. 110°, DMF, 24 hr 2. CH₃ONa, CH₃OH	(sugar, Cl) (52)	248, 201
	Methyl-α-D-mannopyranoside	PF₆⁻ [acetylated (88)]	(CH₃)₄NCl	1. 95°, DMF, 8 hr 2. CH₃ONa, CH₃OH	(sugar, Cl) (76)	201
	"		(C₂H₅)₄NBr	1. 90°, DMF, 2 hr 2. CH₃ONa, CH₃OH	(sugar, Br) (82)	201

137

TABLE X. PREPARATION OF ALKYL HALIDES USING ALKOXY(TRISDIMETHYLAMINO)PHOSPHONIUM SALTS (Continued)

Alcohol	Anion for Salt Isolation and Yield (%)	Nucleophile	Conditions	Products and Yields from Alcohol (%)	Ref.
C$_7$ Methyl-α-D-galactopyranoside (Contd.)	PF$_6^-$ [acetylated (85)]	(CH$_3$)$_4$NCl	1. 110°, DMF, 1 hr 2. CH$_3$ONa, CH$_3$OH	(76)	210
	"	(C$_2$H$_5$)$_4$NBr	1. 90°, DMF, 6 hr 2. CH$_3$ONa, CH$_3$OH	(75)	201
	"	KI	1. 90°, DMF, 20 hr 2. CH$_3$ONa, CH$_3$OH	(60)	201
Methyl-α-D-glucopyranoside	BF$_4^-$ (100) (acetylated)	"	1. 80°, DMF, 24 hr 2. CH$_3$ONa, CH$_3$OH	(64)	248, 201

138

	Alcohol	Anion	Reagent	Conditions	Product (yield)	Refs.
	Methyl-α-D-mannopyranoside	PF₆⁻ [acetylated (88)]	"	1. 90°, DMF, 3 hr 2. CH₃ONa, CH₃OH	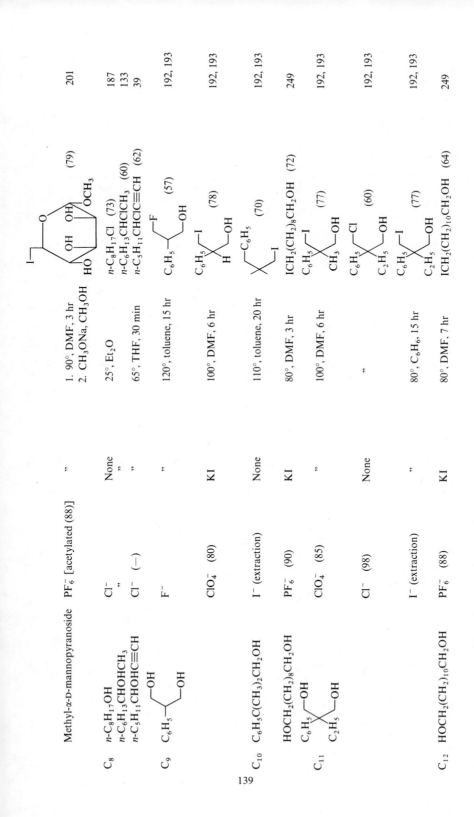 (79)	201
C₈	n-C₈H₁₇-OH	Cl⁻	None	25°, Et₂O	n-C₈H₁₇Cl (73)	187
	n-C₆H₁₃CHOHCH₃	"	"		n-C₆H₁₃CHClCH₃ (60)	133
	n-C₅H₁₁CHOHC≡CH	Cl⁻ (—)	"	65°, THF, 30 min	n-C₅H₁₁CHClC≡CH (62)	39
C₉	C₆H₅ —OH —OH	F⁻	"	120°, toluene, 15 hr	(57)	192, 193
C₁₀	C₆H₅C(CH₃)₂CH₂OH	ClO₄⁻ (80)	KI	100°, DMF, 6 hr	(78)	192, 193
		I⁻ (extraction)	None	110°, toluene, 20 hr	(70)	192, 193
C₁₁	HOCH₂(CH₂)₈CH₂OH	PF₆⁻ (90)	KI	80°, DMF, 3 hr	ICH₂(CH₂)₈CH₂OH (72)	249
	C₆H₅ —OH	ClO₄⁻ (85)	"	100°, DMF, 6 hr	(77)	192, 193
	C₆H₅ C₂H₅ —OH	Cl⁻ (98)	None	"	(60)	192, 193
		I⁻ (extraction)	"	80°, C₆H₆, 15 hr	(77)	192, 193
C₁₂	HOCH₂(CH₂)₁₀CH₂OH	PF₆⁻ (88)	KI	80°, DMF, 7 hr	ICH₂(CH₂)₁₀CH₂OH (64)	249

139

Alcohol	Anion for Salt Isolation and Yield (%)	Nucleophile	Conditions	Products and Yields from Alcohol (%)	Ref.
C$_{12}$ (*Contd.*) α,α'-Trehalose	PF_6^- (88)	Cl$^-$	25°, Et$_2$O	(65)	187
"	PF_6^- (acetylated) (85)	(CH$_3$)$_4$NCl	70°, DMF, 2 hr	6-Chloro-6-desoxy-α,α'-trehalose heptaacetate (37) + 6,6'-dichloro derivative (19)	35
	PF_6^- (acetylated) (85)	(CH$_3$)$_4$NCl	1. 80°, DMF, 10 hr 2. CH$_3$ONa, CH$_3$OH	(72)	201
	PF_6^- (acetylated)	(C$_2$H$_5$)$_4$NBr	90°, DMF, 4 hr	6-Bromo-6-desoxy-α,α'-trehalose heptaacetate (39) + 6,6'-dibromo derivative (19)	35
	PF_6^- (acetylated) (85)	"	1. 85°, DMF, 10 hr 2. CH$_3$ONa, CH$_3$OH	(75)	201
	PF_6^- (acetylated)	KI	90°, DMF, 3 hr	6-Iodo-6-desoxy-α,α'-trehalose heptaacetate (40) + 6,6'-diiodo derivative (14)	35

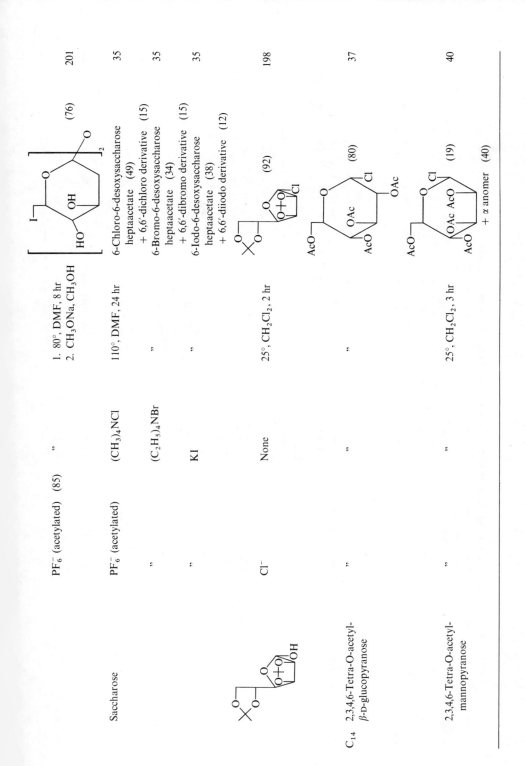

	PF$_6^-$ (acetylated) (85)	"	1. 80°, DMF, 8 hr 2. CH$_3$ONa, CH$_3$OH	(76)	201
Saccharose	PF$_6^-$ (acetylated)	(CH$_3$)$_4$NCl	110°, DMF, 24 hr	6-Chloro-6-desoxysaccharose heptaacetate (49) + 6,6'-dichloro derivative (15)	35
	"	(C$_2$H$_5$)$_4$NBr	"	6-Bromo-6-desoxysaccharose heptaacetate (34) + 6,6'-dibromo derivative (15)	35
	"	KI	"	6-Iodo-6-desoxysaccharose heptaacetate (38) + 6,6'-diiodo derivative (12)	35
	Cl$^-$	None	25°, CH$_2$Cl$_2$, 2 hr	(92)	198
C$_{14}$ 2,3,4,6-Tetra-O-acetyl-β-D-glucopyranose	"	"	"	(80)	37
2,3,4,6-Tetra-O-acetyl-mannopyranose	"	"	25°, CH$_2$Cl$_2$, 3 hr	(19) + α anomer (40)	40

141

TABLE XI. FORMATION OF CARBON–NITROGEN BONDS USING ALKOXY(TRISDIMETHYLAMINO)PHOSPHONIUM SALTS

	Alcohol	Anion Used for Salt Isolation (%)	Nucleophile	Conditions	Products and Yields from Alcohol (%)	Ref.
C_1	CH_3OH	ClO_4^- (100)	$C_6H_5NH_2$	130°, DMF, 10 hr	$C_6H_5NHCH_3$ (70)	189
C_3	$HC{\equiv}CCH_2OH$	"	CH_3NH_2	70°, DMF, 3 hr	$HC{\equiv}CCH_2NHCH_3$ (60)	189
		"	$HOCH_2CH_2NH_2$	80°, DMF, 3 hr	$HC{\equiv}CCH_2NHCH_2CH_2OH$ (57)	189
		"	Pyrrolidine	"	N-Propargylpyrrolidine (83)	189
		"	Morpholine	"	N-Propargylmorpholine (80)	189
		"	Cyclohexylamine	"	N-Propargylcyclohexylamine (77)	189
		"	$C_6H_5NH_2$	130°, DMF, 10 hr	$C_6H_5NHCH_2C{\equiv}CH$ (68)	189
C_4	$CH_3CH{=}CHCH_2OH$	ClO_4^- (not isolated)	CH_3NH_2	70°, DMF, 3 hr	$CH_3CH{=}CHCH_2NHCH_3$ (53)	189
		"	$HOCH_2CH_2NH_2$	80°, DMF, 3 hr	$CH_3CH{=}CHCH_2NHCH_2CH_2OH$ (52)	189
			Pyrrolidine	"	N-Crotylpyrrolidine (60)	189
			Morpholine	"	N-Crotylmorpholine (60)	189
			Cyclohexylamine	"	N-Crotylcyclohexylamine (51)	189
			$C_6H_5NH_2$	130°, DMF, 3 hr	$C_6H_5NHCH_2CH{=}CHCH_3$ (60)	189
C_5	$n\text{-}C_5H_{11}OH$	ClO_4^- (100)	$H_2NCH_2CH_2OH$	80°, DMF, 3 hr	$n\text{-}C_5H_{11}NHCH_2CH_2OH$ (55)	189
		"	Morpholine	"	N-n-Pentylmorpholine (75)	189
		"	Cyclohexylamine	"	N-n-Pentylcyclohexylamine (68)	189
		None	NaN_3	130°, DMF, 10 hr	$n\text{-}C_5H_{11}NHC_5H_{11}\text{-}n$ (60)	189
		ClO_4^- (100)	"	30°, THF, H_2O, 4 hr	$n\text{-}C_5H_{11}N_3$ (70)	189
			Pyrrolidine	80°, DMF, 3 hr	N-n-Pentylpyrrolidine (81)	189
	(Cl·····OH structure)	N_3^- (extraction)	None	80°, C_6H_6, 10 hr	(Cl·····N_3 structure) (66)	192, 193
	(OH structure)	N_3^- (extraction)	"	80°, C_6H_6, 4 hr	(N_3 structure) (80)	192, 193
	(OH structure)	PF_6^- (80)	NaN_3	100°, DMF, 10 hr	(N_3·····OH structure) (68)	192, 193
C_6	$n\text{-}C_6H_{13}OH$	ClO_4^- (100)	CH_3NH_2	70°, DMF, 3 hr	$n\text{-}C_6H_{13}NHCH_3$ (60)	189
		—	NaN_3	30°, THF, H_2O, 4 hr	$n\text{-}C_6H_{13}N_3$ (70)	188

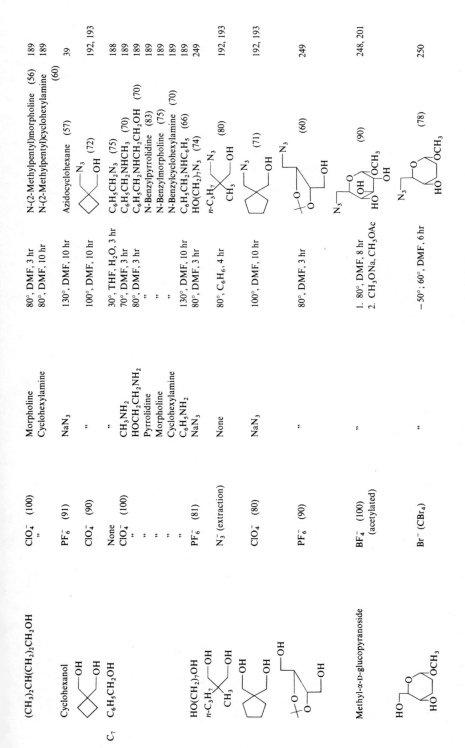

Substrate	Salt (yield)	Reagent	Conditions	Product (yield)	Refs.
(CH₃)₂CH(CH₂)₂CH₂OH	ClO₄⁻ (100)	Morpholine	80°, DMF, 3 hr	N-(2-Methylpentyl)morpholine (56)	189
	"	Cyclohexylamine	80°, DMF, 10 hr	N-(2-Methylpentyl)cyclohexylamine (60)	189
Cyclohexanol	PF₆⁻ (91)	NaN₃	130°, DMF, 10 hr	Azidocyclohexane (57)	39
[structure, OH/OH]	ClO₄⁻ (90)	"	100°, DMF, 10 hr	[structure, N₃/OH] (72)	192, 193
C₇ C₆H₅CH₂OH	None	CH₃NH₂	30°, THF, H₂O, 3 hr	C₆H₅CH₂N₃ (75)	188
	ClO₄⁻ (100)	HOCH₂CH₂NH₂	70°, DMF, 3 hr	C₆H₅CH₂NHCH₃ (70)	189
	"	Pyrrolidine	80°, DMF, 3 hr	C₆H₅CH₂NHCH₂CH₂OH (70)	189
	"		"	N-Benzylpyrrolidine (83)	189
	"	Morpholine	"	N-Benzylmorpholine (75)	189
	"	Cyclohexylamine		N-Benzylcyclohexylamine (70)	189
	"	C₆H₅NH₂	130°, DMF, 10 hr	C₆H₅CH₂NHC₆H₅ (66)	189
	PF₆⁻ (81)	NaN₃	80°, DMF, 3 hr	HO(CH₂)₇N₃ (74)	249
HO(CH₂)₇OH	N₃⁻ (extraction)	None	80°, C₆H₆, 4 hr	[structure, n-C₃H₇/N₃/CH₃/OH] (80)	192, 193
[cyclopentane diol structure]	ClO₄⁻ (80)	NaN₃	100°, DMF, 10 hr	[cyclopentane N₃/OH structure] (71)	192, 193
[bicyclic diol structure]	PF₆⁻ (90)	"	80°, DMF, 3 hr	[bicyclic N₃/OH structure] (60)	249
Methyl-α-D-glucopyranoside	BF₄⁻ (100) (acetylated)	"	1. 80°, DMF, 8 hr 2. CH₃ONa, CH₃OAc	[glucopyranoside N₃ structure] (90)	248, 201
[pyranoside structure]	Br⁻ (CBr₄)	"	−50°; 60°, DMF, 6 hr	[pyranoside N₃ structure] (78)	250

143

TABLE XI. FORMATION OF CARBON–NITROGEN BONDS USING ALKOXY(TRISDIMETHYLAMINO)PHOSPHONIUM SALTS (Continued)

Alcohol	Anion Used for Salt Isolation (%)	Nucleophile	Conditions	Products and Yields from Alcohol (%)	Ref.
C_7 Methyl-α-D-mannopyranoside (Contd.)	PF_6^- [acetylated (88)]	NaN_3	90°, DMF, 3 hr	[pyranose structure: N_3, OAc, OAc, OAc, OCH_3] (87)	201
Methyl-α-D-galactopyranoside	PF_6^- [acetylated (85)]	"	90°, DMF, 15 hr	[pyranose structure: N_3, AcO, OAc, OAc, OCH_3] (80)	201
C_8 (−)-n-C_6H_{13}CHOHCH$_3$	PF_6^- (97) (diisopropylchloramine as oxidant)	"	130°, DMF, 5 hr	(+)-n-C_6H_{13}CH(N$_3$)CH$_3$ (75)	39
HO(CH$_2$)$_8$OH	PF_6^- (77)	"	80°, DMF, 3 hr	HO(CH$_2$)$_8$N$_3$ (70)	249
C_6H_5NHCH$_2$CH$_2$OH	PF_6^- (97)	"	"	C_6H_5NHCH$_2$CH$_2$N$_3$ (87)	251
C_9 HO(CH$_2$)$_9$OH	PF_6^- (77)	"	80°, DMF, 3 hr	HO(CH$_2$)$_9$N$_3$ (70)	249
C_6H_5CH(OH)CH$_2$CH(OH) [structure]	N_3^- (extraction)	None	80°, C_6H_6, 4 hr	C_6H_5CH(N$_3$)CH$_2$CH(OH) [structure] (70)	192, 193
C_{10} 4-t-Butylcyclohexanol (cis–trans = 30:70)	PF_6^- (75)	NaN_3	130°, DMF, 10 hr	1-Azido-4-t-butylcyclohexane (67) (cis–trans = 90:10)	39
C_6H_5C(CH$_3$)$_2$CH$_2$OH	ClO_4^-	None	100°, DMF, 20 hr	[t-butyl-N_3 structure] (67)	192, 193
(−)-Menthol	PF_6^- (86)	NaN_3	130°, DMF, 10 hr	(+)-Azidomenthane (52)	39
C_6H_5C(CH$_3$)(OH)CH(OH) [structure]	N_3^- (extraction)	—	80°, C_6H_6, 4 hr	C_6H_5C(CH$_3$)(N$_3$)CH(OH) [structure] (71)	192, 193

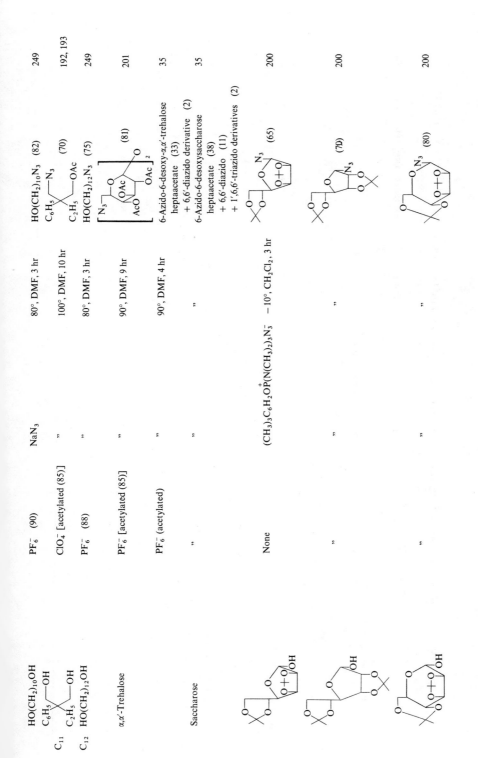

Substrate	Anion (yield)	Reagent	Conditions	Product (yield)	Refs.
$HO(CH_2)_{10}OH$	PF_6^- (90)	NaN_3	80°, DMF, 3 hr	$HO(CH_2)_{10}N_3$ (82)	249
C_{11} C_6H_5—OH C_2H_5—OH	ClO_4^- [acetylated (85)]	"	100°, DMF, 10 hr	C_6H_5—N_3 / C_2H_5—OAc (70)	192, 193
C_{12} $HO(CH_2)_{12}OH$	PF_6^- (88)	"	80°, DMF, 3 hr	$HO(CH_2)_{12}N_3$ (75)	249
α,α'-Trehalose	PF_6^- [acetylated (85)]	"	90°, DMF, 9 hr	(structure) (81)	201
Saccharose	PF_6^- (acetylated)	"	90°, DMF, 4 hr	6-Azido-6-desoxy-α,α'-trehalose heptaacetate (33) + 6,6'-diazido derivative (2)	35
	"	"	"	6-Azido-6-desoxysaccharose heptaacetate (38) + 6,6'-diazido (11) + 1',6,6'-triazido derivatives (2)	35
(structure)	None	$(CH_3)_3C_6H_2OP(N(CH_3)_2)_3 N_3^-$	$-10°$, CH_2Cl_2, 3 hr	(structure, N_3) (65)	200
(structure)	"	"	"	(structure) (70)	200
(structure)	"	"	"	(structure, N_3) (80)	200

145

TABLE XI. FORMATION OF CARBON–NITROGEN BONDS USING ALKOXY(TRISDIMETHYLAMINO)PHOSPHONIUM SALTS (*Continued*)

Alcohol	Anion Used for Salt Isolation (%)	Nucleophile	Conditions	Products and Yields from Alcohol (%)	Ref.
C$_{26}$	None	$(CH_3)_3C_6H_2OP(N(CH_3)_2)_3N_3^-$	$-10°$, CH_2Cl_2, 3 hr	(70) + α anomer (15)	200
C$_{27}$	"	"	"	(75)	200
C$_{37}$	"	NaN_3	$-50°$, $+80°$, DMF, 8 hr	(70)	252

TABLE XII. FORMATION OF ETHERS AND GLYCOSIDES USING ALKOXY(TRISDIMETHYLAMINO)PHOSPHONIUM SALTS

	Alcohol	Nucleophile	Anion Used for Salt Isolation	Conditions	Products and Yields from Salt (%)	Ref.
C_3	$CH_2{=}CHCH_2OH$	C_6H_5OK	PF_6^-	150°, DMF, 15 hr	$C_6H_5OCH_2CH{=}CH_2$ (79)	194
C_5	$(CH_3)_3CCH_2OH$	[2-nitro-2′-biphenylyl OK]	"	"	[2-nitrobiphenyl-2′-yl $OCH_2C(CH_3)_3$] (90)	194
		$p\text{-}O_2NC_6H_4OK$	"	"	$p\text{-}O_2NC_6H_4OCH_2C(CH_3)_3$ (42), $p\text{-}O_2NC_6H_4N(CH_3)_2$ (29)	194
		$p\text{-}CH_3OC_6H_4OK$	"	"	$p\text{-}CH_3OC_6H_4OCH_2C(CH_3)_3$ (66)	194
		C_6H_5OK	"	"	$C_6H_5OCH_2C(CH_3)_3$ (75)	194
C_8	$n\text{-}C_6H_{13}CHOHCH_3$	[2-nitro-2′-biphenylyl OK]	PF_6^-	"	[2-nitrobiphenyl-2′-yl $OCH(CH_3)C_6H_{13}\text{-}n$]	194
		$p\text{-}O_2NC_6H_4OK$	"	"	$n\text{-}C_6H_{13}CH(CH_3)OC_6H_4NO_2\text{-}p$ (80)	194
		$p\text{-}CH_3OC_6H_4OK$	"	"	$n\text{-}C_6H_{13}CH(CH_3)OC_6H_4OCH_3\text{-}p$ (68)	194
		C_6H_5OK	"	"	$n\text{-}C_6H_{13}CH(CH_3)OC_6H_5$ (99)	194
C_{10}	1,10-Decanediol	$NaOCH_3$	"	60°, CH_3OH, 4 hr	10-Methoxy-1-decanol (76)	249
C_{12}	[glycoside structure, OH]	$CH_3OH/N(C_2H_5)_3$	$CH_3C_6H_4SO_3^-$	25°, CH_2Cl_2, 4 hr	[glycoside structure, OCH_3] (34) + β isomer (41)	198

TABLE XII. FORMATION OF ETHERS AND GLYCOSIDES USING ALKOXY(TRISDIMETHYLAMINO)PHOSPHONIUM SALTS (*Continued*)

Alcohol	Anion Used for Salt Isolation	Nucleophile	Conditions	Products and Yields from Salt (%)	Ref.
C_{12} (*Contd.*) [sugar structure with OH]	$CH_3C_6H_4SO_3^-$	$C_6H_5CH_2OH/N(C_2H_5)_3$	"	[sugar structure] $OCH_2C_6H_5$ (52) + β isomer (22)	198
	"	$(CH_3)_2CHOH/N(C_2H_5)_3$	"	[sugar structure] $OCH(CH_3)_2$ (72)	198
	"	$(CH_3)_3COH/N(C_2H_5)_3$	"	[sugar structure] $OC(CH_3)_3$ (90)	198
	"	$CH_2=CHCH_2OH/N(C_2H_5)_3$	"	[sugar structure] $OCH_2CH=CH_2$ (41) + β isomer (34)	198
	$CH_3C_6H_4SO_3^-$	$C_6H_5OH/N(C_2H_5)_3$	$25°, CH_2Cl_2$, 4 hr	[sugar structure] OC_6H_5 (80)	198

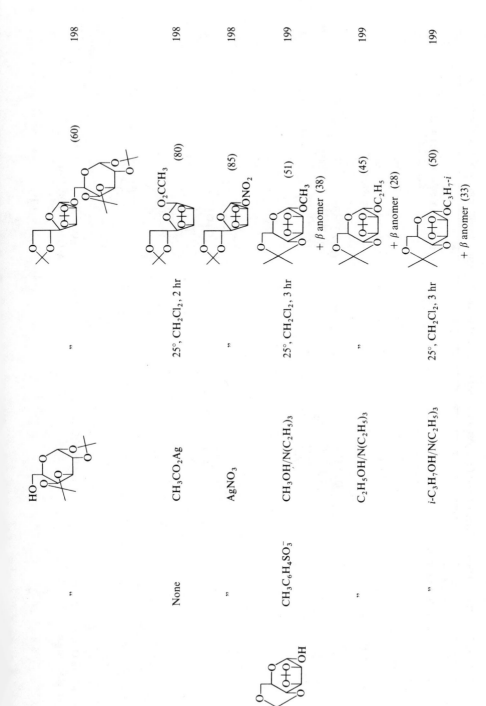

"	"	HO structure	(60)	198
None	CH₃CO₂Ag	O₂CCH₃ structure (80)	25°, CH₂Cl₂, 2 hr	198
"	AgNO₃	ONO₂ structure (85)	"	198
$CH_3C_6H_4SO_3^-$	$CH_3OH/N(C_2H_5)_3$	OCH₃ structure (51) + β anomer (38)	25°, CH₂Cl₂, 3 hr	199
"	$C_2H_5OH/N(C_2H_5)_3$	OC₂H₅ structure (45) + β anomer (28)	"	199
"	$i\text{-}C_3H_7OH/N(C_2H_5)_3$	OC₃H₇-i structure (50) + β anomer (33)	25°, CH₂Cl₂, 3 hr	199

149

TABLE XII. FORMATION OF ETHERS AND GLYCOSIDES USING ALKOXY(TRISDIMETHYLAMINO)PHOSPHONIUM SALTS (*Continued*)

Alcohol	Anion Used for Salt Isolation	Nucleophile	Conditions	Products and Yields from Salt (%)	Ref.
C_{12} (*Contd.*) [structure] OH	$CH_3C_6H_5SO_3^-$	$t\text{-}C_4H_9OH/N(C_2H_5)_3$	25°, CH_2Cl_2, 4 hr	[structure] $OC_4H_9\text{-}t$ (50)	199
	"	[structure] HO— /$N(C_2H_5)_3$	"	[structure] (45)	199
	"	[structure] OH /$N(C_2H_5)_3$	"	[structure] (40)	199
C_{14} 2,3,4,6-Tetra-O-acetyl-β-D-glucopyranose	None	$i\text{-}C_3H_7OH/N(C_2H_5)_3$	"	[structure] AcO AcO OAc $OOC_3H_7\text{-}i$ CH_3 (80)	37

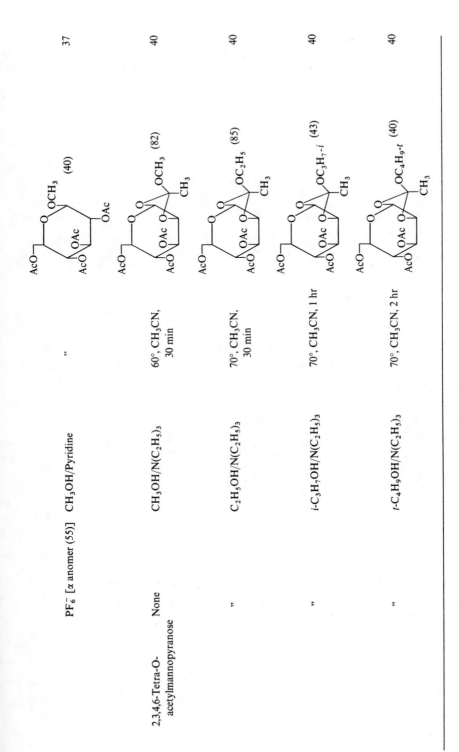

			Product	Ref.
PF$_6^-$ [α anomer (55)]	CH$_3$OH/Pyridine	"	(40)	37
2,3,4,6-Tetra-O-acetylmannopyranose	None	CH$_3$OH/N(C$_2$H$_5$)$_3$	60°, CH$_3$CN, 30 min	OCH$_3$ (82)	40
	"	C$_2$H$_5$OH/N(C$_2$H$_5$)$_3$	70°, CH$_3$CN, 30 min	OC$_2$H$_5$ (85)	40
	"	i-C$_3$H$_7$OH/N(C$_2$H$_5$)$_3$	70°, CH$_3$CN, 1 hr	OC$_3$H$_{7}$-i (43)	40
	"	t-C$_4$H$_9$OH/N(C$_2$H$_5$)$_3$	70°, CH$_3$CN, 2 hr	OC$_4$H$_9$-t (40)	40

TABLE XIII. FORMATION OF CARBON-SULFUR BONDS USING ALKOXY(TRISDIMETHYLAMINO)PHOSPHONIUM SALTS

Alcohol	Anion for Salt Isolation and Yield (%)	Nucleophile	Conditions	Products and Yields from Alcohol	Ref.
C_3 CH_2=CHCH$_2$OH	PF_6^-	p-CH$_3$C$_6$H$_4$SK	150°, DMF, 15 hr	p-CH$_3$C$_6$H$_4$SCH$_2$CH=CH$_2$ (97)	194
	''	C$_6$H$_5$SH/KOH	''	C$_6$H$_5$SCH$_2$CH=CH$_2$ (90)	
C_5 n-C$_5$H$_{11}$OH	None	NH$_4$SCN	30°, THF, 2 hr	n-C$_5$H$_{11}$SCN (70)	188
	''	C$_6$H$_5$SH/N(C$_2$H$_5$)$_3$	−60°, THF, 20 min	n-C$_5$H$_{11}$SC$_6$H$_5$ (70)	188
(see structure)	ClO$_4^-$ [acetylation (80)]	CH$_3$COSH/N(C$_2$H$_5$)$_3$	1. 100°, DMF, 10 hr 2. Ac$_2$O	SCOCH$_3$ (70) OCOCH$_3$	192, 193
(see structure)	ClO$_4^-$ (95)	C$_6$H$_5$SH/N(C$_2$H$_5$)$_3$	100°, DMF, 10 hr	SC$_6$H$_5$ (76) OH	192, 193
C_6 n-C$_6$H$_{13}$OH	None	NH$_4$SCN	30°, THF, 2 hr	n-C$_6$H$_{13}$SCN (70)	188
	''	C$_6$H$_5$SH/N(C$_2$H$_5$)$_3$	−60°, THF, 20 min	n-C$_6$H$_{13}$SC$_6$H$_5$ (70)	188
C_7 C$_6$H$_5$CH$_2$OH	''	NH$_4$SCN	30°, THF, 90 min	C$_6$H$_5$CH$_2$SCN (80)	188
	''	C$_6$H$_5$SH/N(C$_2$H$_5$)$_3$	−60°, THF, 20 min	C$_6$H$_5$CH$_2$SC$_6$H$_5$ (85)	188
	PF_6^-	C$_6$H$_5$SK	150°, DMF, 15 hr	(83)	194
	''	p-CH$_3$C$_6$H$_4$SK	''	C$_6$H$_5$CH$_2$SC$_6$H$_4$CH$_3$-p (98)	194
(see structure)	PF_6^- (90)	C$_6$H$_5$SH/N(C$_2$H$_5$)$_3$	80°, DMF, 5 hr	SC$_6$H$_5$ (50) OH	249
Methyl-α-D-glucopyranoside	BF$_4^-$ [acetylation (100)]	CH$_3$COSH/N(C$_2$H$_5$)$_3$	1. 80°, DMF, 8 hr 2. CH$_2$ONa, CH$_3$OH	(54)	248, 201
Methyl-α-D-mannopyranoside	PF_6^- [acetylation (88)]	NH$_4$SCN	90°, DMF, 10 hr	(83)	201

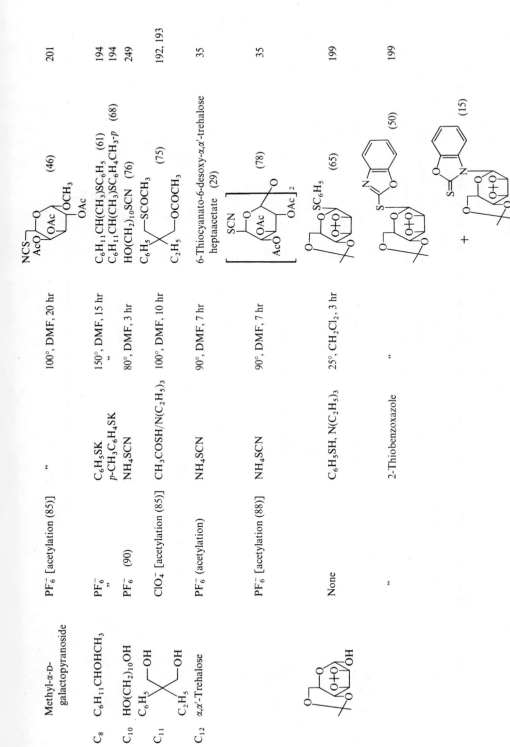

	Substrate	Reagent	Conditions	Product	Ref.
	Methyl-α-D-galactopyranoside	PF$_6^-$ [acetylation (85)]	"	100°, DMF, 20 hr	(structure) NCS AcO OAc OCH$_3$ OAc (46)	201
C$_8$	C$_6$H$_{11}$CHOHCH$_3$	C$_6$H$_5$SK	150°, DMF, 15 hr	C$_6$H$_{11}$CH(CH$_3$)SC$_6$H$_5$ (61)	194
	PF$_6^-$	p-CH$_3$C$_6$H$_4$SK	"	C$_6$H$_{11}$CH(CH$_3$)SC$_6$H$_4$CH$_3$-p (68)	194
C$_{10}$	HO(CH$_2$)$_{10}$OH	NH$_4$SCN	80°, DMF, 3 hr	HO(CH$_2$)$_{10}$SCN (76)	249
	PF$_6^-$ (90)				
C$_{11}$	C$_6$H$_5$—C(C$_2$H$_5$)—OH (structure with OH)	CH$_3$COSH/N(C$_2$H$_5$)$_3$	100°, DMF, 10 hr	C$_6$H$_5$—C(SCOCH$_3$)(C$_2$H$_5$)—OCOCH$_3$ (75)	192, 193
	ClO$_4^-$ [acetylation (85)]				
C$_{12}$	α,α′-Trehalose	NH$_4$SCN	90°, DMF, 7 hr	6-Thiocyanato-6-desoxy-α,α′-trehalose heptaacetate (29)	35
	PF$_6^-$ (acetylation)				
	(trehalose acetonide structure with OH)	NH$_4$SCN	90°, DMF, 7 hr	(structure) SCN AcO OAc (78)	35
	PF$_6^-$ [acetylation (88)]				
		C$_6$H$_5$SH, N(C$_2$H$_5$)$_3$	25°, CH$_2$Cl$_2$, 3 hr	(structure) SC$_6$H$_5$ (65)	199
	None				
		2-Thiobenzoxazole	"	(structure) (50) + (structure) (15)	199
	"				

153

TABLE XIII. FORMATION OF CARBON-SULFUR BONDS USING ALKOXY(TRISDIMETHYLAMINO)PHOSPHONIUM SALTS (*Continued*)

Alcohol	Nucleophile	Conditions	Anion for Salt Isolation and Yield (%)	Products and Yields from Alcohol	Ref.
C_{12} (*Contd.*)	2-Thiobenzothiazole	25°, CH_2Cl_2, 3 hr	None	(65)	199
C_{14} 2,3,4,6-Tetra-O-acetyl-β-D-glucopyranose	$C_6H_5SH/N(C_2H_5)_3$	25°, CH_2Cl_2, 4 hr	"	(65) + α anomer (15)	39
	"	"	PF_6^-, α anomer (59)	(38)	39
2,3,4,6-Tetra-O-acetylmannopyranose	$C_6H_5SH/N(C_2H_5)_3$	25°, CH_2Cl_2, 1 hr	None	(58)	40

154

TABLE XIV. FORMATION OF CARBON–HYDROGEN AND CARBON–CARBON BONDS USING ALKOXY(TRISDIMETHYLAMINO)PHOSPHONIUM SALTS

	Alcohol	Anion for Salt Isolation and Yield (%)	Nucleophile	Conditions	Products and Yields from Alcohol (%)	Ref.
C$_6$	n-C$_6$H$_{13}$OH	None	KCN	30°, CH$_2$Cl$_2$, H$_2$O 3 hr	n-C$_6$H$_{13}$CN (65)	188
C$_7$	Methyl-α-D-glucopyranoside	None (acetylation)	LiAlH$_4$	1. 65° THF, 25 hr 2. CH$_3$ONa, CH$_3$OH	(60)	248, 201
C$_{10}$	HO(CH$_2$)$_{10}$OH	PF$_6^-$ (90)	KCN	80°, DMF, 3 hr	HO(CH$_2$)$_{10}$CN (70)	249
		PF$_6^-$	LiBH(C$_2$H$_5$)$_3$	70° THF, 3 hr	(90)	196
C$_{12}$		ʺ	ʺ	ʺ	(92)	196

TABLE XIV. FORMATION OF CARBON–HYDROGEN AND CARBON–CARBON BONDS USING ALKOXY(TRISDIMETHYLAMINO)PHOSPHONIUM SALTS (*Contd.*)

Alcohol	Anion for Salt Isolation and Yield (%)	Nucleophile	Conditions	Products and Yields from Alcohol (%)	Ref.
C$_{13}$	None	"	"	(92)	196
C$_{21}$	PF$_6^-$ (96)	"	"	(91)	254, 196

TABLE XV. FORMATION OF OXETANES FROM 1,3-GLYCOLS USING 3-HYDROXY(TRISDIMETHYLAMINO)PHOSPHONIUM CHLORIDES

	Salt from Glycol	Conditions	Product and Yield (%)	Ref.
C_5	CH_3—C(—OH)(—OH)—CH_3 (gem-dimethyl 1,3-diol)	150°, Na/Methyl carbitol	3,3-dimethyloxetane (63)	203
	(cyclopropyl 1,3-diol)	"	(10)	203
C_6	C_2H_5, CH_3 1,3-diol	"	(70)	203
	CH_3, CH_3, CH_3 1,3-diol	"	(65)	203
	(cyclobutyl 1,3-diol)	"	(55)	203
C_7	(cyclopentyl 1,3-diol)	"	(60)	203
C_9	$C_6H_5CH(CH_2OH)_2$	70°, Na/CH_3OH	3-Phenyloxetane (35)	203
C_{10}	$C_6H_5C(CH_3)(CH_2OH)_2$	"	3-Methyl-3-phenyloxetane (70)	203
C_{11}	$C_6H_5C(C_2H_5)(CH_2OH)_2$	"	3-Ethyl-3-phenyloxetane (60)	203

REFERENCES

[1] A. Michaelis and R. Kaehne, *Ber.*, **31**, 1048 (1898).

[2] A. E. Arbuzov, *J. Russ. Phys. Chem. Soc.*, **38**, 687 (1906) [*Chem. Zentralbl.*, **1906**, II, 1639].

[3] J. I. G. Cadogan and R. K. Mackie, *Chem. Soc. Rev.*, **3**, 87 (1974).

[4] R. Appel, *Angew. Chem., Int. Ed. Engl.*, **14**, 801 (1975).

[5] T. Mukaiyama, *Angew. Chem., Int. Ed. Engl.*, **15**, 94 (1976).

[6] R. K. Mackie, in *Organophosphorus Reagents in Organic Chemistry*, J. I. G. Cadogan, Ed., Academic Press, London, 1979.

[7] O. Mitsunobu, *Synthesis*, **1981**, 1.

[8] O. Mitsunobu, M. Yamada, and T. Mukaiyama, *Bull. Chem. Soc. Jpn.*, **40**, 935 (1967).

[9] E. Brunn and R. Huisgen, *Angew. Chem., Int. Ed. Engl.*, **8**, 513 (1969).

[10] O. Mitsunobu and M. Iguchi, *Bull. Chem. Soc. Jpn.*, **44**, 2327 (1971).

[11] H. Loibner and E. Zbiral, *Helv. Chim. Acta*, **59**, 2100 (1976).

[12] B. A. Arbuzov, N. A. Polezhaeva, and V. S. Vinogradova, *Izv. Akad. Nauk SSSR, Ser. Khim.*, **1968**, 2525 [*C.A.*, **70**, 87682r (1969)].

[13] G. Alfredsson and P. J. Garegg, *Acta Chem. Scand., Ser. B*, **27**, 724 (1973).

[14] G. Grynkiewicz and H. Burynska, *Tetrahedron*, **32**, 2109 (1976).

[15] M. Gligorijevic and M. J. Gasic, *Bull. Soc. Chim. Beograd*, **41**, 353 (1976).

[16] R. Aneja, A. P. Davies, A. Harkes, and J. A. Knaggs, *J. Chem. Soc., Chem. Commun.*, **1974**, 963.

[17] R. Aneja, A. P. Davies, and J. A. Knaggs, *Tetrahedron Lett.*, **1975**, 1033.

[18] R. Appel, F. Knoll, W. Michel, W. Morbach, H. D. Wihler, and H. Veltmann, *Chem. Ber.*, **109**, 58 (1976).

[19] A. J. Speziale and K. W. Ratts, *J. Am. Chem. Soc.*, **84**, 857 (1962).

[20] R. Appel and K. Warning, *Chem. Ber.*, **108**, 606 (1975).

[21] R. Appel and H. Veltmann, *Tetrahedron Lett.*, **1977**, 399.

[22] L. Gruber, I. Tömösközi, and L. Radics, *Synthesis*, **1975**, 708.

[23] I. M. Downie, J. B. Holmes, and J. B. Lee, *Chem. Ind. (Lond.)*, **28**, 900 (1966).

[24] R. G. Weiss and E. I. Snyder, *J. Org. Chem.*, **35**, 1627 (1970).

[25] L. A. Jones, C. E. Summer, Jr., B. Franzus, T. T. S. Huang, and E. I. Snyder, *J. Org. Chem.*, **43**, 2821 (1978).

[26] R. Aneja, A. P. Davies, and J. A. Knaggs, *Tetrahedron Lett.*, **1974**, 67.

[27] S. J. Cristol, R. M. Strom, and D. P. Stull, *J. Org. Chem.*, **43**, 1150 (1978).

[28] R. G. Weiss and E. I. Snyder, *J. Org. Chem.*, **36**, 403 (1971).

[29] C. Georgoulis and G. Ville, *Bull. Soc. Chim. Fr.*, **1975**, 607.

[30] R. M. Magid, O. S. Fruchey, W. L. Johnson, and T. G. Allen, *J. Org. Chem.*, **44**, 359 (1979).

[31] J. B. Lee and T. J. Nolan, *Tetrahedron*, **23**, 2789 (1967).

[32] B. Castro, R. Burgada, G. Lavielle, and J. Villieras, *C. R. Hebd. Seances Acad. Sci., Ser. C.*, **268**, 1067 (1969).

[33] B. Castro, R. Burgada, G. Lavielle, and J. Villieras, *Bull. Soc. Chim. Fr.*, **1969**, 2770.

[34] B. Castro, J. Villieras, R. Burgada, and G. Lavielle, in *Chimie Organique du Phosphore*, Ed. du CNRS, Paris, 1970, p. 235.

[35] B. Castro, Y. Chapleur, and B. Gross, *Carbohydr. Res.*, **36**, 412 (1974).

[36] B. Castro, M. Nacro, and C. Selve, *Tetrahedron*, **35**, 481 (1979).

[37] B. Castro, Y. Chapleur, and B. Gross, *Bull. Soc. Chim. Fr.*, **1975**, 875.

[38] R. A. Boigegrain, B. Castro, and B. Gross, *Tetrahedron Lett.*, **1975**, 3947.

[39] Y. Chapleur, B. Castro, and B. Gross, *Tetrahedron*, **33**, 1609 (1977).

[40] Y. Chapleur, B. Castro, and B. Gross, *Synth. Commun.*, **7**, 143 (1977).

[41] D. Neibecker and B. Castro, *J. Organomet. Chem.*, **134**, 105 (1977).

[42] D. C. Morrison, *J. Org. Chem.*, **23**, 1072 (1958).

[43] V. A. Ginsburg, M. N. Vasil'eva, S. S. Dubov, and A. Yakubovitch, *Zh. Obshch. Khim.*, **30**, 2854 (1960) [*C.A.*, **55**, 17477b (1961)].

[44] G. Grynkiewicz, J. Jurczak, and A. Zamojski, *Bull. Acad. Pol. Sci.*, **24**, 83 (1976).

[45] G. Grynkiewicz and J. Jurczak, *Carbohydr. Res.*, **43**, 188 (1975).

[46] G. Grynkiewicz, *Bull. Acad. Pol. Sci.*, **26**, 537 (1978).

[47] G. Grynkiewicz, J. Jurczak, and A. Zamojski, *J. Chem. Soc., Chem. Commun.*, **1974**, 413.

[48] O. Mitsunobu, M. Wada, and T. Sano, *J. Am. Chem. Soc.*, **94**, 679 (1972).

[49] A. Zamojski, W. A. Szarek, and K. N. Jones, *Carbohydr. Res.*, **23**, 460 (1972).

[50] M. Wada, T. Sano, and O. Mitsunobu, *Bull. Chem. Soc. Jpn.*, **46**, 2833 (1973).

[51] J. Jurczak, G. Grynkiewicz, and A. Zamojski, *Carbohydr. Res.*, **39**, 147 (1975).

[52] A. Banaszek, B. Szechner, J. Mieczkowski, and A. Zamojski, *Rocz. Chem.*, **50**, 105 (1976).

[53] O. Mitsunobu, S. Takizawa, and H. Morimoto, *J. Am. Chem. Soc.*, **98**, 7858 (1976).

[54] M. Wada and O. Mitsunobu, *Tetrahedron Lett.*, **1972**, 1279.

[55] H. Morimoto, T. Furukawa, K. Miyazima, and O. Mitsunobu, *Chem. Lett.*, **1973**, 821.

[56] W. A. Szarek, C. Depew, and J. K. N. Jones, *J. Heterocycl. Chem.*, **13**, 1131 (1976).

[57] W. A. Szarek, C. Depew, H. C. Jarrell, and J. K. N. Jones, *J. Chem. Soc., Chem. Commun.*, **1975**, 648.

[58] H. Loibner and E. Zbiral, *Helv. Chim. Acta*, **60**, 417 (1977).

[59] H. Loibner and E. Zbiral, *Justus Liebigs Ann. Chem.*, **1978**, 78.

[60] H. Loibner and E. Zbiral, *Tetrahedron*, **34**, 713 (1978).

[61] J. Schweng and E. Zbiral, *Tetrahedron Lett.*, **1978**, 119.

[62] W. Reischi and E. Zbiral, *Justus Liebigs Ann. Chem.*, **1978**, 745.

[63] H. Brandstetter and E. Zbiral, *Helv. Chim. Acta*, **61**, 1832 (1978).

[64] H. Brandstetter and E. Zbiral, *Helv. Chim. Acta*, **63**, 327 (1980).

[65] B. Lal, B. N. Pramanik, M. S. Manhas, and A. K. Bose, *Tetrahedron Lett.*, **1977**, 1977.
[66] O. Mitsunobu and M. Yamada, *Bull. Chem. Soc. Jpn.*, **40**, 2380 (1967).
[67] A. K. Bose, B. Lal, W. A. Hoffman, and M. S. Manhas, *Tetrahedron Lett.*, **1973**, 1619.
[68] S. Shimokawa, J. Kimura, and O. Mitsunobu, *Bull. Chem. Soc. Jpn.*, **49**, 3357 (1976).
[69] L. P. L. Piacenza, *J. Org. Chem.*, **42**, 3778 (1977).
[70] G. Grynkiewicz, *Rocz. Chem.*, **50**, 1449 (1976).
[71] H. Kunz and O. Schmidt, *Z. Naturforsch.*, *B*, **33**, 1009 (1978).
[72] J. Mulzer and G. Brüntrup, *Angew. Chem., Int. Ed. Engl.*, **16**, 255 (1977).
[73] J. Mulzer, A. Pointner, A. Chucholowski, and G. Brüntrup, *J. Chem. Soc., Chem. Commun.*, **1979**, 52.
[74] J. Mulzer, G. Brüntrup, and A. Chucholowski, *Angew. Chem., Int. Ed. Engl.*, **18**, 622 (1979).
[75] T. Kurihara, Y. Nakajima, and O. Mitsunobu, *Tetrahedron Lett.*, **1976**, 2455.
[76] B. Seuring and D. Seebach, *Justus Liebigs Ann. Chem.*, **1978**, 2044.
[77] Y. Fukuyama, C. L. Kirkemo, and J. D. White, *J. Am. Chem. Soc.*, **99**, 646 (1977).
[78] T. A. Hase and E. L. Nylund, *Tetrahedron Lett.*, **1979**, 2633.
[79] A. I. Myers and R. A. Amos, *J. Am. Chem. Soc.*, **102**, 870 (1980).
[80] O. Mitsunobu, J. Kimura, and Y. Fujisawa, *Bull. Chem. Soc. Jpn.*, **45**, 245 (1972).
[81] E. Grochowski and J. Jurczak, *Synthesis*, **4**, 277 (1977).
[82] B. Castro, J. R. Dormoy, G. Evin, and C. Selve, *J. Chem. Res.*, (*S*), **1977**, 182.
[83] G. Grynkiewicz, J. Jurczak, and A. Zamojski, *Tetrahedron*, **31**, 1411 (1975).
[84] M. Furukawa, T. Okawara, Y. Noguchi, and M. Nishikawa, *Synthesis*, **1978**, 441.
[85] O. Mitsunobu, K. Kato, and J. Kimura, *J. Am. Chem. Soc.*, **91**, 6510 (1969).
[86] B. Mlotkowska and A. Zwierzak, *Pol. J. Chem.*, **53**, 359 (1979).
[87] E. Grochowski and J. Jurczak, *Carbohydr. Res.*, **1976**, C15.
[88] E. Grochowski and J. Jurczak, *Synthesis*, **10**, 682 (1976).
[89] E. Grochowski, E. Falent, and J. Jurczak, *Pol. J. Chem.*, **52**, 335 (1978).
[90] E. Grochowski and E. Falent-Kwastowa, *J. Chem. Res.*, (*S*), **1978**, 300.
[91] R. S. Glass and R. J. Swedo, *J. Org. Chem.*, **43**, 2291 (1978).
[92] S. Bittner and Y. Assaf, *Chem. Ind.* (*Lond.*), **1975**, 281.
[93] M. S. Manhas, W. H. Hoffman, B. Lal, and A. K. Bose, *J. Chem. Soc., Perkin Trans. 1*, **1975**, 461.
[94] J. T. Carlock and M. P. Mack, *Tetrahedron Lett.*, **1978**, 5153.
[95] G. Grynkiewicz, *Carbohydr. Res.*, **53**, C11 (1977).
[96] P. J. Garregg, T. Iverson, and T. Norberg, *Carbohydr. Res.*, **73**, 313 (1979).
[97] W. A. Szarek, H. D. Jarrell, and J. K. N. Jones, *Carbohydr. Res.*, **57**, C13 (1977).
[98] G. Grynkiewicz and A. Zamojski, *Synth. Commun.*, **8**, 491 (1978).
[99] J. Kimura, A. Kawashima, M. Sugizaki, N. Nemoto, and O. Mitsunobu, *J. Chem. Soc., Chem. Commun.*, **1979**, 303.
[100] H. Wojciechowska, R. Pawlowicz, R. Andruszkiewicz, and J. Grybowska, *Tetrahedron Lett.*, **1978**, 4063.
[101] F. DiNinno, *J. Am. Chem. Soc.*, **100**, 3251 (1978).
[102] H. Kunz and P. Schmidt, *Tetrahedron Lett.*, **1979**, 2123.
[103] P. J. Garregg and B. Samuelsson, *J. Chem. Soc., Chem. Commun.*, **1979**, 978.
[104] C. R. Haylock, L. D. Melton, K. N. Slessor, and A. S. Tracey, *Carbohydr. Res.*, **16**, 375 (1971).
[105] N. K. Kochetkov and A. I. Usov, *Tetrahedron*, **19**, 973 (1963).
[106] K. Kato and O. Mitsunobu, *J. Org. Chem.*, **35**, 4227 (1970).
[107] O. Mitsunobu, J. Kimura, K. Inzumi, and N. Yanagida, *Bull. Chem. Soc. Jpn.*, **49**, 510 (1976).
[108] R. Mengel and M. Bartke, *Angew. Chem., Int. Ed. Engl.*, **17**, 679 (1978).
[109] R. Boigegrain and B. Castro, *Tetrahedron Lett.*, **1965**, 3459.
[110] J. Kimura, Y. Fujisawa, T. Sawada, and O. Mitsunobu, *Chem. Lett.*, **1974**, 691.
[111] J. Kimura, Y. Hashimoto, and O. Mitsunobu, *Chem. Lett.*, **1974**, 1473.
[112] J. Kimura, K. Yagi, H. Suzuki, and O. Mitsunobu, *Bull. Chem. Soc. Jpn.*, **53**, 3670 (1980).

[113] H. R. Hudson, R. C. Rees, and J. E. Weekes, *J. Chem. Soc., Chem. Commun.*, **1971**, 1297.

[114] S. R. Landauer and H. N. Rydon, *Chem. Ind. (Lond.)*, **1951**, 313.

[114a] A. Campbell and H. N. Rydon, *Chem. Ind. (Lond.)*, **1951**. 312.

[115] S. R. Landauer and H. N. Rydon, *J. Chem. Soc.*, **1953**, 2224.

[116] C. S. L. Baker, P. D. Landor, S. R. Landor, and A. N. Patel, *J. Chem. Soc.*, **1965**, 4348.

[117] N. K. Kochetkov, L. Z. Kudryasov, and A. Z. Uzov, *Dokl. Akad. Nauk SSSR*, **133**, 1091 (1960) [*C.A.*, **55**, 27066i (1961)].

[118] N. K. Kochetkov and A. I. Usov, *Tetrahedron Lett.*, **1963**, 519.

[119] K. S. Adamyants, N. K. Kochetkov, and A. I. Usov, *Bull. Acad. Sci. USSR, Div. Chem. Sci. (Engl. Transl.)*, **1967**, 1262.

[120] J. P. H. Verheyden and J. G. Moffatt, *J. Org. Chem.*, **35**, 2319 (1970).

[121] J. P. H. Verheyden and J. G. Moffatt, *J. Org. Chem.*, **35**, 2868 (1970).

[122] J. P. H. Verheyden and J. G. Moffatt, *J. Am. Chem. Soc.*, **86**, 2094 (1964).

[123] J. P. H. Verheyden and J. G. Moffatt, *J. Am. Chem. Soc.*, **88**, 5684 (1966).

[124] J. P. H. Verheyden and J. G. Moffatt, *J. Org. Chem.*, **37**, 2289 (1972).

[125] J. P. H. Verheyden and J. G. Moffatt, *J. Org. Chem.*, **39**, 3573 (1974).

[126] S. D. Dimitrijevich, J. P. H. Verheyden, and J. G. Moffatt, *J. Org. Chem.*, **44**, 400 (1979).

[127] N. Kornblum and D. C. Iffland, *J. Am. Chem. Soc.*, **77**, 6653 (1955).

[128] J. B. Lee and M. M. El Sawi, *Chem. Ind. (Lond.)*, **1960**, 839.

[129] K. Kerfurt, J. Jary, and Z. Samek, *J. Chem. Soc., D*, **1969**, 213.

[130] E. Santaniello, *Gazz. Chim. Ital.*, **108**, 131 (1978).

[131] R. O. Hutchins, M. G. Hutchins, and C. A. Milewski, *J. Org. Chem.*, **37**, 4190 (1972).

[132] E. S. Lewis, B. J. Walker, and L. M. Ziurys, *J. Chem. Soc., Chem. Commun.*, **1978**, 424.

[133] D. Brett, I. M. Downie, J. B. Lee, and M. F. S. Matought, *Chem. Ind. (Lond.)*, **1969**, 1017.

[134] D. G. Coe, S. R. Landauer, and H. N. Rydon, *J. Chem. Soc.*, **1954**, 2281.

[135] D. G. Coe, H. N. Rydon, and B. L. Tonge, *J. Chem. Soc.*, **1957**, 323.

[136] A. W. Franck and C. F. Baranaukas, *J. Org. Chem.*, **31**, 872 (1966).

[137] G. S. Harris and D. S. Payne, *J. Chem. Soc.*, **1956**, 3038.

[138] H. N. Rydon and B. L. Tonge, *J. Chem. Soc.*, **1956**, 3043.

[139] G. A. Wiley, R. L. Hershkowitz, B. M. Rein, and B. C. Chung, *J. Am. Chem. Soc.*, **86**, 964 (1964).

[140] L. Horner, H. Oediger, and H. Hoffmann, *Justus Liebigs Ann. Chem.*, **626**, 26 (1959).

[141] G. A. Wiley, B. M. Rein, and R. L. Hershkowitz, *Tetrahedron Lett.*, **1964**, 2509.

[142] L. Kaplan, *J. Org. Chem.*, **31**, 3454 (1966).

[143] M. E. Herr and R. A. Johnson, *J. Org. Chem.*, **37**, 310 (1972).

[144] R. A. Arain and M. K. Hargreaves, *J. Chem. Soc., C*, **1970**, 67.

[145] K. Haga, M. Yoshikawa, and T. Kato, *Bull. Chem. Soc. Jpn.*, **43**, 3922 (1970).

[146] J. P. Schaeffer and J. Higgins, *J. Org. Chem.*, **32**, 1607 (1967).

[147] E. Piers and I. Nagakura, *Synth. Commun.*, **5**, 193 (1975).

[148] R. K. Boeckman, Jr., and B. Ganem, *Tetrahedron Lett.*, **1974**, 913.

[149] T. Dahl and R. Stevenson, *J. Org. Chem.*, **36**, 3243 (1971).

[150] P. Caubère and M. S. Mourad, *Tetrahedron*, **30**, 3439 (1974).

[151] I. Okada, K. Ichimura, and R. Sudo, *Bull. Chem. Soc. Jpn.*, **43**, 1185 (1970).

[152] H. M. Relles and R. W. Schluenz, *J. Am. Chem. Soc.*, **96**, 6469 (1974).

[153] D. Brett, I. M. Downie, and J. B. Lee, *J. Org. Chem.*, **32**, 855 (1967).

[154] R. D. Birkenmeyer and F. Kagan, *J. Med. Chem.*, **13**, 616 (1970).

[155] E. H. Axelrod, G. M. Milne, and E. E. Van Tamelen, *J. Am. Chem. Soc.*, **92**, 2139 (1970).

[156] M. Anisuzzaman and R. L. Whistler, *Carbohydr. Res.*, **61**, 511 (1978).

[157] L. M. Beacham, *J. Org. Chem.*, **44**, 3100 (1979).

[158] S. David and G. de Sennyer, *Carbohydr. Res.*, **77**, 79 (1979).

[159] S. L. Regen and D. P. Lee, *J. Org. Chem.*, **40**, 1669 (1975).

[160] P. Hodge and G. Richardson, *J. Chem. Soc., Chem. Commun.*, **1975**, 622.

[161] C. R. Harrison and P. Hodge, *J. Chem. Soc., Chem. Commun.*, **1978**, 813.

[162] R. M. Magid, O. S. Fruchey, and W. L. Johnson, *Tetrahedron Lett.*, **1977**, 2999.

[163] P. J. Garregg and B. Samuelsson, *Synthesis*, **1979**, 813.

[164] T. Hata, I. Yamamoto, and M. Sekine, *Chem. Lett.*, **1975**, 977.

[165] J. C. Florent, C. Monneret, and Q. Khuong-Huu, *Tetrahedron*, **34**, 909 (1978).

[166] A. V. Azhayev, A. A. Krayevsky, and J. Smrt, *Coll. Czech. Chem. Commun.*, **44**, 792 (1979).

[167] I. Yamamoto, M. Sekine, and T. Hata, *J. Chem. Soc., Perkin Trans.* 1, **1980**, 306.

[168] T. Adachi, Y. Arai, and I. Inoue, *Carbohydr. Res.*, **78**, 66 (1980).

[169] R. Appel and R. Kleinstück, *Chem. Ber.*, **107**, 5 (1974).

[170] H. Fukase, N. Mizokami, and S. Horii, *Carbohydr. Res.*, **60**, 289 (1978).

[171] T. Kametani, Y. Rigawa, and M. Ihara, *Tetrahedron*, **35**, 313 (1979).

[172] V. Stoilava, L. S. Trifonov, and A. S. Orahovats, *Synthesis*, **1979**, 105.

[173] D. Barbry, D. Couturier, and G. Ricart, *Synthesis*, **1980**, 387.

[174] G. W. Erickson and J. L. Fry, *J. Org. Chem.*, **45**, 970 (1980).

[175] A. Mizumo, Y. Hamada, and T. Shioiri, *Synthesis*, **1980**, 1007.

[176] R. Appel and H. D. Wihler, *Chem. Ber.*, **109**, 3446 (1976).

[177] Y. Kashman and A. Groweiss, *J. Org. Chem.*, **45**, 3814 (1980).

[178] E. E. Schweizer, W. S. Creazy, K. K. Light, and E. T. Schaffer, *J. Org. Chem.*, **34**, 212 (1969).

[179] M. M. Ponpipom and S. Hanessian, *Carbohydr. Res.*, **18**, 342 (1971).

[180] S. Hanessian, M. M. Ponpipom, and P. Lavallée, *Carbohydr. Res.*, **24**, 45 (1972).

[181] S. Hanessian, R. Massé, and M. L. Capmeau, *J. Antibiot.*, **30**, 893 (1977).

[182] S. Hanessian and P. Lavallée, *Carbohydr. Res.*, **28**, 303 (1973).

[183] S. Hanessian and P. Lavallée, *J. Antibiot.*, **25**, 683 (1972).

[184] S. Hanessian, D. Ducharme, R. Massé, and M. L. Capmau, *Carbohydr. Res.*, **63**, 265 (1978).

[185] A. K. Bose and B. Lal, *Tetrahedron Lett.*, **1973**, 3937.

[186] W. Ried and H. Appel, *Chem. Ber.*, **679**, 51 (1964).

[187] I. M. Downie, J. B. Lee, and M. F. S. Matough, *J. Chem. Soc., Chem. Commun.*, **1968**, 1350.

[188] B. Castro and C. Selve, *Bull. Soc. Chim. Fr.*, **1971**, 2296.

[189] B. Castro and C. Selve, *Bull. Soc. Chim. Fr.*, **1971**, 4368.

[190] B. Castro, M. Ly, and C. Selve, *Tetrahedron Lett.*, **1973**, 4455.

[191] B. Castro, Y. Chapleur, and B. Gross, *Tetrahedron Lett.*, **1974**, 2313.

[192] B. Castro and C. Selve, *Bull. Soc. Chim. Fr., Part 2*, **1974**, 3004.

[193] B. Castro and C. Selve, *Bull. Soc. Chim. Fr., Part 2*, **1974**, 3009.

[194] I. M. Downie, H. Heaney, and G. Kemp, *Tetrahedron Lett.*, **1975**, 3951.

[195] R. E. Ireland, S. Thaisrivongs, N. Vanier, and C. S. Wilcox, *J. Org. Chem.*, **45**, 48 (1980).

[196] P. Simon, J. C. Ziegler, and B. Gross, *Synthesis*, **1979**, 951.

[197] F. Chrétien, Thesis, University of Nancy, 1977.

[198] R. A. Boigegrain, B. Castro, and B. Gross, *Bull. Soc. Chim. Fr.*, **1974**, 2623.

[199] F. Chrétien, Y. Chapleur, B. Castro, and B. Gross, *J. Chem. Soc., Perkin Trans.* 1, **1980**, 381.

[200] F. Chrétien, B. Castro, and B. Gross, *Synthesis*, **1979**, 937.

[201] B. Castro, Y. Chapleur, and B. Gross, *Bull. Soc. Chim. Fr., Part 2*, **1973**, 3034.

[202] R. Toubiana, C. Pizza, Y. Chapleur, and B. Castro, *J. Carbohydr., Nucleosides, Nucleotides*, **5**, 127 (1978).

[203] B. Castro and C. Selve, *Tetrahedron Lett.*, **1973**, 4459.

[204] R. Boigegrain and B. Castro, *Tetrahedron*, **32**, 1283 (1976).

[205] R. Boigegrain and B. Castro, *J. Carbohydr., Nucleosides, Nucleotides*, **3**, 335 (1976).

[206] R. B. Silverman and M. A. Levy, *J. Org. Chem.*, **45**, 815 (1980).

[207] W. Schörkhuber and E. Zbiral, *Justus Liebigs Ann. Chem.*, **1980**, 1455.

[208] P. G. Mattingly, J. F. Kerwin, Jr., and M. J. Miller, *J. Am. Chem. Soc.*, **101**, 3983 (1979).

[209] A. K. Bose, D. P. Sahu, and M. S. Manhas, *J. Org. Chem.*, **46**, 1229 (1981).

[210] G. Grynkiewicz, *Pol. J. Chem.*, **53**, 2501 (1979).

[211] H. Redlich and W. Francke, *Angew. Chem., Int. Ed. Engl.*, **19**, 630 (1980).

[212] H. Redlich, B. Schneider, and W. Francke, *Tetrahedron Lett.*, **1980**, 3013.

[213] K. Tomioka and K. Koga, *Tetrahedron Lett.*, **1980**, 2321.

[214] M. Asaoka, N. Yanagida, and H. Takei, *Tetrahedron Lett.*, **1980**, 4611.

[215] R. Hohlbrugger and W. Klötzer, *Chem. Ber.*, **112**, 3486 (1979).

[216] R. J. Bass, B. J. Banks, and M. Snarey, *Tetrahedron Lett.*, **1980**, 769.

[217] K. Akerfelot, P. J. Garregg, and T. Iverson, *Acta Chem. Scand.*, *Ser. B*, **1979**, 1815.

[218] O. Mitsunobu, T. Kudo, M. Nishida, and N. Tsuda, *Chem. Lett.*, **1980**, 1614.

[219] A. Campbell and H. N. Rydon, *Chem. Ind.* (*Lond.*), **1951**, 312.

[220] M. McCoss, E. K. Ryu, R. S. White, and R. L. Last, *J. Org. Chem.*, **45**, 788 (1980).

[220a] P. L. Durette, *Synthesis*, **1980**, 1037.

[221] J. B. Lee and T. J. Nolan, *Can. J. Chem.*, **44**, 1331 (1966).

[222] J. Hooz and S. S. H. Gilani, *Can. J. Chem.*, **46**, 86 (1968).

[223] J. B. Lee and I. M. Downie, *Tetrahedron*, **23**, 359 (1967).

[224] E. I. Snyder, *J. Org. Chem.*, **37**, 1466 (1972).

[225] J. B. Heather, R. S. D. Mittal, and C. J. Sih, *J. Am. Chem. Soc.*, **98**, 3661 (1971).

[226] H. L. Goering and S. L. Trenbeath, *J. Am. Chem. Soc.*, **98**, 5016 (1976).

[227] I. Tomoskozi, L. Gruber, and L. Radics, *Tetrahedron Lett.*, **1975**, 2473.

[228] G. Stork, M. E. Jung, E. Calvin, and Y. Noel, *J. Am. Chem. Soc.*, **96**, 3684 (1974).

[229] A. W. Friederang and D. S. Tarbell, *J. Org. Chem.*, **33**, 7797 (1968).

[230] R. G. Weiss and E. I. Snyder, *J. Chem. Soc.*, *Chem. Commun.*, **1968**, 1358.

[231] P. J. Kocienski, G. Cornigliaro, and G. Feldstein, *J. Org. Chem.*, **42**, 353 (1977).

[232] R. D. Miller, M. Schneider, and D. L. Dolce, *J. Am. Chem. Soc.*, **95**, 8469 (1973).

[233] T. C. Jain, D. Jenkins, A. F. Russel, J. D. H. Verheyden, and J. G. Moffat, *J. Org. Chem.*, **39**, 30 (1974).

[234] L. B. Hunt, D. F. MacSweeny, and R. Ramage, *Tetrahedron*, **27**, 1491 (1971).

[235] E. E. van Tamelen and R. J. Anderson, *J. Am. Chem. Soc.*, **94**, 8225 (1972).

[236] J. J. Baldwin, A. W. Raab, K. Mensler, B. H. Arison, and D. E. McClure, *J. Org. Chem.*, **43**, 4876 (1978).

[237] R. Aneja and A. P. Davies, *J. Chem. Soc.*, *Perkin Trans.* 1, **1974**, 141.

[238] F. G. de la Heras, R. Alonson, and G. Alonso, *J. Med. Chem.*, **22**, 496 (1979).

[234] C. W. Chiu and R. L. Whistler, *J. Org. Chem.*, **38**, 832 (1973).

[240] R. A. Houghten, R. A. Simpson, R. N. Hanson, and H. Rapoport, *J. Org. Chem.*, **44**, 4536 (1979).

[241] B. T. Lawton, W. A. Szarek, and J. K. N. Jones, *Carbohydr. Res.*, **14**, 255 (1970).

[242] R. S. Klein, H. Ohrui, and I. J. Fox, *J. Carbohydr.*, *Nucleosides*, *Nucleotides*, **1975**, 265.

[243] R. Aneja, A. P. Davies, and J. A. Knaggs, *J. Chem. Soc.*, *Chem. Commun.*, **1973**, 110.

[244] S. Trippett, *J. Chem. Soc.*, **1962**, 2337.

[245] J. D. Morrison, J. E. Tomaszewski, H. S. Mosher, J. Dale, D. Miller, and R. L. Elsenbaumer, *J. Am. Chem. Soc.*, **99**, 3167 (1977).

[246] F. Arcamone, A. Bargiotti, G. Cassinelli, S. Penco, and S. Hanessian, *Carbohydr. Res.*, **46**, C3 (1976).

[247] H. E. Hadd, *Steroids*, **31**, 453 (1978).

[248] B. Castro, Y. Chapleur, B. Gross, and C. Selve, *Tetrahedron Lett.*, **1972**, 5001.

[249] R. Boigegrain, B. Castro, and C. Selve, *Tetrahedron Lett.*, **1975**, 2529.

[250] A. Canas-Rodriguez, A. Martinez-Tobed, A. Gomez-Sanchez, and C. Martin-Madero, *Carbohydr. Res.*, **56**, 289 (1977).

[251] R. Boigegrain, Thesis, University of Nancy, 1975.

[252] A. Canas-Rodriguez and A. Martinez-Tobed, *Carbohydr. Res.*, **68**, 43 (1979).

[253] I. M. Downie, H. Heaney, and G. Kemp, *Angew. Chem.*, *Int. Ed. Engl.*, **14**, 370 (1975).

[254] P. Simon, J. C. Ziegler, and B. Gross, *Carbohydr. Res.*, **64**, 257 (1978).

CHAPTER 2

REDUCTIVE DEHALOGENATION OF POLYHALO KETONES WITH LOW-VALENT METALS AND RELATED REDUCING AGENTS

RYOJI NOYORI AND YOSHIHIRO HAYAKAWA

Nagoya University, Nagoya, Japan

CONTENTS

ACKNOWLEDGMENT

The assistance of Mr. Thomas C. Johns of Wilmington, Delaware in searching the literature is gratefully acknowledged.

INTRODUCTION

α-Halo ketones play a significant role in organic synthesis because of their ready availability and the diversity of their chemical reactions, which reflect the various functionalities in the molecule, *viz.*, a reactive halogen–carbon bond, acidic hydrogen atoms, and a carbonyl moiety. An important reaction of halo ketones is removal of the halogen atom(s) by low-valent transition metals and related reducing agents to generate reactive species that are capable of under-going carbon–carbon bond-forming reactions. The cyclocoupling reactions of α,α'-dihalo or more highly halogenated ketones across 1,3-dienic or olefinic substrates to give seven- or five-membered ketones, respectively, are especially noteworthy from the synthetic viewpoint.

The direct preparation of odd-membered carbocycles has remained a per-sistent problem, and until recently only a few methods of synthesis were available. The existing methods for obtaining seven-membered carbocycles have been limited to expansion of six-membered rings and certain cyclization re-actions.[1] The construction of five-membered carbocyclic systems has generally been accomplished only by intramolecular condensation of open-chain di-carbonyl compounds or dicarboxylic acid derivatives,[2,3] ring contraction of six-membered cyclic ketones,[4,5] or ring expansion of four-membered cyclic ketones.[6] These methods, however, are not always useful because of the limited availability of the starting materials in certain reactions. In principle the most direct method for construction of seven-membered carbocycles is the cyclo-condensation between three-carbon and four-carbon units, viz., 1,3-dienes, as shown in Eq. 1; for preparation of five-membered carbocycles, the combination

$$\text{(Eq. 1)}$$

$$\text{(Eq. 2)}$$

$$\text{(Eq. 3)}$$

of three-carbon and two-carbon units, viz., olefinic substances, or one-carbon and four-carbon units, as illustrated in Eqs. 2 and 3, respectively, can be con-sidered. However, these reactions are not achieved easily because of the difficulty of generating reactive three-carbon and one-carbon units that are capable of undergoing cycloaddition across the unsaturated substrates. Reactions of low-valent metals and α,α'-dihalo or more highly halogenated ketones generate such bifunctional three-carbon species that can enter into both types of cycloaddition reactions (Eqs. 1 and 2). Moreover, these reactive species also undergo [3 + 2] cycloadditions across certain unsaturated bonds containing heteroatoms, thereby giving rise to five-membered heterocycles.

Thus these [3 + 4] and [3 + 2] annulations have allowed easy access to a wide range of organic frameworks that are of theoretical or practical importance. The scope and limitations of these cyclocoupling reactions and their synthetic applications[7,8] are the main concern of this chapter, which briefly describes other examples of dehalogenation of α-mono- and α,α'-polyhalo ketones using low-valent transition metals and related reducing agents. Also described are the mechanistic aspects of the reductions and several representative experimental procedures.

MECHANISM

Reduction of α-Monohalo Ketones

For the initial step in the reduction of organic halides by metals or their complexes, three pathways have been proposed and classified by the mode of electron changes, which, in turn, depend on the nature of the reducing agent.[9] *Atom-transfer process*:

$$R-X + M^n \longrightarrow R\cdot + M^{(n+1)}X$$

*Electron-transfer process**:

$$R-X + M^n \longrightarrow R-X^{\overline{\cdot}} + M^{(n+1)}$$

Oxidative-addition process:

$$R-X + M^n \longrightarrow R-M^{(n+2)}X$$

The reduction of α-halo ketones appears to proceed similarly. The atom-transfer or electron-transfer process generating the radical species **1** (Eq. 4) is

$$\underset{1}{\left[\begin{array}{c}O \\ \diagdown\diagup^{\cdot}\end{array}\right]}$$

(Eq. 4)

considered for reactions using one-equivalent reductants such as alkali metals or certain low-valent transition metal complexes containing, e.g., chromium(II),[10] iron(II),[11] cobalt(II),[11] and molybdenum(0),[12] which tend to release one electron to attain a higher oxidation state. Though not general, use of two or

* The initially generated anion radical R–X$^{\overline{\cdot}}$ suffers subsequent elimination of X$^-$, giving rise to radical R·. Thus in practice this process is difficult to distinguish experimentally from the atom-transfer process, since they differ mainly in the timing of the bond-making and bond-breaking sequence.

more equivalents of such reagents drives reduction via enolate intermediate **2**, which is generated by two successive one-electron changes (Eq. 5).

$$1 + 2M^n \longrightarrow \left[\underset{2}{\overset{OM^{(n+1)}}{\diagup\!\!\diagdown}} \rightleftharpoons \overset{O}{\diagup\!\!\diagdown} M^{(n+1)} \right]$$

$$\text{or} \left[\overset{O}{\diagup\!\!\diagdown}^{-} \right] M^{(n+1)} \quad \text{(Eq. 5)}$$

Reduction takes place via the oxidative addition mechanism, directly forming the metal enolate intermediate **3** (Eq. 6), when it is carried out with two-equivalent reductants, e.g., iodide salts or low-valent transition metals capable

$$\overset{O}{\diagup\!\!\diagdown} X + M^n \longrightarrow \left[\underset{3}{\overset{OM^{(n+2)}X}{\diagup\!\!\diagdown}} \rightleftharpoons \overset{O}{\diagup\!\!\diagdown} M^{(n+2)}X \right]$$

$$\text{or} \left[\overset{O}{\diagup\!\!\diagdown}^{-} \right] M^{(n+2)}X \quad \text{(Eq. 6)}$$

of changing the oxidation state either by concerted two-equivalent transformations or by successions of one-electron transfers. A related mechanism has been proposed also for electrochemical reductions of α-halo ketones.[13] A reactive enolate metal complex has been isolated from a reduction promoted by an iron carbonyl in an aprotic solvent.[14] In most reactions, however, the intervention of a metal enolate species is supported chemically by a subsequent nucleophilic reaction, as depicted in Eq. 7. Zinc enolates sometimes undergo

$$\mathbf{3} \xrightarrow{\text{EX}} \overset{O}{\diagup\!\!\diagdown} E \quad \text{or/and} \quad \overset{OE}{\diagup\!\!\diagdown} + M^{(n+2)}X_2 \quad \text{(Eq. 7)}$$

disproportionation with the starting halo ketones to form two equivalents of radical species (Eq. 8), which have actually been detected by chemical and spectroscopic techniques.[15]

$$\underset{3}{\left[\overset{O}{\diagup\!\!\diagdown}^{-} \right]} M^{(n+2)}X + \overset{O}{\diagup\!\!\diagdown} X \longrightarrow 2 \left[\overset{O}{\diagup\!\!\diagdown}^{\cdot} \right] + M^{(n+2)}X_2 \quad \text{(Eq. 8)}$$

Reduction of α,α-Dihalo Ketones

The reduction of geminal dihalo ketones begins in the same manner as that of α-monohalo ketones to form the radical species **4** (Eq. 9) or enolate intermediate **5** or **6** (Eqs. 9 and 10), as indicated by their chemical behavior. Enolate intermediate **5** or **6** further eliminates the remaining halogen atom, giving rise

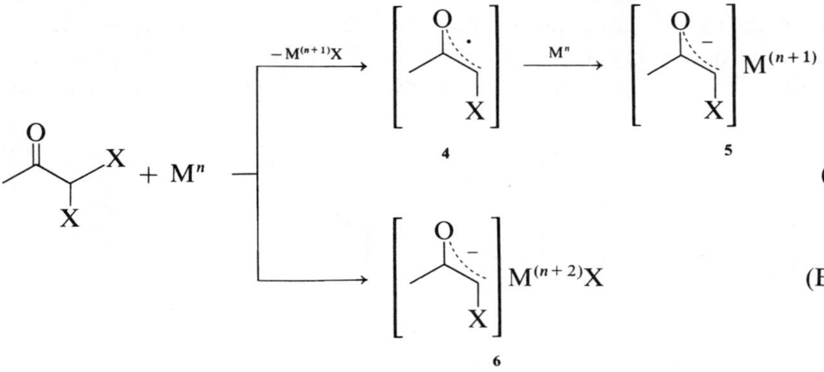

(Eq. 9)

(Eq. 10)

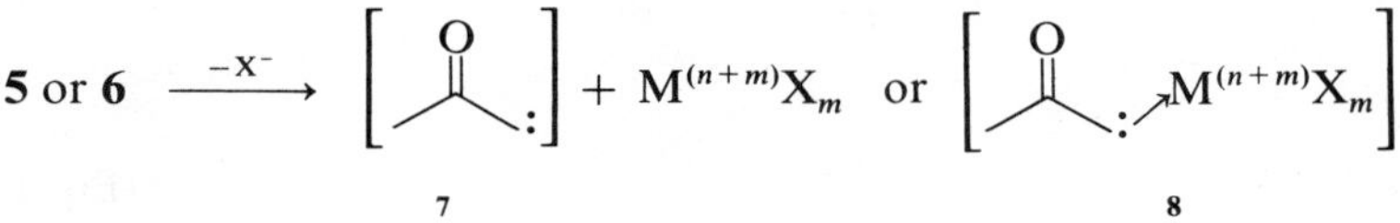

$(m = 1 \text{ or } 2)$

to ketocarbene **7** or its metal complex **8** (Eq. 11). Intervention of such species is convincingly argued for the reduction of **9**, which leads to tricyclic ketone **10** as the result of carbenoid γ C–H bond insertion.[16]

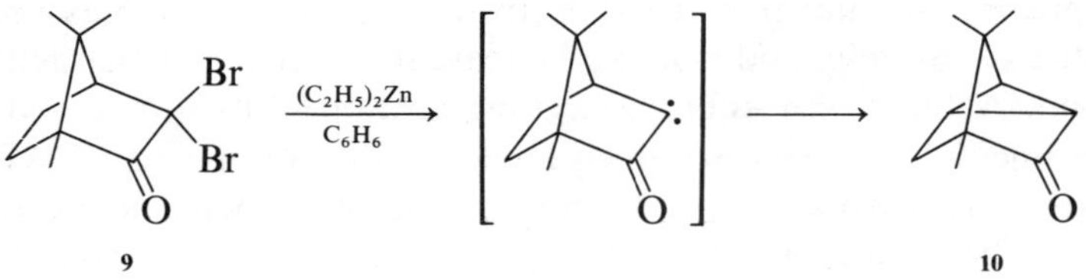

Reduction of α,α′-Dihalo and Polyhalogenated Ketones

Vapor-Phase Reduction.   Vapor-phase reduction of α,α′-dihalo ketones with potassium occurs by the successive atom-transfer mechanism to generate intermediary 2-oxo-1,3-alkadiyl diradical species of type **11**, which decompose to carbon monoxide and olefins through cyclopropanones.[17]

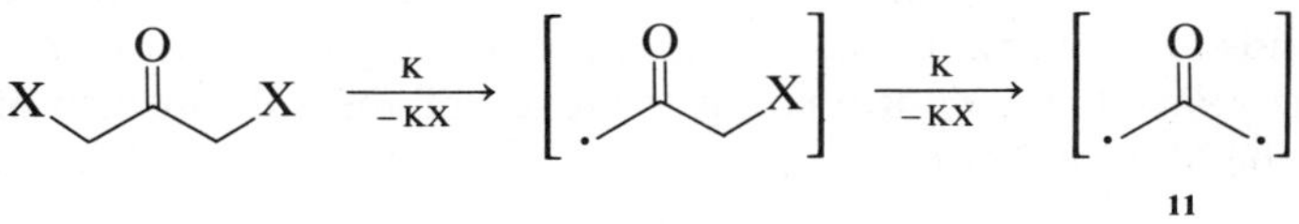

Reduction in Solution.   Reductions of α,α'-dihalo ketones and more highly halogenated ketones in organic solvents have been performed with two-equivalent reducing agents or electrochemically. The reaction is initiated by two-equivalent reduction of the polyhalo ketone to give metal enolate **12**, as ascertained by quenching in protic media to afford the protonated products. Enolate **12** suffers subsequent S_N1-type or Lewis-acid-assisted elimination of the

12

13   **14**   **15**

(Eq. 12)

allylic bromine atom to produce 2-oxyallyl–metal complex **13** or its structural isomers, viz., cyclopropanone **14** and allene oxide **15** (Eq. 12). The relative stabilities of these species are influenced by a variety of factors. Theoretical treatments suggest that the parent 2-oxyallyl dipolar ion in a free form is very labile, isomerizing immediately to **14** and/or **15**.[18,19] The oxyallyl species **13** receives greatest stabilization from electron-releasing substituents present at the C-1 and C-3 termini and from an increase in the covalent character of the M–O bond. Actually, α,α'-dibromoacetone cannot be used as a starting C_3 substance since it leads to the extremely labile oxyallyl intermediate **13** (R = H), which bears no carbocation-stabilizing substituents. On the other hand, $\alpha,\alpha,\alpha',\alpha'$-tetrabromoacetone or secondary and tertiary bis(bromoalkyl) ketones are utilized because the added substituent R (Br or alkyl) increases the stability of the resulting oxyallyl species **13** by mesomeric and inductive effects, respectively. The existence of the 2-oxyallyl species is possible only in the metal-complexed or protonated form (**13**, M = metal or H). The stability of **13** depends strongly on the nature of the M–O bond, which is influenced greatly by the nature of M and the solvents used. The more covalent the character of the metal–oxygen bond, the more the equilibrium lies in the direction of 2-oxyallyl cation formation. When the M–O bond is rather ionic, the equilibrium tends to shift toward the formation of the ring-closed isomers **14** or **15**. The polarity and metal-coordinating tendency of the solvent also affect the nature of the M–O linkage.

Many lines of evidence suggest the intermediacy of oxyallyl species through nucleophilic trapping[20–29] (Eqs. 13–15) and a variety of cationic rearrange-

$$(CH_3)_2CBrCOCBr(CH_3)_2 \xrightarrow[CH_3CO_2D]{Hg} (CH_3)_2C(OCOCH_3)COCD(CH_3)_2$$
16
$$(60\%)$$

$$\text{(Eq. 13)}^{23}$$

$$\text{(Eq. 14)}^{28}$$

$$(61\%, \textit{cis/trans} = 2:1)$$

$$(CH_3)_2CBrCOCHBrC_4H_9\text{-}t$$

$$\xrightarrow[t\text{-}C_4H_9CO_2H/DMF]{2e^-,\ t\text{-}C_4H_9CO_2Na} (CH_3)_2C(OCOC_4H_9\text{-}t)COCH_2C_4H_9\text{-}t \quad \text{(Eq. 15)}^{27}$$
$$(67\%)$$

ments. The skeletal changes include electrocyclization of the 2-oxypentadienyl-type cation (Eq. 16),[28,29] and $[1_a, 4_s]$ sigmatropic (Eq. 17),[28–31] neopentyl (Eq. 18),[28,29] *tert*-butylallyl-cyclopropylcarbinyl (Eq. 19),[28,29] and cationic

$$C_6H_5CHBrCOCHBrC_6H_5 \xrightarrow[C_6H_6]{Fe_2(CO)_9} \left[\text{structure} \right]$$

$$\text{(Eq. 16)}^{28,29}$$

$$(70\%)$$

$$\text{(Eq. 17)}^{28,29}$$

$$(95\%)$$

[3, 4] sigmatropic (Eq. 20)[32] rearrangements. Ene reactions between 2-oxyallyl cations and olefins are also reported (Eq. 21).[33] In some reactions reduction in

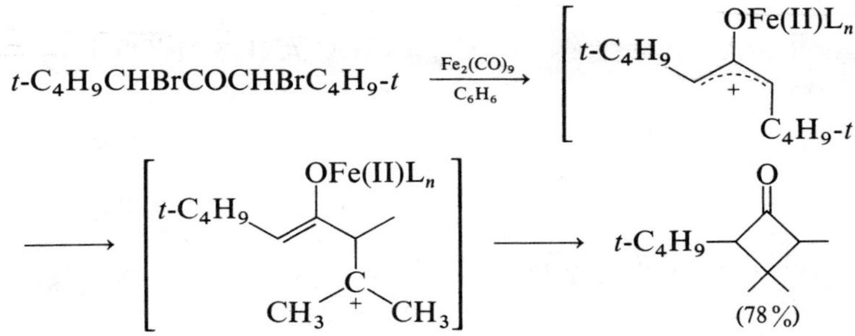

$t\text{-}C_4H_9CHBrCOCHBrC_4H_9\text{-}t \xrightarrow[C_6H_6]{Fe_2(CO)_9}$

(Eq. 18)[28,29]

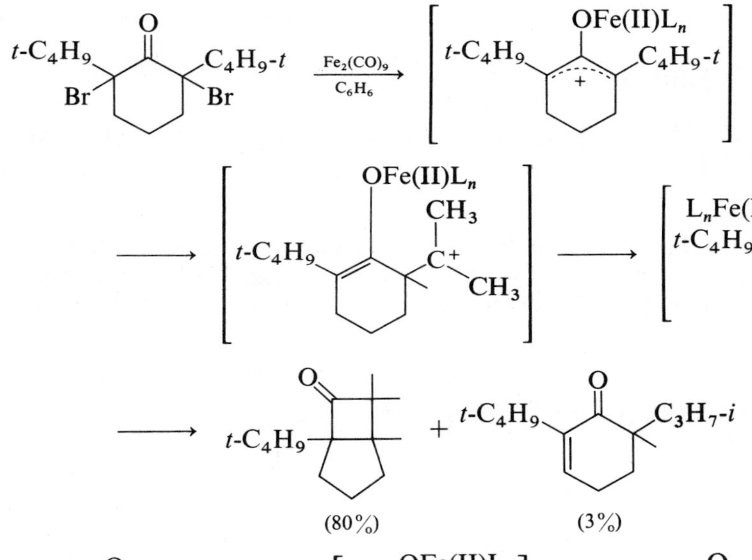

(78%)

(Eq. 19)[28,29]

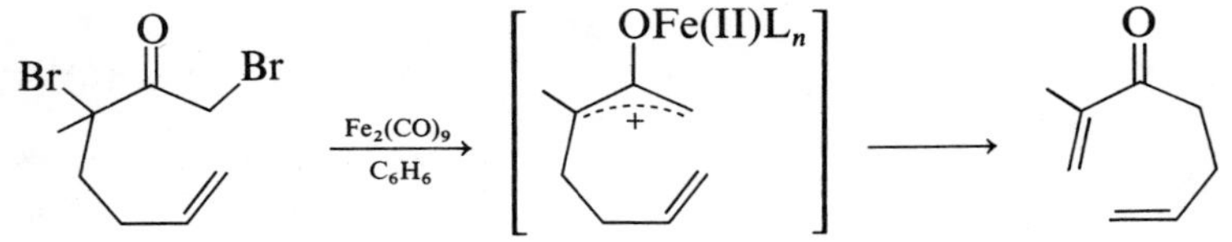

(80%)   (3%)

(Eq. 20)[32]

$C_6H_5CHBrCOCH_2Br \xrightarrow[C_6H_6]{Fe_2(CO)_9}$

$\xrightarrow{(CH_3)_2C=CH_2}$

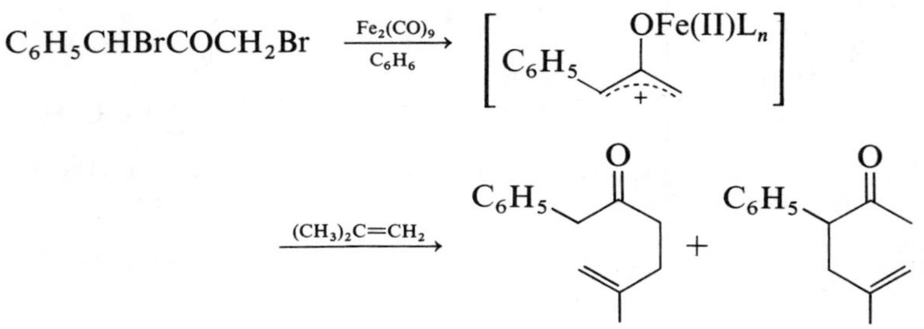

(Eq. 21)[33]

basic media such as tetrahydrofuran (THF) or N,N-dimethylformamide (DMF) yields α,β-unsaturated ketones resulting from prototropy of the 2-oxyallyl intermediate.[20,28,29,34,35] Furthermore, the intervention of an allyl

$$(CH_3)_2CBrCOCBr(CH_3)_2 \xrightarrow[\text{DMF}]{Fe_2(CO)_9} CH_2{=}C(CH_3)COC_3H_7\text{-}i \quad (Eq.\ 22)^{34,35}$$

16   **17** (80%)

cationic species is in accord with the reductive [3 + 4] or [3 + 2] cyclo-coupling reactions between polyhalo ketones and 1,3-dienes or certain olefins, respectively.[33,36] Cyclopropanones as transient intermediates are detected by IR analysis of certain electrochemical reductions in dimethylformamide.[37] Some products derivable in principle from cyclopropanones are isolated from electrolysis[37,38] or zinc/copper couple reduction[39] in protic solvents. Formation of allene oxides as product-determining intermediates has not yet been substantiated.

SCOPE AND LIMITATIONS

Reduction of α-Monohalo Ketones

Formation of the Parent Ketones. When α-monohalo ketones are treated with either one- or two-equivalent reductants in protic or aprotic media, the parent ketones are produced; the latter reaction usually, though not always, requires aqueous workup. Electroreduction proceeds analogously.

Lithium metal, a one-electron reductant, in ethereal solvents is adequate for the reduction of halo ketones lacking active α-hydrogen atoms.[40] Use of

$$C_2H_5C(CH_3)_2COCBr(CH_3)C_2H_5 \xrightarrow[\text{ether}]{Li} C_2H_5C(CH_3)_2COCH(CH_3)C_2H_5$$

(97%)

molybdenum hexacarbonyl–alumina[12] and chromium(II),[41] iron(II),[11,42] and cobalt(II)[42] species also leads to satisfactory reduction; tungsten hexacarbonyl is rather less effective.[43] Usually such reductions require the stoichiometric use

$$p\text{-}C_6H_5C_6H_4COCH_2Br \xrightarrow[\text{THF}]{Mo(CO)_6,\ Al_2O_3} p\text{-}C_6H_5C_6H_4COCH_3$$

18   **19** (73%)

(85%)

of reducing agents. Reduction can be achieved by the use of either catalytic or stoichiometric amounts of iron(II) salts in the presence of thiophenol; the stoichiometric conditions generally provide better yields.[11,42]

$$ \mathbf{18} \quad \xrightarrow[\text{H}_2\text{O–CH}_3\text{OH}]{\text{0.1 eq. Fe(II)-polyphthalocyanine complex, C}_6\text{H}_5\text{SH}} \quad \mathbf{19}_{(67\%)} $$

Equations 23 and 24 show examples of reductions that are assisted by two-equivalent reducing agents. Zinc metal and zinc/copper couple are the most conventional reagents; the addition of acidic substances accelerates the reduction to some extent.[44–49] Iron pentacarbonyl [Fe(CO)$_5$] and diiron enneacarbonyl [Fe$_2$(CO)$_9$] may be employed stoichiometrically;[28,29] the former reagent

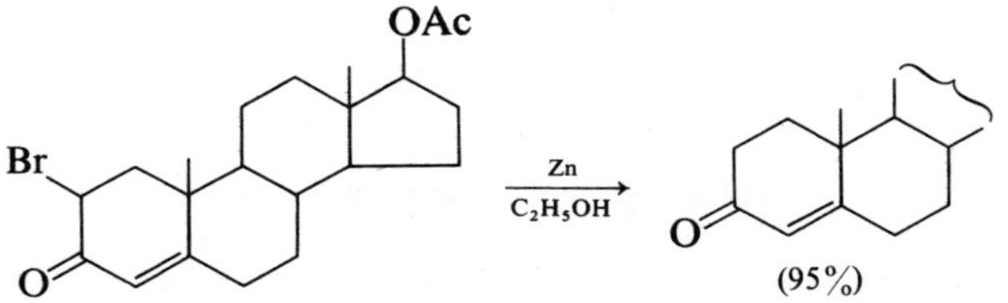

(Eq. 23)

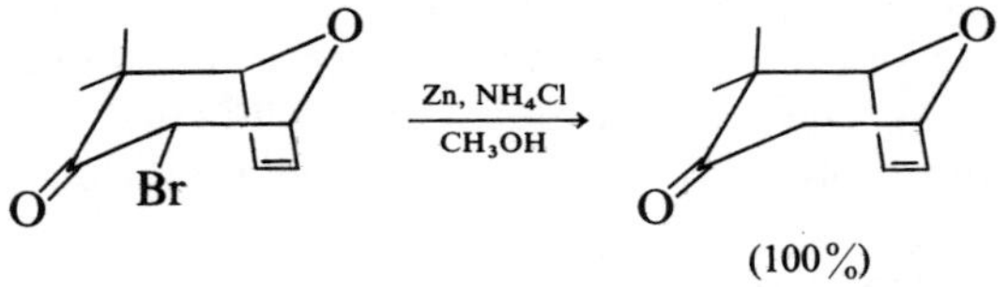

(Eq. 24)

reduces the bromo ketone in an aprotic solvent to give the parent ketone without any aqueous workup.[50] Catalytic reduction using dicobalt octacarbonyl [Co$_2$(CO)$_8$] is a practical and useful process; a phase-transfer catalyst

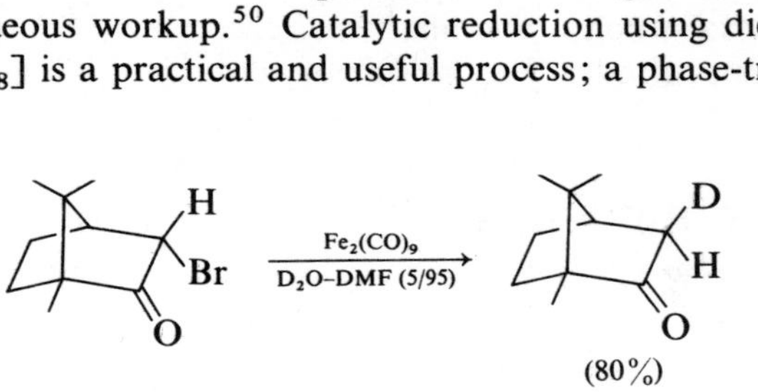

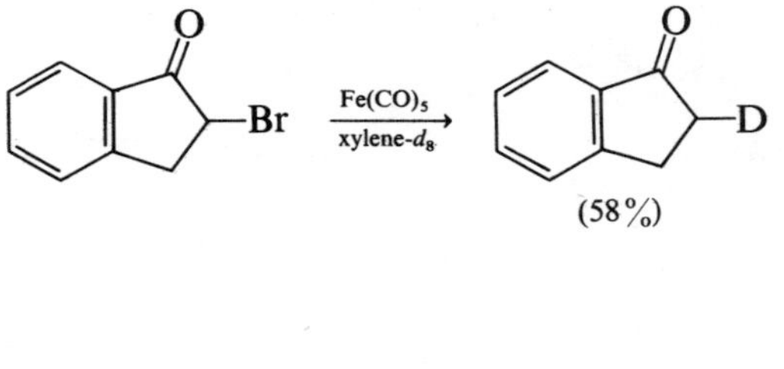

(80%)

(58%)

and sodium hydroxide are required to generate the actual reducing agent, $[Co(CO)_4]^-$.[51]

$$\textbf{18} \xrightarrow[\text{H}_2\text{O–C}_6\text{H}_6]{\text{0.1 eq. Co}_2\text{(CO)}_8, \text{C}_6\text{H}_5\text{CH}_2\text{N(C}_2\text{H}_5)_3\text{Cl, NaOH}} \textbf{19}$$
$$(58\%)$$

Lithium dialkylcuprates perform the reduction under very mild conditions,[52] but undesired alkylations often occur.[53] Use of aqueous titanium(III)[54] or vanadium(II)[55] chlorides is convenient. Reductive dehalogenation with an

$$\xrightarrow[\text{ether}]{(\text{CH}_3)_2\text{CuLi}}$$

(95%)

$$\xrightarrow[\text{aq. CH}_3\text{CN}]{\text{TiCl}_3}$$

(80%)

excess of lithium iodide is promoted to a considerable extent by the combined use of boron trifluoride etherate[56] or hydriodic acid.[31]

$$\xrightarrow[\text{2. H}_2\text{O}]{\text{1. LiI, BF}_3\text{O(C}_2\text{H}_5)_2\text{, THF}}$$

(100%)

A clean reduction, even of fluoro ketones, is effected by electrochemical treatment.[13]

$$C_6H_5COCH_2F \xrightarrow{2e^-, H^+} C_6H_5COCH_3$$

In general, the two-equivalent reduction of bromo ketones, excluding tertiary ones,[7] is strongly accelerated by the addition of sodium iodide,[15] because iodide ion converts the bromo ketones to the more reactive iodo ketones via S_N2 displacement.

1,4-Diketones resulting from reductive dimerization of the starting halo ketones are typical side products in the reaction, particularly of alkyl halomethyl ketones in aprotic media.

Formation of α-Alkyl Ketones.   The reduction of α-monobromo ketones in the presence of electrophilic alkylating agents gives rise to α-alkyl ketones. However, only methyl iodide and certain activated alkyl halides appear to be of synthetic value here; ordinary higher alkyl iodides are not sufficiently reactive. For instance, the lithium-promoted reaction of **20** and methyl iodide in ether affords the methylation product **21** (R = CH$_3$) in 93% yield, while the reaction

$$t\text{-}C_4H_9COCBr(CH_3)_2 \xrightarrow[\text{R = CH}_3,\, \text{C}_2\text{H}_5]{\text{Li, RI, (HMPA)}} t\text{-}C_4H_9COCR(CH_3)_2 \qquad \text{(Eq. 25)}^{40}$$

$$\mathbf{20} \qquad\qquad\qquad\qquad\qquad \mathbf{21}\ (15\text{–}93\%)$$

with ethyl iodide fails to yield the ethylated product (see Eq. 25); the use of hexamethylphosphoramide (HMPA) as a solvent facilitates the ethylation to some extent to give **21** (R = C$_2$H$_5$) in 15% yield.[40] Similar reactivity is observed in alkylations promoted by zinc dust in dimethyl sulfoxide (DMSO)–benzene.[57] Ethylation is achieved very slowly but in moderate to high yields

(66%)

(40%)

in hexamethylphosphoramide–benzene.[58] Efficient alkylations have been reported using activated alkyl halides such as allyl bromide[57] and α-chloro thioethers.[59] Reductive methylations of unsymmetrically substituted bromo ketones take place specifically at the carbon originally attached to bromine, producing single positional isomers.[57] A typical side reaction is the formation

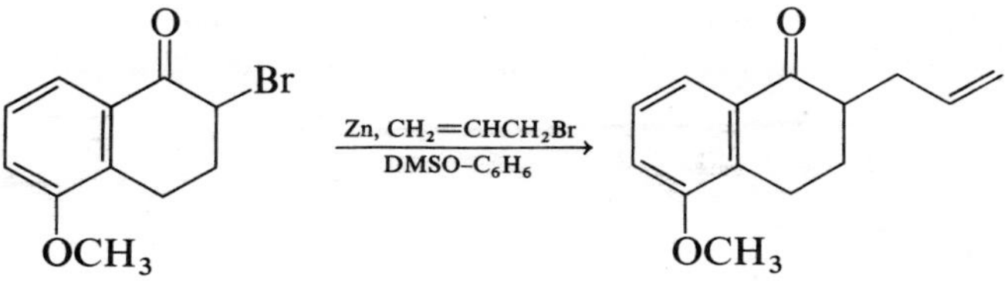

(75%)

$$n\text{-}C_3H_7COCHBrC_2H_5$$

$$\xrightarrow[\text{CH}_3\text{CO}_2\text{C}_2\text{H}_5]{\text{Zn, CH}_3\text{CH(SC}_2\text{H}_5)\text{Cl}} \quad n\text{-}C_3H_7COCH(C_2H_5)CH(SC_2H_5)CH_3$$

(82%)

of furans arising via aldol condensation of an intermediary zinc enolate and an α-bromo ketone, which can be suppressed by lowering the concentration of starting bromide.

The reduction of bromo ketones with lithium dimethylcuprate followed by treatment with methyl iodide affords methylation products.[60]

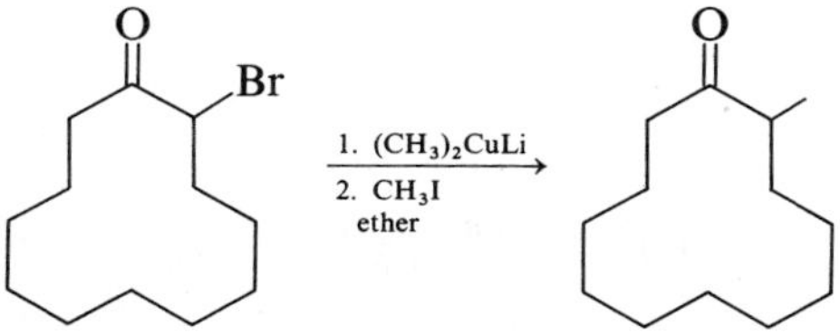

(67%)

The reduction of α-bromo ketones with zinc dust or zinc/copper couple in dimethyl sulfoxide containing sodium iodide and sodium hydrogen carbonate produces a radical species via the mechanism illustrated in Eq. 8 (see p. 168).[15] This intermediate can be trapped by aryl-substituted olefins or methylenecycloalkanes to give substitutive addition products possessing predominantly an endocyclic double bond.[15] Conjugated dienes and enynes undergo similar alkylations.[61]

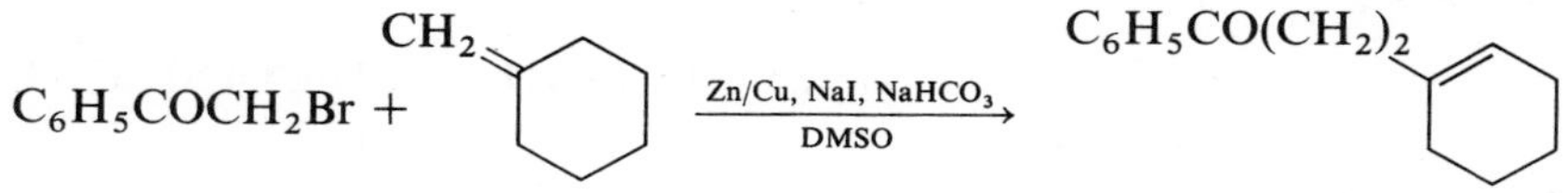

(54%)

Formation of Aldols. A reductive aldol reaction between α-monobromo ketones and aldehydes or ketones can be achieved with zinc dust either alone[57,62] or in combination with diethylaluminum chloride and a catalytic

amount of copper(I) bromide.[63] Regiospecific aldol formation is observed in the
reaction of the bromide derived from an unsymmetrical ketone.[63]

$$(90\%)$$

$$(Eq. 26)^{57}$$

$$n\text{-}C_3H_7COCHBrC_2H_5 + (C_2H_5)_2CO$$

$$\xrightarrow{\text{Zn}} n\text{-}C_3H_7COCH(C_2H_5)C(OH)(C_2H_5)_2 \qquad (Eq. 27)^{62}$$

$$(73\%)$$

$$(94\%)$$

$$(100\%, \textit{erythro/threo} = 4:3)$$

Formation of 1,3-Diketones and Enol Carboxylates.   Acyl chlorides undergo
effective C- and O-acylation reactions with the intermediary enolates generated
by reduction of halo ketones, yielding 1,3-diketones or enol carboxylates,
respectively. Diketones are formed by reduction with zinc dust in ethyl acetate[64]
or lithium dimethylcuprate in ether,[65] whereas enol carboxylates result from
use of disodium tetracarbonylferrate $[Na_2Fe(CO)_4]$ in tetrahydrofuran containing pyridine (Py).[66]

$$n\text{-}C_4H_9COCHBrC_3H_7\text{-}n + CH_3COCl$$

$$\xrightarrow[\text{CH}_3\text{CO}_2\text{C}_2\text{H}_5]{\text{Zn}} n\text{-}C_4H_9COCH(COCH_3)C_3H_7\text{-}n$$

$$(56\%)$$

$$CH_3COCH_2Cl + (CH_3)_2CHCH_2COCl$$

$$\xrightarrow[\text{THF}]{\text{Na}_2\text{Fe(CO)}_4, \text{ Py}} CH_3C[OCOCH_2CH(CH_3)_2]{=}CH_2$$

$$(97\%)$$

Reductive Dimerization to 1,4-Diketones, β-Epoxy Ketones, and Furans.
The following three types of reductive dimerizations are reported (Scheme I):

(1) formation of 1,4-diketone **22** through coupling at α carbons, (2) formation of β-epoxy ketone **23** by aldol reaction of an enolate intermediate with another α-halo ketone molecule followed by intramolecular S_N2 displacement of the resulting aldolate anion, and (3) production of unsymmetrically substituted furan **24**, probably arising from thermal rearrangement of the β-epoxy ketone accompanied by dehydration.[67]

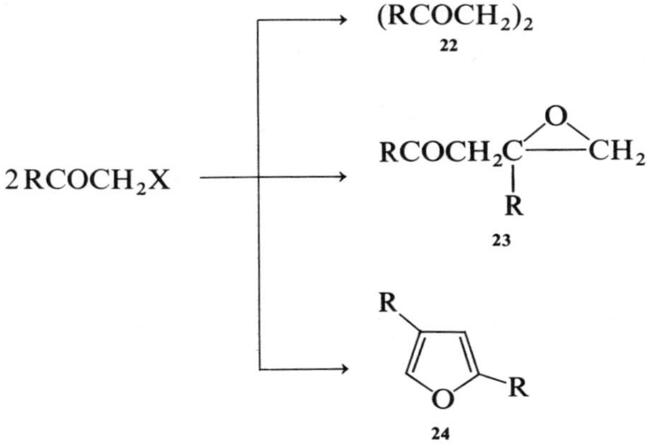

Scheme I

These self-coupling reactions are usually nonselective, resulting in complex mixtures. However, a clean reaction producing 1,4-diarylbutane-1,4-diones is achieved by use of $Fe(CO)_5$ in 1,2-dimethoxyethane (DME).[14] The water–benzene two-phase reaction using $Co_2(CO)_8$ in the presence of sodium hydroxide and a phase-transfer catalyst is usually efficient for the self-coupling of sterically crowded bromo ketones to give 1,4-diketones.[51]

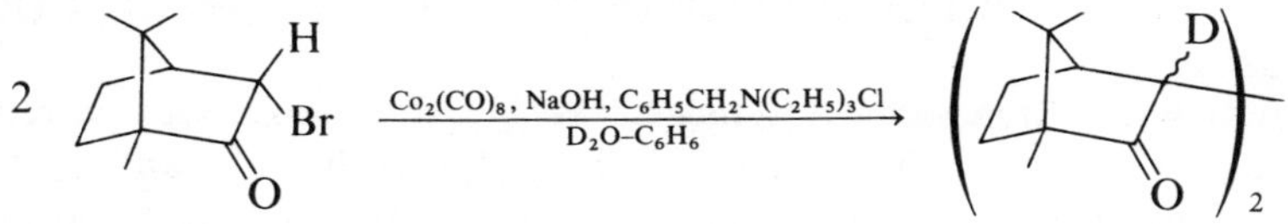

A high-concentration reduction with zinc dust in dimethylsulfoxide–benzene produces furans.[57] Noteworthy is the use of nickel tetracarbonyl $[Ni(CO)_4]$

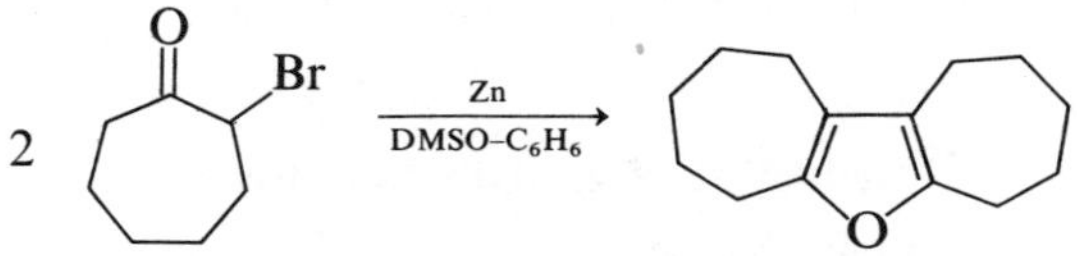

(58%)

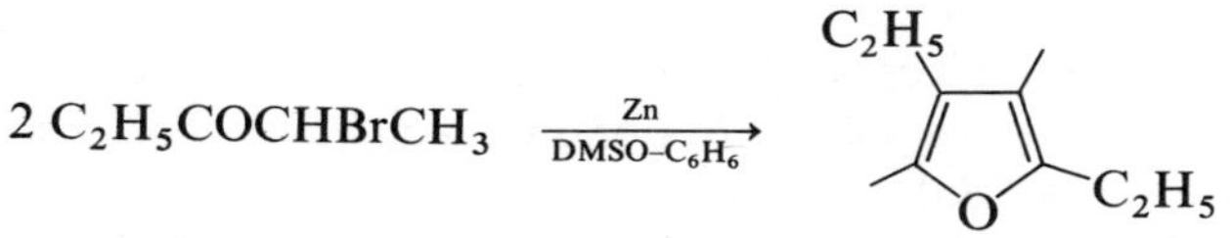

(97%)

for the reaction of aryl bromomethyl ketones, which leads exclusively to 1,4-diketones in tetrahydrofuran or to 2,4-diarylfurans in dimethylformamide, respectively.[68] β-Epoxy ketones are obtained from alkyl (but not aryl) bromo-alkyl ketones in dimethylformamide.[67]

$$2\,p\text{-}RC_6H_4COCH_2Br + Ni(CO)_4 \xrightarrow{}$$
$$(R = H, CH_3, Br)$$

$$\xrightarrow{\text{THF}} (p\text{-}RC_6H_4COCH_2)_2$$
$$(15\text{--}45\%)$$

$$\xrightarrow{\text{DMF}}$$

$$p\text{-}RC_6H_4$$

$$(50\text{--}92\%)\quad C_6H_4R\text{-}p$$

$$2\,t\text{-}C_4H_9COCH_2Br \xrightarrow[\text{DMF}]{\text{Ni(CO)}_4} t\text{-}C_4H_9COCH_2C\overset{\displaystyle O}{\underset{\displaystyle C_4H_9\text{-}t}{{-\!-\!-}}}CH_2$$

$$(61\%)$$

Formation of 1,5-Dicarbonyl Compounds. α-Chloro or α-bromo ketones can undergo reductive Michael addition to α,β-unsaturated ketones or esters, giving rise to 1,5-dicarbonyl products. Reactions with magnesium metal[69] or an organo cuprate[65] in ether are examples.

$$(CH_3)_2CHCH_2COCHBrC_3H_7\text{-}i + C_6H_5CH{=}CHCOC_6H_5$$

$$\xrightarrow[\text{ether}]{\text{Mg}} [(CH_3)_2CHCH_2]COCH(i\text{-}C_3H_7)CH(C_6H_5)CH_2COC_6H_5$$

Formation of α-Arenesulfenyl Ketones. When α-bromo ketones are reduced with zinc dust in ether in the presence of arenesulfenyl chlorides and a catalytic amount of mercury(II) chloride, α-arenesulfenyl ketones are produced in high yields.[70]

$$CH_3COCBr(CH_3)_2 + C_6H_5SCl \xrightarrow[\text{cat HgCl}_2]{\text{Zn}} CH_3COC(SC_6H_5)(CH_3)_2$$

$$(71\%)$$

Reduction of α,α-Dihalo Ketones

Formation of α-Monohalo Ketones and Parent Ketones. α,α-Dihalo ketones are reduced stepwise to form initially the corresponding α-monohalo ketones and subsequently the parent ketones.

Zinc dust in protic solvents is frequently used for this purpose. Glacial acetic acid is recommended as a solvent to obtain a high yield of completely dehalo-genated ketone (Eq. 28a),[71] whereas milder reaction in aqueous acetic acid tends to afford monohalo ketones predominantly (Eq. 28b).[72]

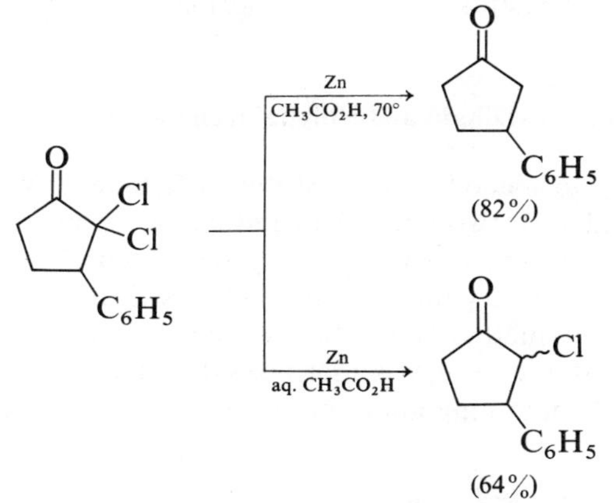

(Eq. 28a)[71]

(82%)

(Eq. 28b)[72]

(64%)

Reduction of α,α-dichloro ketone **25** with lithium dimethylcuprate in tetrahydrofuran at $-78°$ affords α-monochloro ketone **26** exclusively after aqueous workup;[73] this procedure is substantially easier and higher yielding than the zinc reduction. Diethylzinc in benzene is another excellent reducing agent for the preparation of monohalo ketones.[16]

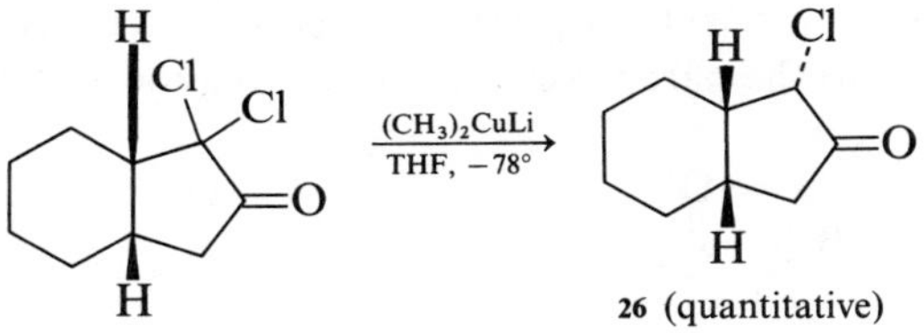

26 (quantitative)

25

Formation of α-Alkyl α-Monohalo Ketones and α-Alkyl α,β-Unsaturated Ketones. Quenching of the reaction mixture of **25** and lithium dimethylcuprate in tetrahydrofuran–ether with alkyl iodides and hexamethylphosphoramide below $-40°$ gives rise to α-alkyl α-chloro ketone **27**;[73] workup at

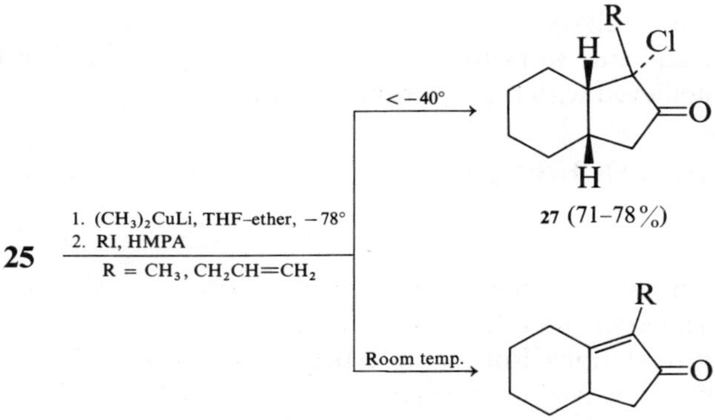

1. $(CH_3)_2CuLi$, THF–ether, $-78°$
2. RI, HMPA
$R = CH_3, CH_2CH=CH_2$

27 (71–78%)

28 (68–86%)

ambient temperature results in dehydrochlorination of **27** and formation of enone **28**.[73]

Reduction of α,α'-Dihalo and Polyhalogenated Ketones in Solution

Formation of α-Monohalo Ketones and Parent Ketones. When α,α'-dibromo ketones are treated with two-equivalent reductants in protic media, stepwise debromination occurs to afford ultimately the parent ketones; the selective preparation of intermediary monobromo ketones is difficult.[28] Zinc dust and zinc/copper couple in methanol are the most commonly employed reductants. Formation of α-methoxy ketones is a serious side reaction, but can be prevented by using methanol containing ammonium chloride as solvent.[74,75] Hydriodic

$$\text{Zn/Cu, NH}_4\text{Cl, CH}_3\text{OH}$$
$$Z = CH_2, O, NCO_2CH_3$$

(100%)

acid also can effect the reduction.[32] Use of two molar equivalents of primary or secondary alkoxides in tetrahydrofuran or alcoholic tetrahydrofuran reduces certain dibromides to the parent ketones in fair to good yields.[76,77]

$$\xrightarrow[\text{THF}]{2\ i\text{-}C_3H_7OK}$$

(60%)

Formation of α-Monoalkyl Ketones. Reduction of open-chain and cyclic dibromo ketones with organo cuprate reagents such as lithium dialkylcuprates and lithium alkyl(alkoxy)- or alkyl(phenylthio)cuprates gives α-monoalkyl ketones after aqueous workup.[60,78] Primary, secondary, and tertiary alkylations can be achieved with the appropriate quantity of cuprate, the structure of

$$n\text{-}C_3H_7CHBrCOCHBrC_3H_7\text{-}n \xrightarrow[\text{ether}]{(CH_3)_2CuLi} n\text{-}C_4H_9COCH(CH_3)C_3H_7\text{-}n$$
(70%)

which has an important influence on the yield of product. Alkyl heterocuprates generally work better than the corresponding dialkyl reagents for α-secondary and α-tertiary alkylations. For example, the *sec*-butylation of 2,6-dibromocyclo-hexanone (**29**) is achieved in 75% yield using $[C_2H_5CH(CH_3)](t\text{-}C_4H_9O)CuLi$ in tetrahydrofuran, but only in 38% yield with $[C_2H_5CH(CH_3)]_2CuLi$ in ether.[60,78] Introduction of primary alkyl groups is performed in comparable yield using dialkyl cuprates or alkyl heterocuprates.[60,78]

(75%)

Grignard reagents are adequately effective only for the preparation of α-methylated ketones; other higher-alkylated derivatives cannot be prepared in this manner.[78] The major side product obtained in these alkylation reactions is the parent ketone; no dialkylated products are formed.

Alkylation of unsymmetrical α,α'-dibromo ketones usually affords a mixture of two possible positional isomers, with the less substituted one predominating. The degree of regioselectivity decreases in the order tertiary alkylation > secondary alkylation > primary alkylation (cf. Eq. 29).[78] β-Substituents also affect the regioselectivity (Eq. 30).[78]

$R = n\text{-}C_4H_9$	48%	16%
$CH(CH_3)C_2H_5$	61%	8%
$t\text{-}C_4H_9$	31%	2%

(Eq. 29)[78]

(33%)   (Eq. 30)[78]

Another type of reductive α-alkylation of dibromo ketone **30** is achieved with propargyl alcohol or its ethers using copper powder in acetonitrile containing sodium iodide. α-Allenyl ketones are produced.[79]

$$CH_3CHBrCOCHBrCH_3 + HC\equiv CCH_2OR$$

30

$$\xrightarrow[\text{R = H, CH}_3]{\text{Cu, NaI, CH}_3\text{CN}} C_2H_5COC(CH_3)\!\!=\!\!C\!\!=\!\!CHCH_2OR$$

(25–35%)

Formation of α,α'-Dialkyl Ketones.   Reaction of α,α'-dibromo ketones with organo cuprates followed by treatment with alkyl iodides affords α,α'-dialkyl ketones.[78,80,81] Use of alkyl bromides or alkyl tosylates results in relatively little alkylation.[81] Addition of N,N,N',N'-tetramethylethylenediamine (TMED)

t-$C_4H_9CHBrCOCHBrC_4H_9$-t

$$t\text{-}C_4H_9CHBrCOCHBrC_4H_9\text{-}t \xrightarrow[\text{2. CH}_3\text{I}]{\text{1. (CH}_3)_2\text{CuLi, ether}} t\text{-}C_4H_9CH(CH_3)COCH(CH_3)C_4H_9\text{-}t$$
$$(80\%)$$

or hexamethylphosphoramide to the reaction mixture facilitates the second alkylation.[81] The formation of α-monoalkyl[80] or α,α-dialkyl ketones is a serious side reaction.

Quenching of cuprate-generated enolates with aldehydes followed by treatment with ammonium chloride yields α-alkyl α'-(E)-alkylidene ketones.[81]

Reductive Dimerization to Form 1,4-Diketones. α,α'-Dibromo ketones undergo two types of reductive dimerizations, giving the corresponding open-chain and cyclic 1,4-diketones. Open-chain diones are obtained by reduction with zinc/copper couple in N-methylformamide[82] or $Fe_2(CO)_9$ in tetrahydrofuran,[83] whereas cyclic diones result from sodium iodide reduction in acetone or acetone–carbon disulfide.[84,85] Reactions of unsymmetrical dibromides are nonregioselective.[82]

$$(CH_3)_2CBrCOCBr(CH_3)_2 \xrightarrow[\text{HCONHCH}_3]{\text{Zn/Cu}} [i\text{-}C_3H_7COC(CH_3)_2]_2$$
$$\mathbf{16} \qquad\qquad\qquad\qquad (71\%)$$

Formation of Seven-Membered Ketones by
Intermolecular [3 + 4] Cyclocoupling

General Aspects. Before the scope and limitations of the reductive [3 + 4] cyclocouplings outlined in Scheme II are discussed, it may be useful to consider some of the general aspects of the reaction. The yields of 4-cycloheptenones appear to depend on the efficiency of the generation of 2-oxyallyl or related

$$\text{(Eq. 31)}$$

$$\text{(Eq. 32)}$$

Scheme II

species and the ease of the trapping by 1,3-dienes. As mentioned in the MECHANISM section, secondary and tertiary dibromides can generate the three-carbon reactive intermediates, while dibromoacetone and other methyl alkyl ketone dibromides are unable to produce such intermediates effectively. However, $\alpha,\alpha,\alpha',\alpha'$-tetrabromoacetone (**33**) reacts with cyclic dienes to give **34**, which on brief treatment with zinc/copper couple in methanol is converted to ketone **35**.[74] The overall process is regarded formally as a cyclocoupling reaction

$$\text{CHBr}_2\text{COCHBr}_2 \quad + \qquad \xrightarrow{\text{Reducing agent}}$$

33

34

$$\xrightarrow[\text{CH}_3\text{OH}]{\text{Zn/Cu, NH}_4\text{Cl}}$$

35

between dibromoacetone and the diene. The modified procedure is widely applicable; **37** is obtained using methyl alkyl ketone tribromides such as **36**.[41]

$$\text{R}^1\text{R}^2\text{CBrCOCHBr}_2 \quad + \qquad \xrightarrow[\text{2. Zn/Cu, NH}_4\text{Cl, CH}_3\text{OH}]{\text{1. Reducing agent}}$$

36

37

The [3 + 4] reaction is a cycloaddition of the intermediary 2-oxyallyl species or a cyclopropanone, depending on the reaction conditions, across the diene.[36] Therefore, as in the Diels-Alder reaction, dienes with a high equilibrium concentration of the *s-cis* conformer serve as efficient four-carbon receptors. Thus in general cyclic dienes undergo the cycloaddition reaction better than conformationally flexible open-chain derivatives.[8]

Discussed below are the scope and limitations of the [3 + 4] cyclocoupling reaction classified according to the structure of the diene component.

Formation of 4-Cycloheptenones and Related Compounds.   When α,α'-dihalo ketones are reduced with iron carbonyls or zinc/copper couple in the presence of open-chain 1,3-dienes, 4-cycloheptenones are obtained. Of the various kinds of dienes, 2,3-dialkylbutadienes usually give the best yields of cycloadducts, because of their high equilibrium concentration of *s-cis* conformer. For instance, the $Fe_2(CO)_9$-promoted cyclocoupling of **16** and 2,3-dimethylbuta-diene (**40**, $R^1 = R^2 = CH_3$) proceeds in a higher yield (71%) than reactions using butadiene (33%) or isoprene (**40**, $R^1 = H$, $R^2 = CH_3$, 47%).[74] The 1,2-bismethylenecycloalkane **42**, in which the *s-cis* conformation is frozen by inclusion of the C-2 and C-3 positions of the diene in the cycloalkane system, is among the most reactive four-carbon units.[74] Diiron enneacarbonyl is conventionally employed as the reducing agent. When dieneiron tricarbonyl complexes of type **43** are used in place of the free dienes and $Fe_2(CO)_9$, forcing reaction conditions are required, but the result is still a remarkable increase in yields of cycloheptenones. For example, the reaction of **16** and butadiene with $Fe_2(CO)_9$ at 60° for 38 hours produces **41** ($R^1 = R^2 = H$) in only 33% yield, while use of **43** at 80° for 4 hours gives the same adduct in 90% yield.[74] Iron pentacarbonyl gives a lower yield.[74,86] Zinc/copper couple as the reducing agent affords in addition to 4-cycloheptenones many undesirable side products such as five-membered ketones and 1,4-diketones.[87,88]

$$RCHBrCOCHBrR + CH_2{=}CHCH{=}CH_2 \xrightarrow[R = CH_3,\, i\text{-}C_3H_7]{Fe_2(CO)_9,\, C_6H_6}$$

38

39 (30–44%)

$$(CH_3)_2CBrCOCBr(CH_3)_2 + CH_2{=}CR^1CR^2{=}CH_2 \xrightarrow[C_6H_6]{Fe_2(CO)_9}$$

16   **40**

41 (33–71%)

$$CH_3CHBrCOCHBrCH_3 + \text{(structure 42)} \xrightarrow[C_6H_6]{Fe_2(CO)_9} \text{(structure, 80\%)}$$

30

42

(80%)

$$(CH_3)_2CBrCOCBr(CH_3)_2 + \text{(structure 43)} \xrightarrow[C_6H_6]{Fe(CO)_3} \textbf{41} (R^1 = R^2 = H)$$

16

43

(90%)

$$(CH_3)_2CBrCOCH_2I + CH_2{=}C(CH_3)CH{=}CH_2$$

44

$$\xrightarrow[CH_3CN\text{–}DME]{Zn/Cu} \text{(structure 45)} + \text{(structure)} + \text{(structure)}$$

45

(total 5–12%)

Synthetic Applications. 2,7-Dialkylated 4-cycloheptenones such as **39** are utilized as key intermediates in the preparation of various kinds of troponoid compounds including the 2,7-dialkylated tropone **46**,[86,89] γ-tropolone **47**,[86,89] and 4,5-homotropone **48**, which exists as the hydrohomotropylium ion **49** in concentrated sulfuric acid.[89,90] A single-step synthesis of karahanaenone (**45**),

46   47   48   49

a monoterpene, was achieved by the reaction between **44** and isoprene with zinc/copper couple or $Fe_2(CO)_9$, though in very poor yield.[87]

Formation of Bicyclo[3.2.n]alkenones and Related Compounds. Reductive cyclocoupling of dibromo ketones across cyclopentadiene gives rise to bicyclo[3.2.1]oct-6-en-3-ones in generally high yields. Copper–sodium iodide in acetonitrile,[91,92] zinc dust–triethyl borate in tetrahydrofuran,[47] zinc/copper couple in 1,2-dimethoxyethane or acetone,[93] $Fe_2(CO)_9$ in benzene,[74,87] and sodium iodide in acetone or acetonitrile[84,94] are reducing agents and solvents of choice. The combination of zinc powder and a trialkyl borate effects reduction of the dihalo derivatives of methyl alkyl ketone **44**.[47] The 2,4-disubstituted

$$CH_3CHBrCOCHBrCH_3 \ + \ \text{(cyclopentadiene)} \xrightarrow[CH_3CN]{Cu, \ NaI}$$

30

$(82\%, \ \alpha,\alpha/\beta,\beta = 6.4:1)$

$$(CH_3)_2CBrCOCBr(CH_3)_2 \ + \ \text{(cyclopentadiene)} \xrightarrow[DME]{Zn/Cu}$$

16

(65%)

$$(CH_3)_2CBrCOCH_2I \ + \ \text{(cyclopentadiene)} \xrightarrow[THF]{Zn, \ (C_2H_5O)_3B}$$

44

50 (76%)

$$i\text{-}C_3H_7CHBrCOCHBrC_3H_7\text{-}i \ + \ \text{(cyclopentadiene)} \xrightarrow[C_6H_6]{Fe_2(CO)_9}$$

$(90\%, \ \alpha,\alpha/\beta,\beta = 1:1)$

$$C_6H_5CHBrCOCHBrC_6H_5 \ + \ \text{(cyclopentadiene)} \xrightarrow[CH_3CN]{NaI}$$

31

$(99\%, \ \alpha,\alpha/\alpha,\beta = 4:6)$

product obtained is usually a mixture of two *cis* isomers[74,91] except for the diphenyl adduct.[94] The reaction of 2,5-dibromocyclopentanone yields tricyclic ketone **51**.[95]

$$Br\text{-}(\text{cyclopentanone})\text{-}Br \ + \ \text{(cyclopentadiene)} \xrightarrow{Fe_2(CO)_9}$$

51

Tri- and tetrabromo ketones also react with cyclopentadiene with the assistance of Fe(CO)$_5$ in benzene[96] or tetrahydrofuran–benzene;[48,74] alternatively, cyclopentadieneiron tricarbonyl (**52**) is a good reducing agent.[97] The extra bromine atoms in the resulting cycloadducts are quantitatively removed by treatment with zinc/copper couple in ammonium chloride–saturated methanol.[48,74] Poor results are obtained with $Fe_2(CO)_9$[48] or zinc/copper

$$(CH_3)_2CBrCOCHBr_2 \ + \ \text{(cyclopentadiene)} \quad \xrightarrow[\text{2. Zn/Cu, NH}_4\text{Cl, CH}_3\text{OH}]{\text{1. Fe(CO)}_5\text{, THF–C}_6\text{H}_6}$$

50 (83%)

$$CHBr_2COCHBr_2 \ + \ \underset{\underset{\mathbf{52}}{\text{Fe(CO)}_3}}{\text{(cyclopentadiene·Fe(CO)}_3)} \quad \xrightarrow[\text{2. Zn/Cu, NH}_4\text{Cl, CH}_3\text{OH}]{\text{1. Room temp., ether}}$$

33

(40–60%)

couple–triethyl borate.[47] 5,5-Dimethylcyclopentadiene fails to give a cyclo-adduct.[92]

Fulvenes also enter into cyclocoupling with dibromo ketones, aided by copper–sodium iodide in acetonitrile,[92] zinc/copper couple in 1,2-dimethoxy-ethane,[92,98] or sodium iodide in acetonitrile,[98] giving 8-alkylidenebicyclo-[3.2.1]oct-6-en-3-ones. The reactions with 6,6-dialkylfulvenes generally give

$$CH_3CHBrCOCHBrCH_3 \ + \ \text{(6-OCOCH}_3\text{-fulvene)} \quad \xrightarrow[\text{CH}_3\text{CN}]{\text{Cu, NaI}}$$

30

OCOCH$_3$

(30–40%, $\alpha,\alpha/\beta,\beta = 5:1$)

$$\mathbf{30} \ + \ \text{(6-methyl-6-}t\text{-butylfulvene)} \quad \xrightarrow[\text{DME}]{\text{Zn/Cu}}$$

C$_4$H$_9$-t

$\alpha,\alpha/\alpha,\beta/\beta,\beta = 5:1:1$

$$C_6H_5CHBrCOCHBrC_6H_5 \ + \ \text{(6,6-dimethylfulvene)} \quad \xrightarrow[\text{CH}_3\text{CN}]{\text{NaI}}$$

31

C$_6$H$_5$ ... C$_6$H$_5$

(90%)

good results; hetero substituents at the C-6 position depress the yields. 6-(Dimethylamino)fulvene acts as a 6π-electron component, giving a different type of cyclic adduct, **53**, in the reaction with $Fe_2(CO)_9$.[99]

$$(CH_3)_2CBrCOCBr(CH_3)_2 \quad + \qquad \xrightarrow[\text{C}_6\text{H}_6]{\text{Fe}_2(\text{CO})_9}$$

16

53 (17%)

Cyclopentadienones are fair to good C_4 substrates, giving bicyclo[3.2.1]oct-6-ene-3,8-dione **54** by reaction of **16** with $Fe_2(CO)_9$ in benzene.[99] Diiron

$$\textbf{16} \quad + \qquad \xrightarrow[R = CH_3,\, C_2H_5,\, n\text{-}C_3H_7,\, n\text{-}C_4H_9]{\text{Fe}_2(\text{CO})_9,\, \text{C}_6\text{H}_6}$$

54 (22–35%)

enneacarbonyl assists the cyclocoupling of **30** and 1,3-cyclohexadiene to form bicyclo[3.2.2]non-6-en-3-one **55**.[100]

$$\textbf{30} \quad + \qquad \xrightarrow[\text{C}_6\text{H}_6]{\text{Fe}_2(\text{CO})_9}$$

55 (50%, $\alpha,\alpha/\beta,\beta = 1.5:1$)

Synthetic Applications. Carbocamphenilone (**56**), a terpenic α-diketone, and its oxidized product, camphenic acid (**57**), are prepared via the cycloadduct **50** as a key intermediate.[48]

56

$$HO_2C \qquad \qquad \text{--}CO_2H$$

57

Formation of 8-Azabicyclo[3.2.1]oct-6-en-3-ones and Related Compounds.
Pyrrole and its N-alkyl derivatives can be employed as diene substrates for the copper–sodium iodide-promoted [3 + 4] cyclocoupling reaction of open-chain dibromo ketones.[92,101,102] Reaction with cyclic dibromo ketones such

as 2,6-dibromocyclohexanone affords only substitution products. Treatment of both open-chain and cyclic dibromides with zinc or iron carbonyls results only in substitution.[74,102,103]

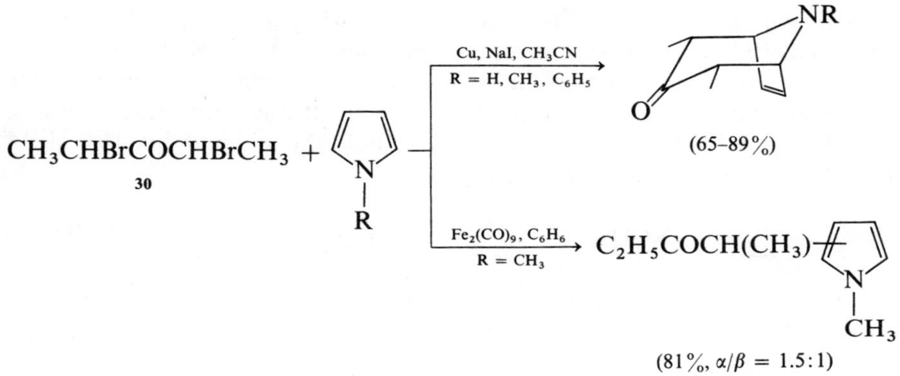

$CH_3CHBrCOCHBrCH_3$ + (pyrrole, N–R)

30

Cu, NaI, CH_3CN
R = H, CH_3, C_6H_5
→ (product with NR)
(65–89%)

$Fe_2(CO)_9$, C_6H_6
R = CH_3
→ $C_2H_5COCH(CH_3)$–(pyrrole, N–CH_3)

(81%, α/β = 1.5:1)

Pyrrole derivatives bearing an electron-withdrawing substituent such as acetyl or methoxycarbonyl on the nitrogen undergo the cyclocoupling reaction with dibromo[74] or more highly brominated ketones.[75] The reducing agent of

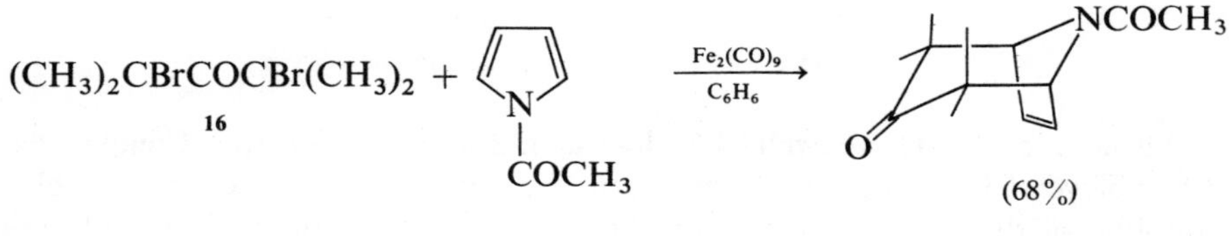

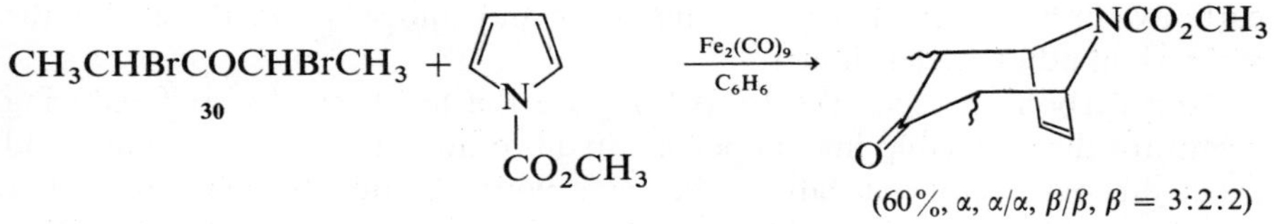

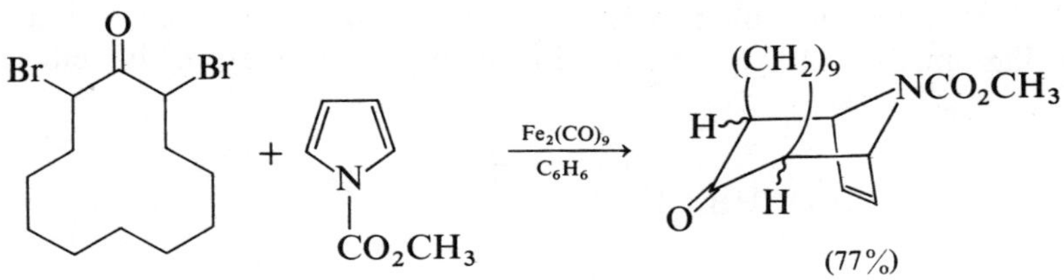

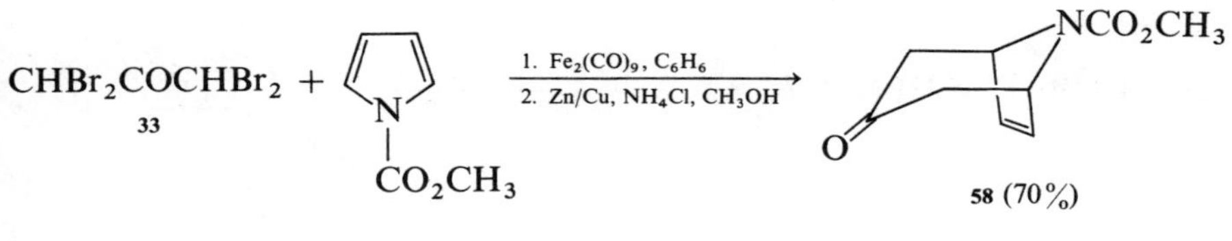

$(CH_3)_2CBrCOCBr(CH_3)_2$ + (pyrrole, N–$COCH_3$)

16

$Fe_2(CO)_9$
C_6H_6
→ (product with $NCOCH_3$)
(68%)

$CH_3CHBrCOCHBrCH_3$ + (pyrrole, N–CO_2CH_3)

30

$Fe_2(CO)_9$
C_6H_6
→ (product with NCO_2CH_3)
(60%, α, α/α, β/β, β = 3:2:2)

+ (pyrrole, N–CO_2CH_3)

$Fe_2(CO)_9$
C_6H_6
→ (product with $(CH_2)_9$, NCO_2CH_3)
(77%)

$CHBr_2COCHBr_2$ + (pyrrole, N–CO_2CH_3)

33

1. $Fe_2(CO)_9$, C_6H_6
2. Zn/Cu, NH_4Cl, CH_3OH
→ (product with NCO_2CH_3)
58 (70%)

choice is $Fe_2(CO)_9$ in benzene. Use of $Fe(CO)_5$ or zinc/copper couple gives less satisfactory results. Attempts to use copper–sodium iodide have not succeeded.

Synthetic Applications. The synthesis of a variety of tropane alkaloids has been achieved via 6,7-dehydrotropine (**59**), derived from **58** by diisobutyl-aluminum hydride reduction in tetrahydrofuran; subsequent appropriate modification of the double bond leads to several naturally occurring products, including tropine (**60**), scopine (**61**), tropanediol (**62**), and teloidine (**63**).[75,104] Unnatural analogs can also be prepared.[101–103]

Formation of 8-Oxabicyclo[3.2.1]oct-6-en-3-ones and Related Compounds.
8-Oxabicyclo[3.2.1]oct-6-en-3-ones can be prepared by reductive cyclo-coupling of dibromo or polybromo ketones with furan or its alkyl, alkoxy-carbonyl, or halo derivatives; the ester group and halogen atom attached to the sp^2-carbon atom are left intact.

Iron carbonyls, especially $Fe_2(CO)_9$, are the most widely used reducing agents for the cyclocoupling of open-chain dibromo or polybromo ketones and cyclic dibromo ketones to afford the corresponding adducts, generally in fair to high yields;[74,89] cyclic dibromides undergo somewhat erratic transformations accompanied by α-substitution reactions to give **69**.[74] Reactions involving methyl alkyl ketone dibromides are reasonably efficient.[36]

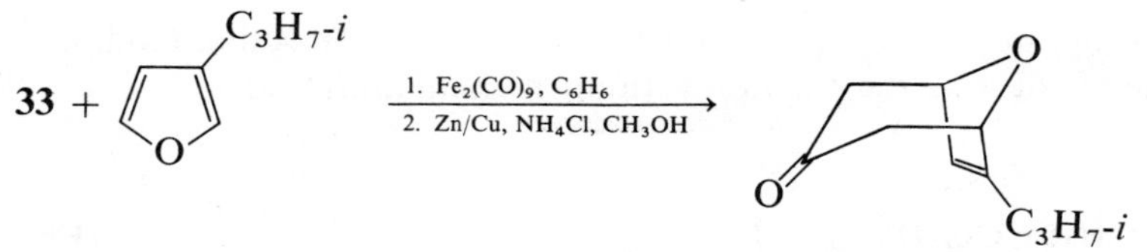

65 (71%)

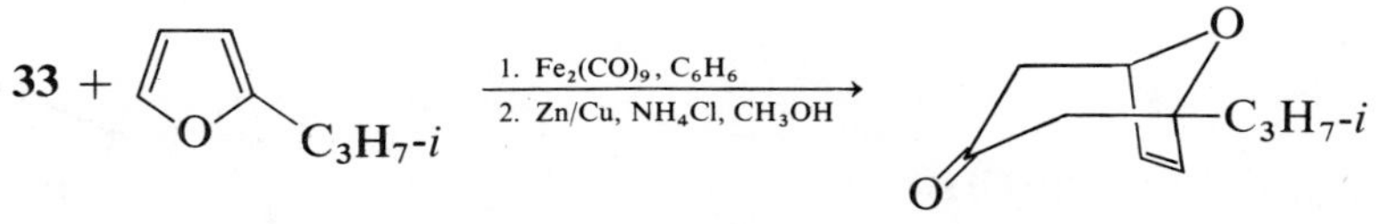

66 (47%)

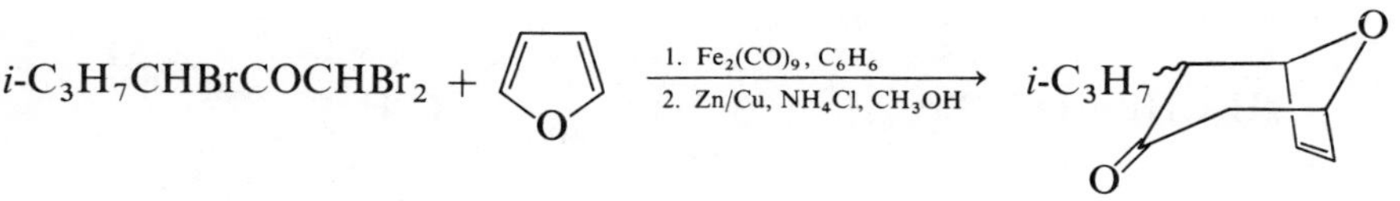

$i\text{-}C_3H_7CHBrCOCHBr_2$ +

67 (35%)

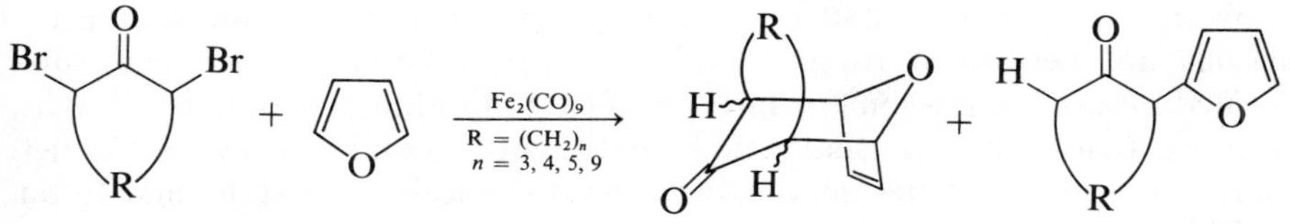

68 (35–54%)   **69** (10–35%)

Zinc/copper[92,93,102,105,106] or zinc/silver couple[45] also effects the reactions. In addition, Grignard reagents,[78] the copper–sodium iodide reagent,[92,102,107]

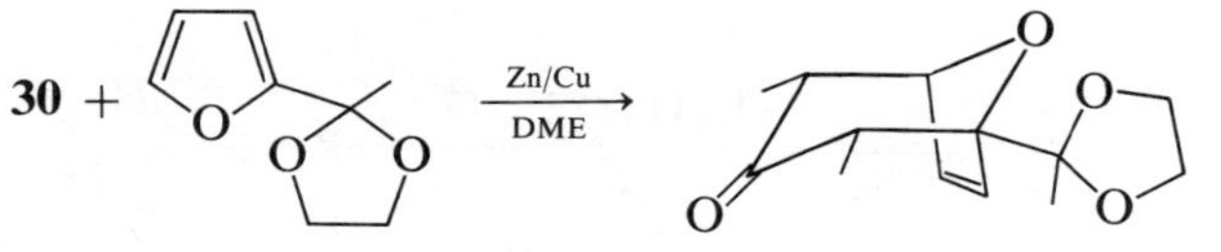

(Eq. 33)[105]

70 (53%)

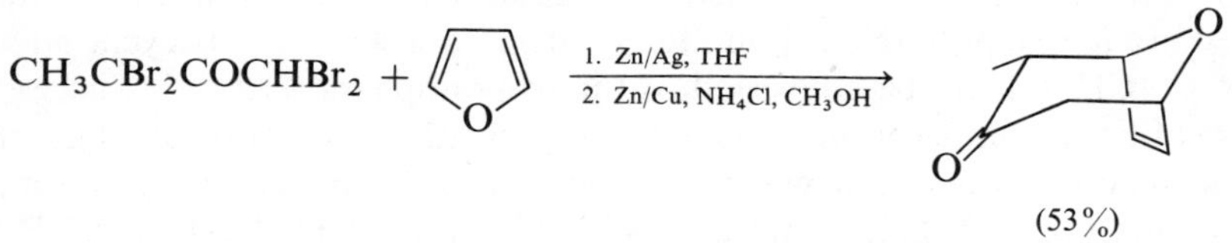

$CH_3CBr_2COCHBr_2$ +

(53%)

organo cuprates,[78] a copper/isonitrile complex,[108] and sodium iodide[84,94] are capable of effecting the reaction with open-chain dibromides.

$$CH_3CHBrCOCHBrCH_3 + \text{(furan)} \xrightarrow[CH_3CN]{Cu, NaI} \text{(product)} \qquad (Eq. \ 34)^{107}$$
30

(40–48%)

$$(CH_3)_2CBrCOCBr(CH_3)_2 + \text{(methylfuran)} \xrightarrow[C_6H_6]{Cu, \ t\text{-}C_4H_9NC} \text{(product)}$$
16

(30–40%)

$$C_6H_5CHBrCOCHBrC_6H_5 + \text{(furan)} \xrightarrow[Acetone]{NaI} \ C_6H_5 \text{(product)} \ C_6H_5$$
31

(65%)

When the $Fe_2(CO)_9$- and zinc/copper-couple-promoted cyclocoupling reactions are performed using 1,3-dibromo-1-phenylpropan-2-one and substituted furans, a regioisomeric mixture of [3 + 4] adducts is obtained.[36] The regioselectivity, which is usually moderately good, is controlled by the frontier molecular orbitals of the intermediary oxyallyl species (LUMOs) and furan (HOMOs).

$$C_6H_5CHBrCOCH_2Br + \text{(substituted furan with } R^1, R^2, R^3, R^4 \text{)}$$

$$\xrightarrow[C_6H_6]{Fe_2(CO)_9} \ C_6H_5 \text{(product with } R^1, R^4, R^2, R^3 \text{)} + C_6H_5 \text{(product with } R^4, R^1, R^3, R^2 \text{)}$$

Synthetic Applications. The resulting oxabicyclic ketones are readily transformed in a few steps to the tropone **46** or γ-tropolone **47**.[89] The tricyclic adduct **68** [R = $(CH_2)_9$] can be converted to the troponophane **71**.[89,109] Success in the use of tetrabromoacetone and tribromo derivatives of methyl alkyl ketones in these reactions opens a new route to various naturally occurring troponoid compounds, viz., nezukone (**72**) from **65**,[89,110] α-thujaplicin (**73**) from **67**,[89,111] and hinokitiol (β-thujaplicin, **74**) from **66**,[89,111] respectively. The reaction in

Eq. 33 giving **70** has been utilized for the synthesis of nonactic acid (**75**).[106] The Prelog-Djerassi lactone (**76**), a pivotal intermediate in Masamune's methymycin synthesis,[112] has been prepared through the key reaction shown in

Eq. 34.[105] The readily available oxabicyclic ketone **64** is converted to **77**, which provides an efficient entry into the C-nucleoside family. Thus there is accomplished the stereocontrolled synthesis of natural products such as pseudouridine (**78**), pseudocytidine (**79**), and showdomycin (**80**), as well as a number of unnatural analogs such as pseudoisocytidine (**81**) and 2-thiopseudouridine

(82).[113,114] Construction of the basic skeleton of lolium alkaloids **83** has also been achieved by way of **64.**[115]

83

Attempted Cyclocoupling between Dibromo Ketones and Thiophene. Several cyclocoupling reactions between dibromo ketones and thiophene have been tried with copper–sodium iodide[102] or $Fe_2(CO)_9$[74] without success, resulting only in the α-substitution product.

30 + $\overset{\displaystyle}{\underset{S}{\text{[thiophene]}}}$ $\xrightarrow[C_6H_6]{Fe_2(CO)_9}$ $C_2H_5COCH(CH_3)\overset{\displaystyle}{\underset{S}{\text{[thiophene]}}}$

(37%)

Formation of 6,7;8,9-Dibenzobicyclo[3.2.2]nonan-3-ones. Reduction of α,α′-dibromo ketones with zinc/copper couple in dioxane containing anthracene affords the 6,7;8,9-dibenzobicyclo[3.2.2]nonan-3-one [3 + 4] cyclocoupling products in 3–25% yields.[116] Addition of chlorotrimethylsilane (TMSCl) to the system and changing the solvent to benzene result in substantially increased yields of products (71–97%).[116]

$(CH_3)_2CBrCOCHBrCH_3$ +

$\xrightarrow[C_6H_6]{Zn/Cu, \text{ TMSCl}}$

(97%)

**Intramolecular [3 + 4] Cyclocoupling Leading to
11-Oxatricyclo[5.3.0.1^{1,4}]undec-2-en-6-ones (Oxidoperhydroazulenes)**

The $Fe_2(CO)_9$ reduction of dibromide **84** in benzene permits the direct construction of oxatricyclic ketone **85**, whose oxidoperhydroazulene structure is the basic skeleton of naturally occurring daucon, ambrosic acid, and germacrol.[117]

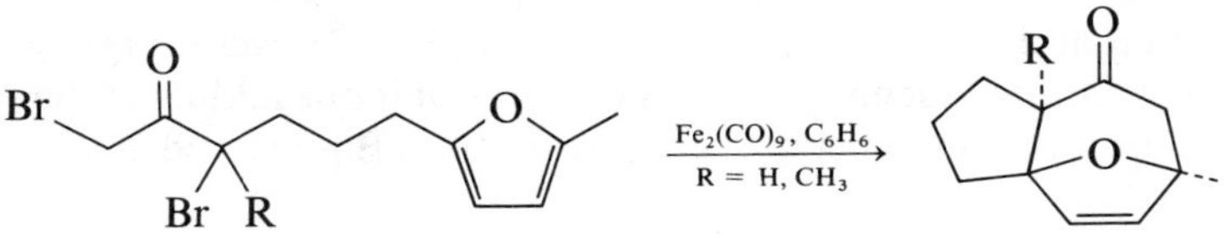

Formation of Five-Membered Ketones and Heterocycles by Intermolecular [3 + 2] Cyclocoupling

General Aspects.   When α,α'-dibromo ketones are reduced in the presence of certain olefins or carbonyl compounds, the [3 + 2] cyclocoupling reactions proceed to give five-membered ketones or oxacycles, respectively; the reaction course depends on the nature of the reducing agents and the structures of the dibromo ketones or olefins employed.

Scheme III outlines the general reaction pathways. Actually the C_3 units must be the dibromides derived from alkyl or aryl ketones rather than acetone, and the C_2 components must be the substances which, through addition of **86**, form the intermediate **87** stable enough to shift the equilibrium to the right, thereby completing the cyclocoupling reaction. Therefore the choice of cation-stabilizing R group in the C_2 units is crucial for the success of the reaction. Only a few examples have been reported using polybromo ketones or cyclic dibromo ketones.

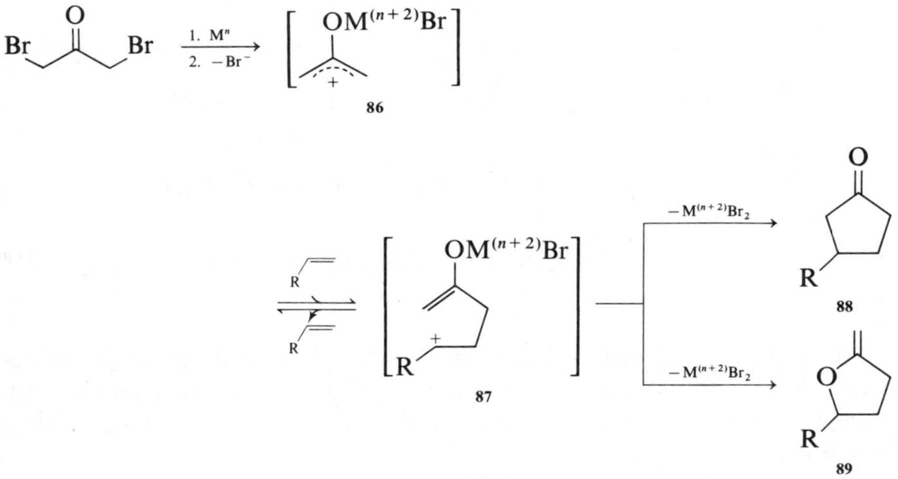

Scheme III

Below are described the scope and limitations of the [3 + 2] cyclocoupling reaction classified by the kinds of receptors and reducing agents.

Formation of 3-Arylcyclopentanones. The [3 + 2] cyclocoupling reaction with aryl-substituted olefins gives 3-arylcyclopentanone adducts. The products are mixtures of diastereomers, where possible (see Eqs. 35 and 36).[118,119]

$$CH_3CHBrCOCHBrCH_3 + C_6H_5CR=CH_2 \xrightarrow[\substack{R = H, CH_3, \\ c\text{-}C_3H_5, C_6H_5}]{Fe_2(CO)_9,\ C_6H_6}$$

30

$(60-95\%)$

(Eq. 35)[118]

$$30 + \quad \text{(ferrocenyl-vinyl)} \quad \xrightarrow[C_6H_6]{Fe_2(CO)_9} \quad \text{(ferrocenyl cyclopentanone)} \qquad (\text{Eq. 36})^{118}$$

90

(30%)

$$(CH_3)_2CBrCOCBr(CH_3)_2 + C_6H_5C(CH_3)=CH_2$$

16

$$\xrightarrow[C_6H_6]{Fe_2(CO)_9}$$

91 (5%) 92 (16%)

$+ (E)\text{-}i\text{-}C_3H_7COC(CH_3)_2CH=C(CH_3)C_6H_5$

93 (37%)

$+ i\text{-}C_3H_7COC(CH_3)_2CH_2C(C_6H_5)=CH_2$ (Eq. 37)[118]

94 (11%)

Most phenylated olefins, even ferrocenylethylene (**90**), undergo the reaction. Placement of a carbocation-stabilizing group, particularly cyclopropyl, phenyl, or methoxyl, at the arylated olefin carbon or in the aromatic ring, results in a smooth reaction.

As for the reducing agent, $Fe_2(CO)_9$ gives a satisfactory result; $Fe(CO)_5$ is less effective. Use of aromatic olefin–$Fe(CO)_4$ complexes does not improve the yield.[118] Reaction using zinc/copper couple in benzene or 1,2-dimethoxyethane gives rise to no cyclopentanones.[118]

In general, reaction of tertiary dibromides and aromatic olefins proceeds rather sluggishly, probably because of steric factors, and tends to afford various kinds of byproducts, including substitution products like **93** and **94**, and 2-alkylidenetetrahydrofurans like **92** (see Eq. 37).[118]

An experiment using the α,α'-dibromo ketone **16** and (Z)-β-deuteriostyrene (**95**) demonstrates that the cyclocoupling reaction giving **96** proceeds stereospecifically, whereas the olefinic substitution giving **97** occurs in a nonstereospecific fashion.[118–120]

A high degree of regioselectivity is observed in the reaction using styrene as the C_2 component, which is best interpreted in terms of relative stabilities of the possible zwitterionic intermediates of type **87** in Scheme III.[36] The regioselective, single-step synthesis of α-cuparenone (**98**), a sesquiterpenic ketone, has been performed according to Eq. 38.[121]

Formation of 2-Cyclopentenones and Related Compounds. Enamines are two-carbon substrates capable of entering into the reductive [3 + 2] reaction.

The $Fe_2(CO)_9$ reduction of tertiary dibromo ketone **16** in the presence of enamine **99** produces the β-morpholinocyclopentanone derivative **100**.[122,123]

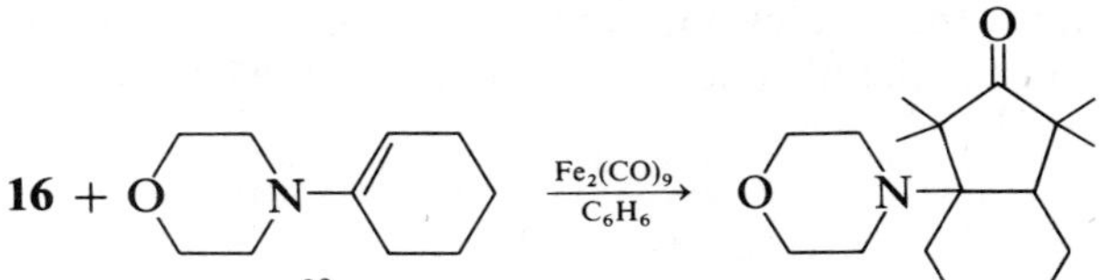

16 + [structure] **99** $\xrightarrow[C_6H_6]{Fe_2(CO)_9}$ [structure] **100** (87%)

In the reaction of secondary dibromide **38**, the initially formed labile morpholino adduct **101** suffers elimination of morpholine on brief treatment with silica gel[124] or dilute ethanolic sodium hydroxide, or even spontaneously in certain cases, to give the 2-cyclopentenone **102**, as depicted in Eq. 39. Thus the

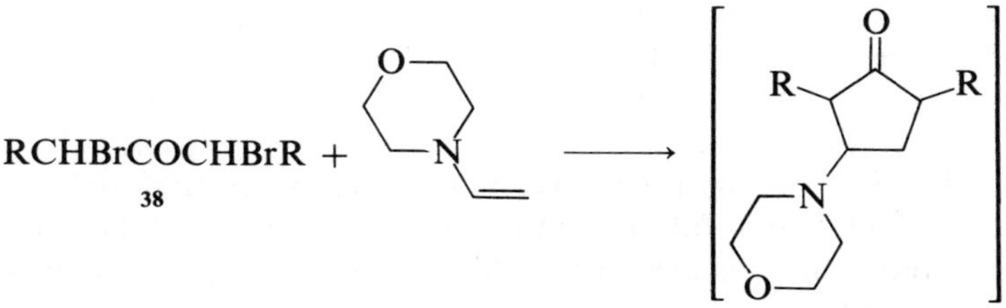

RCHBrCOCHBrR + [structure] ⟶ [structure **101**]

38

$\xrightarrow{- \text{Morpholine}}$ [structure] **102**   (Eq. 39)

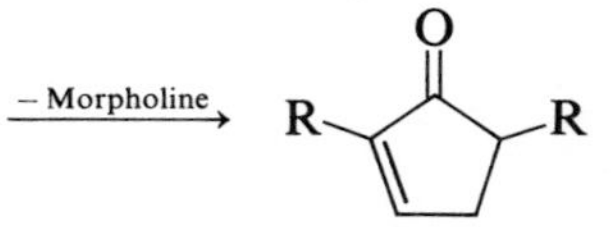

overall transformation provides a single-pot procedure for the preparation of substituted cyclopentenones. Several representative reactions are shown in Eqs. 40–43.[122]

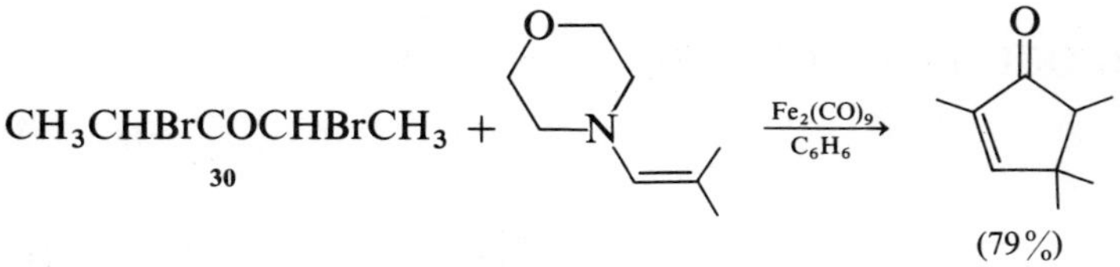

$CH_3CHBrCOCHBrCH_3$ + [structure] $\xrightarrow[C_6H_6]{Fe_2(CO)_9}$ [structure]   (Eq. 40)

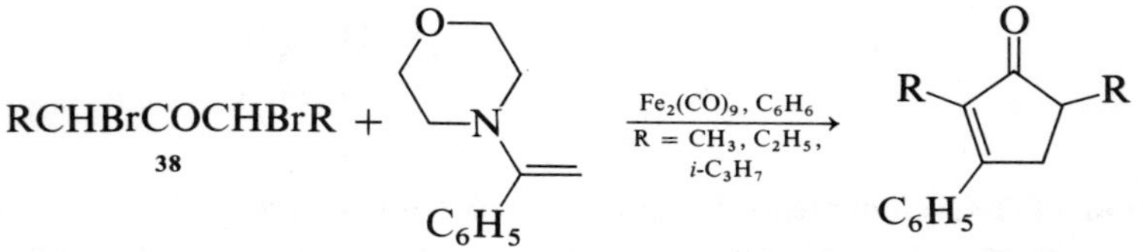

30

(79%)

RCHBrCOCHBrR + [structure C_6H_5] $\xrightarrow[\substack{R = CH_3, C_2H_5, \\ i\text{-}C_3H_7}]{Fe_2(CO)_9, C_6H_6}$ [structure C_6H_5]   (Eq. 41)

38

(64–94%)

$$RCHBrCOCHBrR + \text{(103)} \xrightarrow[C_6H_6]{Fe_2(CO)_9} \text{(104)}$$

38 · 103 · 104 (R = CH$_3$, R' = (CH$_2$)$_4$; 100%)

(Eq. 42)

$$CH_3CHBrCOCHBrCH_3 + \text{(105)} \xrightarrow[\substack{R = (CH_2)_5,\\(CH_2)_{11}}]{Fe_2(CO)_9, C_6H_6} \text{(106)}$$

30 · 105 · 106 (71–90%)

(Eq. 43)[122]

   This cyclocoupling reaction is achieved using enamine derivatives of either open-chain or cyclic ketones, and of aldehydes. Diiron enneacarbonyl is the best reducing agent; Fe(CO)$_5$ gives less satisfactory results. Benzene is recommended as the solvent; tetrahydrofuran or dimethylformamide gives little or no cyclocoupling products. The reduction with zinc/copper couple in 1,2-dimethoxyethane drastically diminishes the yield of adduct, and sodium iodide is not usable.

   Equation 44 illustrates some reactions of unsymmetrically substituted C$_2$ and C$_3$ components, giving mixtures of regioisomeric [3 + 2] adducts. The regiochemistry is controlled primarily by the stability of zwitterionic intermediates of type **87** (see p. 197).[36]

$$(CH_3)_2CBrCOCHBrR + \text{(99)}$$

99

$$\xrightarrow[\text{2. NaOH, C}_2\text{H}_5\text{OH}]{\text{1. Fe}_2(CO)_9, C_6H_6}$$

(Eq. 44)[36]

R = CH$_3$ (41:59, total 86%)
C$_6$H$_5$ (75:25, total 61%)

   Some enol ethers and ketene diphenyl acetal undergo a similar reaction, but much less effectively.[83]

Synthetic Utility and Applications. This cyclocoupling reaction using secondary dibromo ketones opens a new route to α,α′-dialkylcyclopentenones. The attractive features include the ready availability of starting materials, operational simplicity, and wide generality. The advantages may be offset to some extent by the incapability of preparing α-unsubstituted cyclopentenones. This feature, however, can be a virtue of this method in view of the difficulty of introducing alkyl groups, particularly bulky ones such as isopropyl, into α positions of five-membered ketones.[125]

The reaction with cycloalkanone enamine **103** and secondary dibromo ketones leads efficiently to the bicyclic cyclopentenone **104**, where the second ring size can be changed *ad libitum*. The 5/7-fused derivatives [**104**, R′ = $(CH_2)_5$] are potential intermediates for the synthesis of 1,3-dialkylazulenes of type **107**.

107

Another utility is the convenient formation of the spiro[n,4]alkenone system **106**, accomplished by reaction between secondary dibromo ketones and enamines of cycloalkanecarboxaldehydes (**105**) (Eq. 43). This spiroannelation finds wider generality in comparison with existing methods[126] because one can prepare derivatives bearing optional alkyl substituents at the α and α′ positions to the carbonyl group or a desired second ring size by suitable choice of starting materials.

Formation of 3(2H)-Furanones and Dihydro Derivatives. *N,N*-Dimethyl-carboxamides undergo the hetero [3 + 2] cyclocoupling reaction with dibromo ketones, providing a tool for the preparation of the title five-membered hetero-cyclic ketones. The $Fe_2(CO)_9$-assisted reaction of **38** and **108** first leads to the labile dimethylamino cycloadduct **109**, which in turn suffers facile elimination of dimethylamine to product **110** in moderate to good yield.[34,35] The overall

$$RCHBrCOCHBrR + R'CON(CH_3)_2$$

38   108

conversion can be viewed as the construction of a carbon–oxygen bridge between the α and α' positions of the parent dialkyl ketones. Other examples are shown in Eqs. 45–48. The reaction of dibromides with bulky alkyl substituents like isopropyl and *tert*-butyl gives amino ketones of type **109** in isolable form; the complete deamination requires brief heating at elevated temperature. The di-*tert*-butyl derivative **112** (Eq. 47) is quite stable on heating.[34] Use of tertiary dibromo ketones of type **16** generally affords little of the cyclocoupling product; instead, enones of type **17** predominate (Eq. 48, see also Eq. 22, p. 173).[28,34]

$(CH_3)_2CBrCOCHBrC_6H_5 + HCON(CH_3)_2$

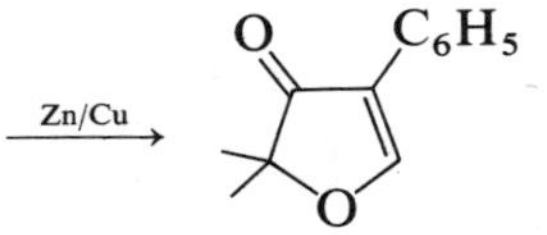

$$\xrightarrow{\text{Zn/Cu}}$$

(Eq. 45)[127]

$i\text{-}C_3H_7CHBrCOCHBrC_3H_7\text{-}i +$

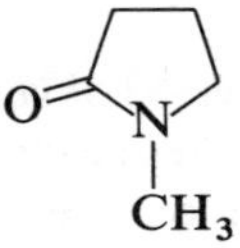

111

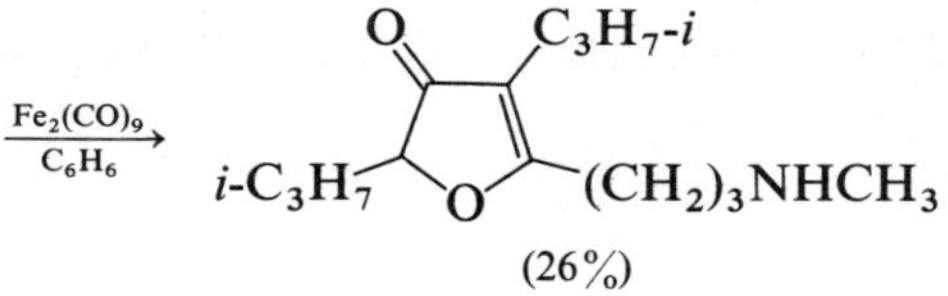

$$\xrightarrow[C_6H_6]{Fe_2(CO)_9}$$

$i\text{-}C_3H_7$ $(CH_2)_3NHCH_3$

(26%)

(Eq. 46)

$t\text{-}C_4H_9CHBrCOCHBrC_4H_9\text{-}t + HCON(CH_3)_2$

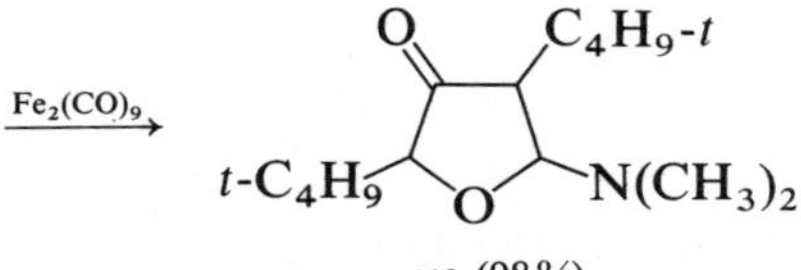

$$\xrightarrow{Fe_2(CO)_9}$$

$t\text{-}C_4H_9$ $N(CH_3)_2$

112 (98%)

(Eq. 47)[34]

$(CH_3)_2CBrCOCBr(CH_3)_2 + HCON(CH_3)_2$
16

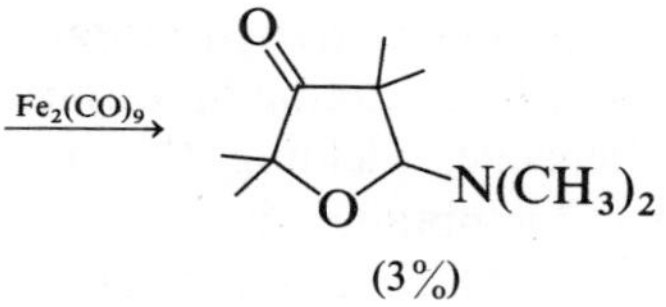

$$\xrightarrow{Fe_2(CO)_9}$$

$N(CH_3)_2$

(3%)

(Eq. 48)[28,34]

This operationally simple cyclocoupling procedure has wide applicability. The employable nucleophilic substrates include dimethylformamide, *N,N*-dimethylacetamide, and lactam **111**.[34]

The reaction with $Fe_2(CO)_9$ occurs most successfully.[34,35] Use of $Fe(CO)_5$ requires irradiation by visible light to achieve reaction.[34] Zinc/copper couple can be used only in limited cases such as the reaction shown in Eq. 45;[127] usually another type of 1:1 cyclocoupling product with a 1,3-dioxolane structure is formed (see p. 205).[128]

Synthetic Applications. 3(2*H*)-Furanones obtained by this reaction are structurally related to muscarine alkaloids,[129] and the synthesis of 4-methyl-muscarine iodide (**113**) has been achieved via **110** (R = CH_3, R' = H).[35]

113

Formation of Other Ketonic Five-Membered Compounds. The reduction of **31** with sodium iodide in acetone or acetonitrile containing certain dipolarophiles such as tetracyanoethylene or diethyl azodicarboxylate forms five-membered ketone **114**[85] and pyrazolidone **115**,[85] respectively.

$$C_6H_5CHBrCOCHBrC_6H_5 + (NC)_2C{=}C(CN)_2$$
31

114 (35%)

115 (15%)

Formation of Alkylidenetetrahydrofurans. Diiron enneacarbonyl promotes the hetero cyclocoupling reaction of tertiary dibromo ketone **16** and α-morpholinostyrene in benzene, yielding the alkylidenetetrahydrofuran **116**.[118] A similar product **117** is obtained in the reaction of **30** and 1,1-dimethoxyethene with copper–sodium iodide in acetonitrile.[79]

$(CH_3)_2CBrCOCBr(CH_3)_2$ **16** + [morpholine-N-C(=CH_2)C_6H_5] $\xrightarrow[C_6H_6]{Fe_2(CO)_9}$ [structure] **116** (86%)

$CH_3CHBrCOCHBrCH_3$ **30** + $CH_2{=}C(OCH_3)_2$ $\xrightarrow[CH_3CN]{Cu,\ NaI}$ [structure with OCH_3, OCH_3] **117** (90%)

Formation of 1,3-Dioxolanes. Reaction of **16** and ultrasonically dispersed mercury in ketonic solvents (but not aldehydes) allows a simple preparation of the 4-isopropylidene-5,5-dimethyl-1,3-dioxolane ring system.[130] Zinc–copper

$(CH_3)_2CBrCOCBr(CH_3)_2$ **16** + $CH_3COC_2H_5$ $\xrightarrow{Hg}$ [structure with O, O, C_2H_5] (59%)

couple reduction of dibromo or bromo iodo ketones in dimethylformamide or N,N-dimethylacetamide also affords the 1,3-dioxolane **118**,[128] which is readily converted to the above-mentioned 3(2H)-furanone by acid treatment. Of

$(CH_3)_2CBrCOCXR_2^1$ **16** + $R^2CON(CH_3)_2$ $\xrightarrow[\substack{R^1 = H, CH_3 \\ R^2 = H, CH_3 \\ X = Br, I}]{Zn/Cu}$ [structure: R^1, O, R^2, R^1, O, $N(CH_3)_2$] **118** (11–50%)

greater importance, these products react with cyclopentadiene or furan under acid catalysis to give bicyclic ketones of type **32** (Y = CH$_2$ or O, respectively, see Eq. 32, p. 185) on elimination of dimethylformamide.[93]

Formation of 5-Alkylidene-2-oxazolines. When dibromo or bromo iodo ketones are reduced with zinc/copper couple[87] or Fe$_2$(CO)$_9$[83] in acetonitrile, 5-alkylidene-2-oxazolines are produced in generally low yields.

$(CH_3)_2CBrCOCH_2Br$ + CH_3CN $\xrightarrow[DME]{Zn/Cu}$ [structure with N, O]

$(CH_3)_2CBrCOCBr(CH_3)_2$ **16** + CH_3CN $\xrightarrow{Fe_2(CO)_9}$ [structure with N, O] (50%)

Intramolecular [3 + 2] Cyclocoupling to Form Bicyclo[2.2.1]heptan-2-ones

Simple olefins such as isobutylene do not smoothly undergo the intermolecular [3 + 2] cyclocoupling reaction;[33] rather, the intramolecular version occurs to give bicyclic products. Thus reduction of dibromide **119** with $Fe_2(CO)_9$ in benzene affords a mixture of C_{10} products with camphor (**120**) as the major

component.[117] In a similar fashion reaction of dibromo ketone (E)-**121** and $Fe(CO)_5$ in benzene affords a 2:1 mixture of campherenone (**122**) and epicampherenone (**123**).[117] The reaction using the stereoisomer, (Z)-**121**, produces a 1:2 mixture of **122** and **123**.[117] The ease with which intramolecular [3 + 2]

reactions take place is affected profoundly by the substitution pattern around the olefinic bond as well as the length of the methylene chain that links the double bond and the dibromo ketone moiety. Attempted reactions of **124–126** fail to afford the desired bicyclic ketones.

$$R^1R^2C=CH(R^3)(CH_2)_nCBr(R^4)COCH_2Br$$

124, $R^1 = R^2 = R^3 = H; R^4 = CH_3; n = 2$

125, $R^1 = R^2 = R^4 = CH_3; R^3 = H; n = 3$

126, $R^1 = R^2 = R^4 = H; R^3 = CH_3; n = 3$

Miscellaneous Reactions

Formation of 2-(N-Alkylimino)cyclobutanones.   Treatment of a mixture of **16** and an isonitrile with copper in benzene or zinc in pyridine–benzene gives the 2-(N-alkylimino)cyclobutanone **127**.[108] Certain cyclic dibromides undergo an unusual annelation with ring contraction.[108]

Formation of 8-Azabicyclo[5.4.0]undecatrienones. 8-(4-Chlorophenyl)-8-azaheptafulvene serves as an 8π-electron receptor and undergoes $Fe_2(CO)_9$-promoted cyclocoupling with **16** to produce azabicyclic ketone **128**.[131a] In

contrast, its tricarbonyliron complex **129** serves as a 2π-electron receptor and, on reaction with **16**, gives rise to spiro adduct **130**, which is converted to **131** on treatment with trimethylamine N-oxide.[131a]

Formation of 8-Oxabicyclo[5.4.0]undeca-2,4,6-trien-10-ones. Tropone derivatives undergo $Fe_2(CO)_9$-aided cyclocoupling with **16**, yielding the oxabicyclic ketone **132**.[99]

Reaction with Seven-Membered Cyclic Trienes. 1,3,5-Cycloheptatriene reacts with **16** in the presence of $Fe_2(CO)_9$ to give **133** as the sole product.[131b] By contrast, reaction with tricarbonyl(cycloheptatriene)iron gives the isolable

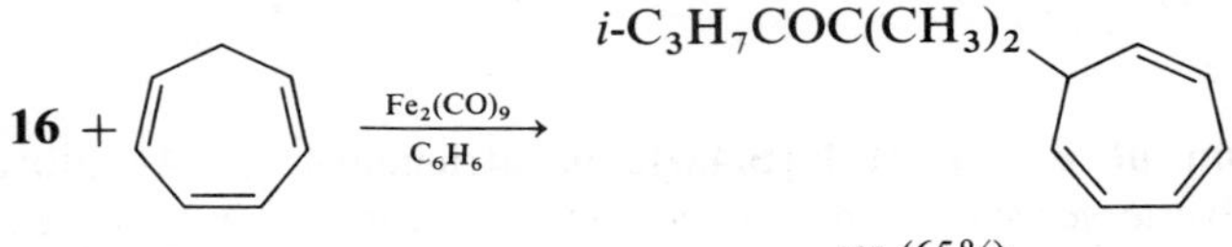

133 (65%)

tricarbonyl complex **134**, whose oxidative degradation with *o*-chloranil affords the tricyclic dione **135**.[131a] When tricarbonyl(*N*-ethoxycarbonylazepine)iron

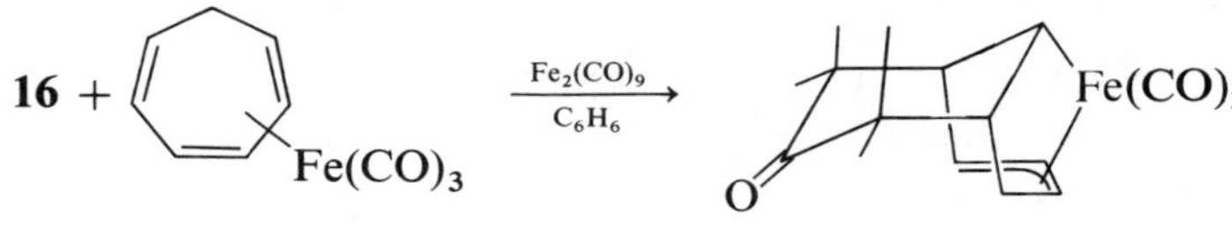

134 (20%)

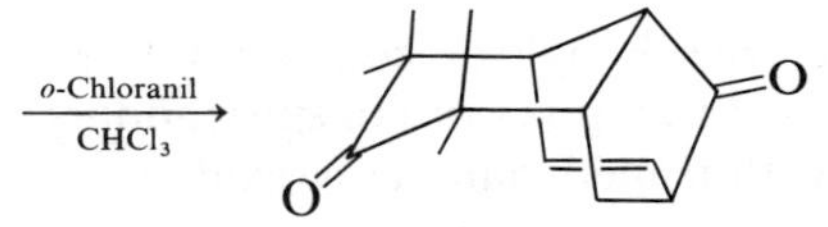

135

is used as a cyclic triene, a mixture of adducts **136** and **137** is obtained.[131a] Successive oxidative treatment of the mixture with *o*-chloranil affords a mixture of **138** and **139**.

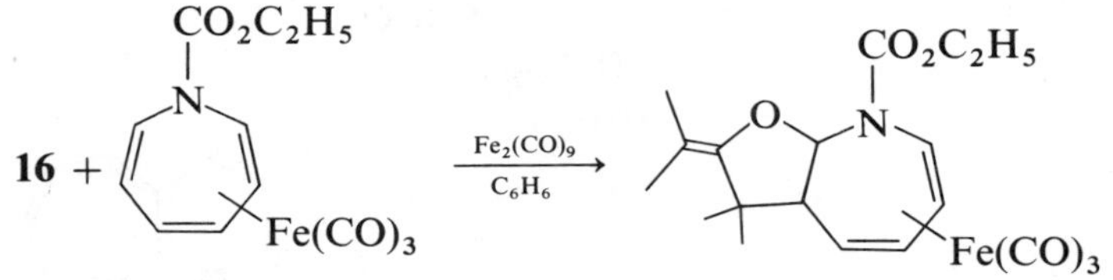

136

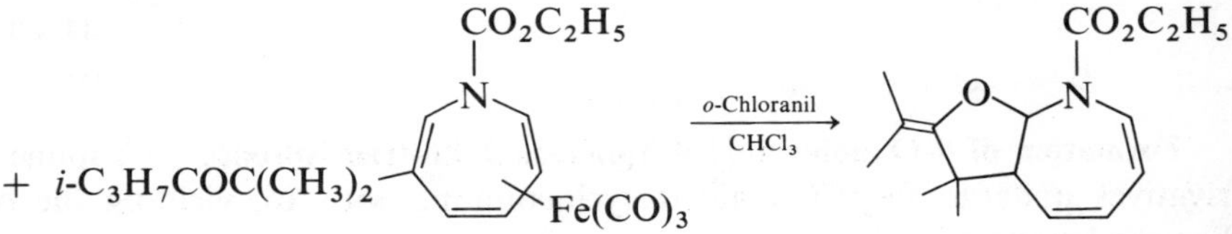

137

138 (21%)

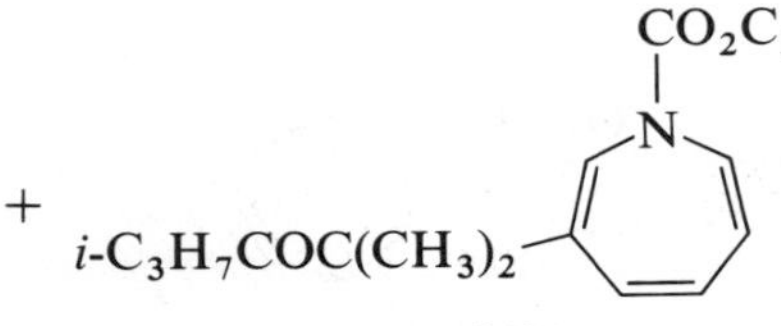

139 (24%)

Formation of Cyclo[3.3.2]azine Derivatives. When dibromide **31** is reduced with $Fe_2(CO)_9$ or copper–sodium iodide in the presence of 2-phenylindoline (**140**), which is an alternative 8π-electron reactant, the tricyclic polyenic ketone **141** is prepared, accompanied by a minor amount of **142**.[132]

31 + C₆H₅—

140

$\xrightarrow[\text{C}_6\text{H}_6]{\text{Fe}_2(\text{CO})_9}$

141 (35%)   +   142 (10%)

EXPERIMENTAL FACTORS

Preparation and Properties of Bromo Ketones

α-Bromo ketones are prepared according to the methods found in the literature.[133] Care should be taken to prevent the bromo derivatives from coming into contact with the skin, since allergic reactions and lachrymatory properties have been observed in several cases. Although certain bromo ketones are obtained as mixtures of diastereomers, no efforts to separate them are usually needed. All substances should be stored under an inert gas in a refrigerator. It is recommended that liquid materials be used immediately after distillation or, if more convenient, after passage of stored material through a short alumina column.

Preparation and Handling of Reducing Agents

Most commercially available reducing agents can be employed directly or after ordinary purification. Reagents that require preparation in the laboratory are: zinc/copper[134] or zinc/silver[135] couple made from zinc metal and copper(II) acetate or silver(I) acetate, respectively; 1,3-diene–Fe(CO)₃[136] complexes from the dienes and Fe(CO)₅ by irradiation with visible light; organo cuprates from alkyllithiums and copper(I) salts.[55,78]

Caution: Reactions with metal carbonyls and the subsequent workup should be carried out in a well-ventilated hood because the carbonyl complexes are extremely toxic and carbon monoxide gas is evolved during the reaction.

EXPERIMENTAL PROCEDURES

The following examples of reactions have been chosen to illustrate useful and general experimental procedures. They are organized into the following six categories: (1) reactions of α-monobromo ketones, (2) alkylations of α,α'-dibromo ketones, (3) [3 + 4] cyclocoupling reactions of α,α'-dibromo ketones and 1,3-dienes, (4) [3 + 4] cyclocoupling reactions of tri- or tetra-bromo ketones, (5) [3 + 2] cyclocoupling reactions of α,α'-dibromo ketones and olefinic substrates, and (6) miscellaneous reactions.

Reactions of α-Monobromo Ketones

π-Bromo-(+)-camphor (Reduction of an α-Monobromo Ketone to the Parent Ketone).[49]   To 155 g (0.50 mol) of α,π-dibromocamphor dissolved in 600 mL of dichloromethane was added 104 g (1.59 g-atoms) of zinc powder. A gentle stream of hydrogen bromide was passed into the mixture with stirring for 2.5–4 hours. Occasionally it was necessary to moderate the reaction by cooling with a water bath. The mixture was filtered, and the dichloromethane solution was washed with water and dried over magnesium sulfate. Removal of dichloromethane by distillation on a steam bath yielded 109 g of crude product, which was recrystallized from hexane. There was obtained 74 g (64%) of π-bromo-(+)-camphor, mp 93–95°, $[\alpha]_D^{26}$ +115° (CHCl$_3$): IR (CCl$_4$) cm^{-1}: 1750 (C=O); UV (C$_2$H$_5$OH), nm max (ε): 289 (58.7). Concentration of the mother liquor yielded an additional 12.7 g (11%) of product, mp 80–90°, $[\alpha]_D^{26}$ +115°. The total yield of ketone was 75%.

1,5-Diphenyl-3-hydroxypent-4-en-1-one (Aldol Reaction of an α-Bromo Ketone with an Aldehyde).[63]   To a stirred slurry of 196 mg (3.0 mg-atoms) of zinc dust and 14 mg (0.1 mmol) of copper(I) bromide in 6 mL of anhydrous tetrahydrofuran at 25° under an argon atmosphere was added via syringe 2.2 mL (2.2 mmol) of a 1 M solution of diethylaluminum chloride in hexane. The resulting mixture was stirred at 25° for 10 minutes and cooled to − 20° with a carbon tetrachloride–dry ice bath. A solution of 398 mg (2.0 mmol) of α-bromo-acetophenone and 290 mg (2.2 mmol) of *trans*-cinnamaldehyde in 10 mL of anhydrous tetrahydrofuran was added dropwise at − 20° over a period of 40 minutes. Stirring was continued for an additional 15 minutes at − 20°, and then the reaction mixture was quenched by the addition of 0.6 mL of pyridine at − 20°. After removal of the cooling bath, the reaction mixture was poured into 20 mL of iced 6% hydrochloric acid and extracted with ether. The ether extract was washed with brine, dried over magnesium sulfate, concentrated *in vacuo*, and purified by preparative TLC on silica gel with 1:5 ether–benzene as eluent to give 464 mg (92%) of the title compound as a yellow oil that was crystallized from ether–pentane, mp 51–53°; IR (neat) cm^{-1}: 3740 (OH), 1675 (C=O); ^{1}H NMR (CDCl$_3$) δ: 2.84–3.61 (broad singlet, 1H, OH), 3.22 (d, J = 6 Hz, 2H, CH$_2$C=O), 4.91 (q, J = 6 Hz, 1H, CHOH), 6.24 (dd, J = 16 and

6 Hz, 1H, CH=CHC$_6$H$_5$), 6.70 (d, J = 16 Hz, 1H, CH=CHC$_6$H$_5$), 6.94–8.31 (m, 10H, aromatic protons); mass spectrum m/e: 234 (M − 18).

Alkylations of α,α′-Dibromo Ketones

2-Methylcyclododecanone (Monoalkylation of an α,α′-Dibromo Ketone).[78] A flame-dried, three-necked, round-bottomed flask equipped with serum stoppers and a nitrogen-filled balloon was charged with 7.76 g (40.8 mmol) of copper(I) iodide and 14 mL of anhydrous ether, and to this at 0° was added 46.5 mL of 1.72M (80 mmol) methyllithium in ether via syringe. To the resulting clear solution at −78° was added 3.44 g (10.0 mmol) of solid cis-2,12-dibromo-cyclododecanone through a side arm. After 15 minutes 10 mL of methanol was added to the yellow suspension. The reaction mixture was allowed to warm to room temperature and poured into 200 mL of saturated, aqueous ammonium chloride, and the yellow precipitate thus formed was removed by suction filtration. The aqueous layer was extracted with three 100-mL portions of ether, and the combined ethereal extracts were washed twice with 50 mL of 1N sodium hydroxide and dried over magnesium sulfate. Solvent was removed in vacuo to leave 1.88 g (99%) of a yellow oil. Analytical GC (FFAP on Chromosorb W, 60–80 mesh, 304 × 0.6 cm, 200°) indicated two peaks: **A**, 7 minutes; **B**, 8 minutes. **A**, identified as cyclododecanone by comparing its retention time with that of authentic material, accounted for ca. 1% of the mixture. **B** corresponded to 2-methylcyclododecanone. Distillation at a bath temperature of ca. 100° (0.2 mm) gave a clear colorless oil, semicarbazone mp 211–213°: IR (CCl$_4$) cm^{-1}: 1710 (C=O); ^{1}H NMR (CCl$_4$) δ: 1.02 (d, J = 7.0 Hz, 3H, CH$_3$), 1.3 (m, 18H, 9 CH$_2$), 2.5 (broad multiplet, 3H, methylene and methine protons α to carbonyl group); mass spectrum (70 eV) m/e: 196 (M).

2,12-Dimethylcyclododecanone (Dialkylation of an α,α′-Dibromo Ketone).[78] To 10.0 mmol of lithium dimethylcuprate prepared according to the same procedure as above was added at −78° through a side arm 680 mg (2.0 mmol) of solid cis-2,12-dibromocyclododecanone. After 30 minutes 3.6 mL (50 mmol) of neat methyl iodide was added, and the reaction mixture was allowed to stir at ambient temperature overnight. Workup as in the preceding example was followed by solvent removal, leaving 420 mg of a yellow oil. Analytical GC (5% Silicone SE-30 on Chromosorb G, 100–120 mesh, 213 × 0.3 cm, 180°) indicated three products: **A**, 6 minutes; **B**, 7 minutes; **C**, 8 minutes. Product **A** was identified as 2-methylcyclododecanone by comparing its retention time with that of authentic material. Preparative GC (20% Silicone SE-30 on Chromosorb W, 45–60 mesh, 610 × 0.3 cm, 215°) was used to isolate **B** (30 minutes) and **C** (34 minutes). **B** was 2,12-dimethylcyclododecanone (285 mg, 68%) recrystallized from methanol, mp 43–44°: IR (CCl$_4$) cm^{-1}: 1710 (C=O); ^{1}H NMR (CCl$_4$) δ: 1.00 (d, J = 7 Hz, 6H, 2 CH$_3$), 1.31 (m, 18H, 9 CH$_2$), 2.72 (broad multiplet, 2H, 2 CH); mass spectrum (70 eV) m/e: 210 (M). **C** was an isomeric 2,12-dimethylcyclododecanone (122 mg, 29%): IR (CCl$_4$) cm^{-1}: 1700 (C=O): ^{1}H NMR (CCl$_4$) δ: 1.04 (d, J = 7 Hz, 6H, 2 CH$_3$), 1.33 (m, 18H

9 CH_2), 2.72 (broad multiplet, $2H$, 2 CH); mass spectrum (70 ev) m/e: 210 (M). The isomeric nature of **B** and **C** was confirmed by combining them for microanalysis. They are apparently *cis*- and *trans*-2,12-dimethylcyclododecanones. The total yield of the product was 97%.

[3 + 4] Cyclocoupling of α,α′-Dibromo Ketones and 1,3-Dienes

3,5-Dimethylbicyclo[5.4.0]undec-1(7)-en-4-one ([3 + 4] Cyclocoupling Reaction between an α,α′-Dibromo Ketone and an Open-Chain 1,3-Diene with Diiron Enneacarbonyl).[74]   In a 30-mL, two-necked flask equipped with a serum cap and a three-way stopcock carrying a nitrogen-filled rubber balloon was placed 906 mg (2.49 mmol) of diiron enneacarbonyl. The system was flushed with nitrogen, and to this was added a solution of 108 mg (1.00 mmol) of 1,2-bis-methylenecyclohexane in 2.5 mL of dry benzene followed by 608 mg (2.49 mmol) of 2,4-dibromopentan-3-one. The mixture was stirred magnetically at 62–64° for 46 hours, then cooled to room temperature, and poured into 15 mL of saturated sodium hydrogen carbonate solution. To this was added 60 mL of ethyl acetate, and the solid material was removed by filtration through a pad of Celite 545. The organic layer was separated, and the aqueous layer was extracted with two 10-mL portions of ethyl acetate. The combined extracts were dried over sodium sulfate, and the solvent was evaporated to give 261 mg of a dark brown oil. Purification by silica gel chromatography using 1:1 benzene–hexane as eluent yielded 153 mg (80%) of the title adduct as a colorless oil that was homogeneous on gas chromatography analysis (10% Silicone SE-30 on Chromosorb W AW, 80–100 mesh, 100 × 0.3 cm, 110°): IR (CCl_4) cm^{-1}: 1705 (C=O), 1660 (C=C); ^{1}H NMR (CCl_4) δ: 1.03 (d, J = 6.5 Hz, $6H$, 2 CH_3), 1.4–1.6 (m, $4H$, 2 CH_2), 1.7–2.3 (m, $8H$, 4 =CCH_2), 2.6–3.0 (m, $2H$, 2 $CHCH_3$); mass spectrum (70 eV) m/e: 192 (M).

2,2,4,5,7,7-Hexamethyl-4-cycloheptenone (41, $R^1 = R^2 = CH_3$, p. 186) ([3 + 4] Cyclocoupling Reaction of an α,α′-Dibromo Ketone and a 1,3-Dieneiron Tricarbonyl Complex).[74]   In a 300-mL pressure bottle was placed 3.92 g (20.0 mmol) of iron pentacarbonyl. The system was evacuated and flushed with nitrogen. Then a mixture of 4.93 g (60.0 mmol) of 2,3-dimethyl-1,3-butadiene and 20 mL of dry benzene was introduced with a syringe. The mixture was irradiated at 55° for 6.5 hours with a 200-W high-pressure mercury lamp. The bottle was opened, and an aliquot of the mixture was analyzed by ^{1}H NMR using dioxane as an internal standard to confirm the quantitative formation of the dieneiron tricarbonyl complex. To this was added a solution of 4.08 g (15.0 mmol) of 2,4-dibromo-2,4-dimethylpentan-3-one in a small amount of benzene. The mixture was again covered with nitrogen and heated at 80° for 12 hours. The reaction mixture was filtered through a pad of 2 g of Celite 545, and the filtrate was concentrated on a rotary evaporator to give a pale yellow oil. This oil was dissolved in 20 mL of acetone, mixed with 1.70 g (10 mmol) of copper(II) chloride dihydrate, and stirred at room temperature for 15 minutes to decompose the remaining dieneiron tricarbonyl complex. The slurry was then passed through a short column packed with Celite 545, and the filtrate was concentrated under reduced pressure. The residue was dissolved in 40–50 mL of dichloromethane and washed with four 20-mL portions of a 5% aqueous solution of ethylenediaminetetraacetic acid disodium salt (Na_2H_2EDTA). The organic layer was dried over anhydrous sodium sulfate, and the solvent was evaporated on a rotary evaporator to give 6.20 g of a yellow oil, distillation of which under reduced pressure with a Kugelrohr apparatus yielded 2.90 g (100 %) of the title adduct as a colorless oil; bp 120–130° (0.02 mm). An analytical sample was obtained by preparative GC (33% Apiezon grease L on Neopak 1 A, 60–80 mesh, 213 × 1 cm, 165°); retention time (t_R), 20 minutes: IR (CCl_4) cm^{-1}: 1685 (C=O); ^{1}H NMR (CCl_4) δ: 1.07 (s, 12H, 4 CH$_3$), 1.79 (s, 6H, 2 =CCH$_3$), 2.25 (s, 4H, 2 =CCH$_2$); mass spectrum (70 eV) m/e: 194 (M), 123 (M − 71, base peak), 110 (M − 84), 109 (M − 85).

2,4-Diphenylbicyclo[3.2.1]oct-6-en-3-one ([3 + 4] Cyclocoupling of an α,α′-Dibromo Ketone and Cyclopentadiene with Sodium Iodide).[94]   A mixture of 5.0 g (14 mmol) of 1,3-dibromo-1,3-diphenylpropan-2-one, 25 g (167 mmol) of sodium iodide, 50 mL (40 g, 606 mmol) of cyclopentadiene, and 150 mL of acetonitrile was boiled for 15 minutes. Chloroform was added, and the mixture was washed, first with sodium thiosulfate solution, then water, and dried over sodium sulfate. Concentration gave the crude material, which was chromatographed on a long column of silica gel to afford 7.4 g (99%) of a mixture of 2α,4α- and 2α,4β-diphenylbicyclo[3.2.1]oct-6-en-3-ones in a 40:60 ratio as judged by the IR spectrum. Crystallization from ethanol yielded mainly the *cis* isomer as the first crop of crystals, and crystallization of the material in the mother liquor from acetone gave mainly the *trans* isomer. The *cis* adduct melted at 149.5–151.5° (from ethanol): IR (Nujol) cm^{-1}: 1700, 755, 740, 705. The *trans* adduct melted at 134–136° (from acetone); IR (Nujol) cm^{-1}: 1700, 1665, 750, 740, 700.

**2α,4β,8-Trimethyl-8-azabicyclo[3.2.1]oct-6-en-3-one([3 + 4] Cyclocoupling Reaction of an α,α′-Dibromo Ketone and N-Methylpyrrole with Copper–Sodium Iodide).*[102]   A solution of 0.97 g (12 mmol) of N-methylpyrrole and 2.4 g (10 mmol) of 2,4-dibromopentan-3-one was added during a 30-minute period to a solution of 6.0 g (40 mmol) of sodium iodide in 50 mL of acetonitrile containing 1.9 g (30 mg-atoms) of copper powder reduced by hydrogen, as supplied commercially. The mixture was stirred magnetically and maintained under an atmosphere of nitrogen throughout the preparation. After 8 hours, 25 mL of water and 50 mL of dichloromethane were added, and the mixture was stirred vigorously for 10 minutes to precipitate copper(II) iodide and filtered. The filtrate was washed successively with dilute ammonium hydroxide solution and brine. The organic layer was separated and extracted with cold 2N hydrochloric acid. After neutralization the aqueous layer was extracted again with

---

* See also reference 101, which indicates that the reaction gave the 2α,4α,8-trimethyl-8-aza-bicyclo[3.2.1]oct-6-en-3-one, but not the 2α,4β,8-trimethyl isomer, in 89% yield.

dichloromethane, and the extract was dried over magnesium sulfate. Removal of solvent gave a crude oil, which was purified by fast filtration through silica gel with ether as eluent, giving 0.8–1.0 g (50–60%) of rather unstable title product. A sample of at least 95% purity, obtained by preparative TLC, had the following spectral characteristics: IR (neat) cm^{-1}: 3055, 1705, 707; ^{1}H NMR (CDCl$_3$) δ: 1.0 (d, $J = 7$ Hz, 6H, 2 CH$_3$), 2.3 (s, 3H, NCH$_3$), 2.7 (m, 2H, 2 COCH), 3.5 (d, $J = 4$ Hz, 2H, 2 NCH), 6.1 (d, $J = 0.5$ Hz, 2H, 2 =CH); mass spectrum m/e (relative intensity): 165 (M, 25), 108 (M − 57, 100), 94 (M − 71, 50).

8-Acetyl-2,2,4,4-tetramethyl-8-azabicyclo[3.2.1]oct-6-en-3-one ([3 + 4] Cyclocoupling of an α,α'-Dibromo Ketone and N-Acetylpyrrole with Diiron Enneacarbonyl).[74]   Into a 50-mL, two-necked flask equipped with a serum cap and a nitrogen balloon was placed 1.10 g (3.02 mmol) of diiron enneacarbonyl. After the system was flushed with nitrogen, 10 mL of dry benzene, 1.62 g (5.96 mmol) of 2,4-dibromo-2,4-dimethylpentan-3-one, and 218 mg (2.00 mmol) of N-acetylpyrrole freshly distilled from sodium hydride were successively added through the rubber septum by a syringe. The mixture was stirred at 40–50° for 18.5 hours. The reaction mixture was diluted with 15 mL of ethyl acetate, washed with three 10-mL portions of saturated sodium hydrogen carbonate solution followed by 5 mL of brine, and dried over sodium sulfate. Concentration of the organic layer gave 1.2 g of an orange oil, which was subjected to column chromatography (25 g of silica gel). Elution with 1:3 ethyl acetate–n-hexane followed by evaporation of the solvents gave some unreacted starting dibromo ketone. The fractions eluted with ethyl acetate afforded 302 mg (68%) of the title compound as pale yellow crystals. Recrystallization from hexane gave an analytical sample: IR (CCl$_4$) cm^{-1}: 1720 (C=O), 1660 (NCOCH$_3$); ^{1}H NMR (CCl$_4$) δ: 1.02 (s, 6H, 2 CH$_3$), 1.23 (s, 3H, CH$_3$), 1.30 (s, 3H, CH$_3$), 2.08 (s, 3H, COCH$_3$), 4.32 (broad singlet, 1H, NCH), 4.92 (broad singlet, 1H, NCH), 6.42 (broad singlet, 2H, 2=CH); mass spectrum (70 eV) m/e: 221 (M), 206 (M − 15), 178 (M − 43), 164 (M − 57), 151 (M − 70), 150 (M − 71), 109 (M − 112), 108 (M − 113).

2α,4α-Dimethyl-8-oxabicyclo[3.2.1]oct-6-en-3-one ([3 + 4] Cyclocoupling of an α,α'-Dibromo Ketone and Furan with the Aid of Zinc/Silver Couple).[45]   A mixture of 19.6 g (0.30 g-atom) of zinc/silver couple, 140 mL (2.00 mol) of furan, and 300 mL of tetrahydrofuran was stirred at −10°, and a solution of 48.8 g (0.20 mol) of 2,4-dibromopentan-3-one in 150 mL of tetrahydrofuran was added dropwise over a period of 1 hour. The mixture was allowed to warm to room temperature and stirred for an additional 12 hours. The insoluble precipitate was removed by filtration, and the filtrate was concentrated. Chromatography of the residue on a column using 500 g of silica gel with 1:20 to 1:5 ethyl acetate–n-hexane as eluent afforded 24.2 g (80%) of the title compound as a colorless liquid: IR (CCl$_4$) cm^{-1}: 1715 (C=O); ^{1}H NMR (CCl$_4$) δ: 0.90 (d, $J = 7$ Hz, 6H, 2 CH$_3$), 2.67 (d of q, $J = 5$ and 7 Hz, 2H, 2 COCH),

4.72 (d, $J = 5$ Hz, $2H$, 2 OCH), 6.38 (broad singlet, $2H$, 2 =CH); mass spectrum m/e (relative intensity): 152 (M, 23), 137 (M − 15, 14), 109 (M − 43, 10), 97 (M − 55, 9), 96 (M − 56, 54), 95 (M − 57, 40), 91 (M − 61, 9), 81 (M − 71, 100), 68 (M − 84, 9), 67 (M − 85, 16), 65 (M − 87, 7), 57 (M − 95, 7), 56 (M − 96, 14), 55 (M − 97, 15), 53 (M − 99, 15), 44 (M − 108, 10), 43 (M − 109, 13), 41 (M − 111, 20), 39 (M − 113, 27).

2,2,4,4-Tetramethyl-8-oxabicyclo[3.2.1]oct-6-en-3-one ([3 + 4] Cyclocoupling of an α,α′-Dibromo Ketone and Furan with Diiron Enneacarbonyl).[74]

In a nitrogen-flushed, 100-mL two-necked flask equipped with a serum cap was placed 3.64 g (10.0 mmol) of diiron enneacarbonyl. To this were added 2.18 g (8.00 mmol) of 2,4-dibromo-2,4-dimethylpentan-3-one, 5.44 g (80.0 mmol) of furan, and 20 mL of benzene. The mixture was stirred at 40° for 38 hours. The resulting precipitate was removed by filtration through a Celite 545 pad. The filtrate was washed with 5% aqueous Na_2H_2EDTA solution. The organic layer was dried over sodium sulfate and evaporated to give a yellow-green oil, which was subjected to silica gel column chromatography with 1:10 ether–hexane as eluent to afford 12.8 g (89%) of the title adduct: IR (CCl_4) cm^{-1}: 1710 (C=O); ^{1}H NMR (CCl_4) δ: 0.86 (s, $6H$, 2 CH$_3$), 1.29 (s, $6H$, 2 CH$_3$), 4.23 (s, $2H$, 2 OCH), 6.19 (s, $2H$, 2 =CH); mass spectrum (75 eV) m/e (relative intensity): 180 (M, 23), 110 (M − 70, 85), 95 (M − 85, 100) (major peaks only).

[3 + 4] Cyclocoupling Reactions of Tri- or Tetrabromo Ketones

2,2-Dimethylbicyclo[3.2.1]oct-6-en-3-one (Iron Pentacarbonyl-Assisted [3 + 4] Cyclocoupling Reaction of a Tribromo Ketone and Cyclopentadiene).[48]

To a stirred, hot (80°) mixture of 1.0 mL of freshly distilled cyclopentadiene, 0.78 mL (1.17 g, 5.98 mmol) of iron pentacarbonyl (Strem), 1.0 mL of tetrahydrofuran, and 10.0 mL of dry benzene was added dropwise, under argon, a solution of 1.62 g (5.00 mmol) of 1,1,3-tribromo-3-methylbutan-2-one in 7.5 mL of 1:1 cyclopentadiene–benzene over 25 minutes. The resulting mixture was stirred for an additional 45-minute period at the same temperature, cooled, quenched by addition of 12.0 mL of methanol saturated with ammonium chloride, and shaken vigorously with 3.80 g (57.8 mg-atoms) of zinc/copper couple for 20 minutes. The reaction mixture was diluted with 200 mL of dichloromethane and 100 mL of saturated aqueous Na_2H_2EDTA solution. The insoluble materials were removed by filtration, and the filtrate was extracted twice with 50- and 30-mL portions of dichloromethane. The combined organic layers were dried over sodium sulfate and concentrated to give 2.16 g of an oily residue, whose GC analysis showed the yield of the title bicyclic adduct (t_R 8 minutes) to be 83%. The oil was dissolved in 20 mL of dichloromethane and added dropwise to 200 mL of vigorously stirred hexane. The resulting precipitate was removed by passage through a Celite 545 pad. The filtrate was evaporated to leave 1.10 g of an oil, which was chromatographed on 15.0 g of Merck Kieselgel 60 (70–230 mesh). Elution with 110 mL of 1:1 benzene–n-hexane and 50 mL of 1:10 ethyl acetate–n-hexane in 10-mL fractions gave 520 mg (66%) of the title product (95% pure based on PMR analysis) in fractions 9–15. Bulb-to-bulb distillation of this oil produced 375 mg of an analytical sample, bp (bath temperature of 70°) 120° (2 mm), as colorless crystals, mp 45–48°: IR (CCl$_4$) cm^{-1}: 1714 (C=O); ^{1}H NMR (CCl$_4$) δ: 1.00 (s, 3H, CH$_3$), 1.17 (s, 3H, CH$_3$), 1.75–2.97 (m, 6H, 2 CH$_2$, 2 =CHCH), 5.95–6.25 (multiplet appearing to be an AB quartet centered at δ 6.1 with J_{AB} approximately 6 Hz and $\Delta\nu_{AB}$ ~4.5 Hz, 2H, 2 =CH); mass spectrum (75 eV) m/e (relative intensity): 150 (M, 94), 93 (M − 57, 100), 84 (M − 66, 90.8), 79 (M − 71, 87.8) (major peaks only).

N-**Methoxycarbonyl-2,4-dibromo-8-azabicyclo[3.2.1]oct-6-en-3-ones and *N*-Methoxycarbonyl-8-azabicyclo[3.2.1]oct-6-en-3-one (Diiron Enneacarbonyl-Promoted [3 + 4] Cyclocoupling of α,α,α′α′-Tetrabromoacetone and *N*-Methoxycarbonylpyrrole).**[75] Into a 30-mL, two-necked flask charged with 1.09 g (3.00 mmol) of diiron enneacarbonyl was poured a solution of 1.13 g (3.00 mmol) of α,α,α′,α′-tetrabromoacetone in 7.5 mL of benzene. The mixture was heated at 50° with stirring for 5 minutes. *N*-Methoxycarbonylpyrrole (125 mg, 1.00 mmol) was added, and the mixture was stirred at 50° for 72 hours. The dark brown reaction mixture was diluted with 15 mL of ethyl acetate, and the insoluble material was removed by filtration through a Celite 545 pad. The filtrate was evaporated under reduced pressure, leaving a black, tarry oil. ^{1}H NMR analysis using 1,1,2,2-tetrachloroethane as internal standard showed that the desired 1:1 adducts, *N*-methoxycarbonyl-2α,4α-dibromo-8-azabicyclo[3.2.1]oct-6-en-3-one and *N*-methoxycarbonyl-2α,4β-dibromo-8-azabicyclo[3.2.1]oct-6-en-3-one, were formed in 70% yield in a 2:1 ratio. Preparative TLC of the tar with 1:3 ethyl acetate–hexane afforded 145 mg (52%) of a mixture of the bicyclic adducts (R_f 0.6–0.7). Fractional recrystallization from ethyl acetate–*n*-hexane afforded an analytical sample of the *cis* isomer as colorless crystals, mp 155–157°: IR (CHCl$_3$) cm^{-1}: 1748 (C=O), 1710 (NCO$_2$CH$_3$); ^{1}H NMR (CDCl$_3$) δ: 3.82 (s, 3H, OCH$_3$), 4.80 (d, J = 4.0 Hz, 2H, 2 CHBr), 5.11 (dd, J = 1.0 and 4.0 Hz, 2 NCH), 6.53 (tripletlike multiplet, J = 1.0 Hz, 2H, 2 =CH); mass spectrum (70 eV) m/e: 341, 339, 337 (1:2:1 ratio, M), 310, 308, 306 (1:2:1 ratio, M − OCH$_3$), 282, 280, 278 (1:2:1 ratio, M − CO$_2$CH$_3$), 260, 258 (1:1 ratio, M − Br). The mother liquor was concentrated and subjected to further TLC using 1:3 ethyl acetate–*n*-hexane as eluent, yielding the pure *trans* dibromide (R_f 0.60), mp 112–114° from ethyl acetate–hexane: IR (CHCl$_3$) cm^{-1}: 1740 (C=O), 1710 (NCO$_2$CH$_3$); ^{1}H NMR (CDCl$_3$) δ: 3.82 (s, 3H, OCH$_3$), 4.27 (d, J = 2.0 Hz, 1H, an equatorial methine proton at C-4), 5.11 (d, J = 3.5 Hz, 1H, an axial methine proton at C-2), 5.0–5.3 (m, 2H, 2 NCH), 6.36 (dd, J = 2.0 and 6.0 Hz, 1H, =CH), 6.61 (dd, J = 2.0 and 6.0 Hz, 1H, =CH); mass spectrum (70 eV) m/e: 341, 339, 337 (1:2:1 ratio, M), 310, 308, 306 (1:2:1 ratio, M − OCH$_3$), 282, 280, 278 (1:2:1 ratio, M − CO$_2$CH$_3$), 260, 258 (1:1 ratio, M − Br).

A 300-mg (0.87 mmol) mixture of the dibromo ketones and 750 mg (11.5 mg-atoms) of zinc/copper couple in 15 mL of methanol saturated with ammonium chloride was stirred at room temperature for 10 minutes. To this mixture was added 30 mL of ethyl acetate, and the insoluble material was removed by filtration. The filtrate was concentrated to afford an oil containing some solid, to which was added 20 mL of 1:1 ethyl acetate–n-hexane. The precipitate was removed by decantation and rinsed with 1:1 ethyl acetate–n-hexane. The organic layers were combined, washed with a small amount of water, dried, and evaporated to give 160 mg (100%) of N-methoxycarbonyl-8-azabicyclo[3.2.1]oct-6-en-3-one as colorless crystals. Recrystallizations from hexane produced an analytical specimen, mp 69–70°: IR (CHCl$_3$) cm^{-1}: 3005 (=C—H), 1700–1710 (C=O and NCO$_2$CH$_3$), 1600 (C=C); ^{1}H NMR (CDCl$_3$) δ: 2.40 (dd, J = 16.5 and 1.5 Hz, 2H, equatorial protons at C-2 and C-4), 2.80 (dd, J = 16.5 and 4.5 Hz, 2H, axial protons at C-2 and C-4), 3.84 (s, 3H, OCH$_3$), 4.90 (broad doublet, J = 4.5 Hz, 2H, 2 NCH), 6.27 (tripletlike multiplet, J = 1.0 Hz, 2H, 2 =CH); mass spectrum (70 eV) m/e: 181 (M), 138 (M − 43).

8-Oxabicyclo[3.2.1]oct-6-en-3-one ([3 + 4] Cyclocoupling of $\alpha,\alpha,\alpha',\alpha'$-Tetrabromoacetone and Furan with Zinc/Silver Couple).[45] A mixture of 29.3 g (0.45 g-atom) of zinc/silver couple and 210 mL (3.00 mol) of furan in 450 mL of dry tetrahydrofuran was placed in a 2-L, three-necked flask and cooled to − 10°. A solution of 112 g (0.30 mol) of $\alpha,\alpha,\alpha',\alpha'$-tetrabromoacetone in 200 mL of dry tetrahydrofuran was added dropwise over a period of 2 hours with stirring, and the resulting mixture was allowed to warm to room temperature and stirred for 12 hours. The insoluble materials were removed by filtration, and the filtrate was concentrated. The tarry brown residue was chromatographed on a column packed with 700 g of Merck Kieselgel 60 (70–230 mesh) using 1:20 to 1:5 ethyl acetate–n-hexane as eluent to give 55 g of crude 2α,4α-dibromo-8-oxabicyclo[3.2.1]oct-6-en-3-one as yellow crystals. Recrystallization from hexane gave a pure sample melting at 126.5–127°. The dibromo adduct was dissolved in 1.3 L of saturated ammonium chloride solution in methanol, and to this was added portionwise 120 g (1.85 g-atoms) of zinc/copper couple over 10 minutes at room temperature. The suspension was stirred for an additional hour and filtered. The filtrate was divided into four portions, and each portion was diluted with 200 mL of water and 300 mL of saturated Na$_2$H$_2$EDTA solution. Extraction was performed with 300 mL and then two 200-mL portions of dichloromethane successively. The combined extracts were dried, and the organic solvent was removed by distillation at atmospheric pressure through a 40-cm Vigreux column. Finally the residue was subjected to light suction by an aspirator at room temperature to leave 20.4 g of the title bicyclic compound, contaminated with a small amount of dichloromethane (95% pure by ^{1}H NMR analysis). Overall yield based on tetrabromoacetone was 55%. A pure crystalline sample of the desired product, mp 37–39°, was obtained by bulb-to-bulb distillation at a bath temperature of 50–80° (0.01 mm): IR (CCl$_4$) cm^{-1}: 1720 (C=O); ^{1}H NMR (CCl$_4$) δ: 2.19 (dd, J = 17 and ~ 1.5 Hz, 2H, equatorial methylene protons at C-2 and C-4), 2.61 (dd, J = 5 and 17 Hz, 2H, axial methylene protons at C-2 and C-4), 4.91 (broad doublet,

$J = 5$ Hz, $2H$, 2 OCH), 6.19 (broad singlet, $2H$, 2 =CH); mass spectrum
(70 eV) m/e (relative intensity): 124 (M, 16), 95 (M − 29, 8), 83 (M − 41, 4), 82
(M − 42, 60), 81 (M − 43, 100), 68 (M − 56, 14), 67 (M − 57, 14), 66 (M −
58, 4), 65 (M − 59, 4).

[3 + 2] Cyclocoupling of α,α′-Dibromo Ketones and Olefinic Substrates

**3-Cyclopropyl-2,5-dimethyl-3-phenylcyclopentanone ([3 + 2] Cyclocoupling
of an α,α′-Dibromo Ketone and an Aryl-Substituted Olefin).**[118] In a two-
necked flask equipped with a magnetic stirrer, a rubber septum, and a three-way
stopcock fitted with an argon-filled balloon was placed 437 mg (1.20 mmol) of
diiron enneacarbonyl; the flask was evacuated and flushed with argon. To the
flask were added successively 9.0 mL of benzene, 288 mg (2.00 mmol) of
α-cyclopropylstyrene, and 244 mg (1.00 mmol) of 2,4-dibromopentan-3-one
through the rubber septum via syringe. The resulting mixture was magnetically
stirred at 65° for 24 hours, diluted with 30 mL of ethyl acetate, washed with
saturated sodium hydrogen carbonate solution and potassium nitrate solution,
dried over sodium sulfate, and concentrated on a rotary evaporator under
reduced pressure (60–90 mm) at 25–50° to leave an oil that consisted mainly of
3-cyclopropyl-*cis*-2,5-dimethyl-3-phenylcyclopentanone   and   3-cyclopropyl-
trans-2,5-dimethyl-3-phenylcyclopentanone (ca. 3:1 ratio based on ^{1}H NMR
analysis). The *cis* isomer: IR (CCl$_4$) cm^{-1}: 1736 (C=O); ^{1}H NMR (CCl$_4$)
δ: 0.0–0.5 (m, $4H$, methylene protons of cyclopropyl group), 0.75 (d, $J = 7.5$ Hz,
$3H$, CH$_3$), 1.17 (d, $J = 6.5$ Hz, $3H$, CH$_3$), 1.5–3.0 (m, $5H$, methylene and
methine protons of five-membered ring and a methine proton of cyclopropyl
group), 7.15 (m, $5H$, aromatic protons); mass spectrum (70 eV) m/e: 228
(M). The *trans* isomer: IR (CCl$_4$) cm^{-1}: 1736 (C=O); ^{1}H NMR (CCl$_4$) δ:
0.0–1.0 (m, $4H$, methylene protons of cyclopropyl group), 1.14 (d, $J = 6.5$ Hz,
$3H$, CH$_3$), 1.35 (d, $J = 6.5$ Hz, $3H$, CH$_3$), 1.5–2.9 (m, $5H$, methylene and
methine protons of five-membered ring and a methine proton of cyclopropyl
group), 7.11 (m, $5H$, aromatic protons); mass spectrum (70 eV) m/e: 228 (M).
After long standing, the mixture gave 215 mg (95%) of the crystalline *trans*
ketone as a single epimer. Recrystallization from hexane afforded colorless
prisms: mp 92–94°.

**2,5-Dimethyl-3-phenylpent-2-enone ([3 + 2] Cyclocoupling of an α,α′-Di-
bromo Ketone and an Enamine with Diiron Enneacarbonyl).***[122] A mixture of
4.00 g (11.0 mmol) of diiron enneacarbonyl, 2.44 g (10.0 mmol) of 2,4-dibromo-
pentan-3-one, and 3.78 g (20.0 mmol) of α-*N*-morpholinostyrene in 25 mL of
benzene was maintained at 30° for 20 hours with efficient stirring. The reaction
mixture was diluted with 15–50 mL of ethyl acetate and then washed with 30 mL
of saturated sodium hydrogen carbonate solution and 30 mL of brine. The
organic layer was dried over sodium sulfate and concentrated under reduced
pressure (50–90 mm), giving 3.0 g of an orange liquid. The oil was subjected to

---

* A different procedure on a large scale is cited in *Organic Syntheses*.[124]

column chromatography on 150 g of Merck Kieselgel 60 (70–230 mesh). Elution with 1:10 ethyl acetate–n-hexane yielded 1.68 g (91%) of the title cyclopentenone as a semisolid, which was recrystallized from hexane, giving rise to a pure sample, mp 57–59°: IR (neat) cm^{-1}: 1696 (C=O), 1626 (conjugated C=C); UV (C$_2$H$_5$OH), nm max (ε): 220 (5220), 279 (11200); ^{1}H NMR (CCl$_4$) δ: 1.22 (d, J = 7.0 Hz, 3H, CHCH_3), 1.91 (t, J = 2.0 Hz, 3H, vinyl CH$_3$), 2.1–2.7 (m, 2H, CHCH$_3$ and a methylene proton cis to CH$_3$ group), 3.14 (dd of quartet, J = 18, 7.5, and 2.0 Hz, 1H, a methylene proton $trans$ to CH$_3$ group), 7.38 (m, 5H, aromatic protons); mass spectrum (70 eV) m/e: 186 (M), 171 (M − 15), 158 (M − 28).

Miscellaneous Reactions

2,4-Diisopropyl-5-methyl-3(2H)-furanone (Diiron Enneacarbonyl-Promoted [3 + 2] Cyclocoupling of an α,α'-Dibromo Ketone and an N,N-Dimethylcarboxamide).[34] To a mixture of 874 mg (2.40 mmol) of diiron enneacarbonyl and 2.03 g (6.00 mmol) of Na$_2$H$_2$EDTA was added a solution of 600 mg (2.00 mmol) of 3,5-dibromo-2,6-dimethylheptan-4-one and 26.4 mg (0.20 mmol) of tetralin in 7.0 mL of N,N-dimethylacetamide. The resulting mixture was stirred at room temperature for 12 hours. The reaction mixture was poured into 20 mL of saturated sodium hydrogen carbonate–potassium nitrate solution and extracted with five 8-mL portions of ethyl acetate. The combined organic extracts were washed with three 8-mL portions of water and dried over sodium sulfate. After evaporation of the solvent an oil was obtained. The ^{1}H NMR spectrum of this residue indicated two doublets at δ 3.95 (J = 3.8 Hz) and 3.38 (J = 4.5 Hz) due to the C-2 methine protons of the title 3(2H)-furanone and 5-(dimethylamino)-2,4-diisopropyl-5-methyltetrahydro-3-furanone, respectively. The intensity of signals, as compared with that of a singlet due to the aromatic protons of tetralin added as an internal standard, showed that the 3(2H)-furanone and aminotetrahydro-3-furanone were produced in 49% and 39% yields, respectively. When the crude oil was heated at 110° for 15 minutes under nitrogen, 316 mg (87%) of the 3(2H)-furanone was obtained as an oily product. Distillation at a bath temperature of 90° (3 mm), followed by preparative GC (10% LAC on Chromosorb W AW, 80–100 mesh, 400 × 0.4 cm, 108°), afforded an analytical sample: IR (neat) cm^{-1}: 1691 (C=O), 1631 (C=C); UV (C$_2$H$_5$OH), nm max (ε): 273 (9550); ^{1}H NMR (CCl$_4$) δ: 0.79 [d, J = 6.5 Hz, 3H, CHCH(CH_3)$_2$], 1.05 [d, J = 6.5 Hz, 3H, CHCH(CH_3)$_2$], 1.15 [d, J = 7.0 Hz, 6H, =CCH(CH_3)$_2$], 2.15 (s, 3H, =CCH_3), 2.0–2.8 [m, 2H, 2 CH(CH$_3$)$_2$], 3.95 (d, J = 3.8 Hz, 1H, OCH); mass spectrum (70 eV) m/e: 182.1295 (M).

2-(N-Cyclohexylimino)-3,3,4,4-tetramethylcyclobutanone (Reaction of an α,α'-Dibromo Ketone and a Copper–Isocyanide Complex).[108] A solution of 2.72 g (10 mmol) of 2,4-dibromo-2,4-dimethylpentan-3-one in 4 mL of benzene was added to a mixture of 1.27 g (20 mg-atoms) of metallic copper and 8.7 g (80 mmol) of cyclohexyl isocyanide in 20 mL of benzene at room temperature.

The mixture was stirred at room temperature for 12 hours and treated with *ca.* 25 mL of ether to precipitate the resulting copper(I) cyclohexyl isocyanide complex. The ether solution was filtered and concentrated, and the residue was distilled to give 1.9 g (86%) of 2-(*N*-cyclohexylimino)-3,3,4,4-tetramethylcyclobutanone, bp 98–100° (2 mm): IR (neat) cm^{-1}: 1765 (C=O), 1669 (C=N), 890; ^{1}H NMR (CDCl$_3$) δ: 1.04 (s, 6*H*, 2 CH$_3$), 1.10 (s, 6*H*, 2 CH$_3$), 0.9–2.5 (broad envelope, 10*H*, 5 CH$_2$), 3.9–4.2 (broad multiplet, 1*H*, =NCH); mass spectrum *m/e*: 221 (M).

TABULAR SURVEY

An attempt has been made to include in Tables I–X synthetic applications of reductive dehalogenation of halo ketones using low-valent transition metals and related reducing agents reported through December 1980, but lack of a systematic method of searching the literature for the reaction makes it likely that some examples were overlooked.

The tables are arranged, in general, according to the presentation in the discussion. Table I summarizes the reductions of α,α′-dihalo ketones that form intramolecular skeletal-changed compounds, α,β-unsaturated ketones, etc. Table II gives examples of the reduction of α,α′-dihalo ketones in the presence of oxy or amino nucleophiles producing the parent ketones, α-oxy-substituted ketones, etc. Table III covers the reductive dehalogenation of α-monohalo ketones to the parent ketones. Table IV is concerned with reductive alkylation of monohalo ketones, including examples of reductive aldol reaction, acylation, self-dimerization, etc. Table V comprises reactions of α,α-dihalo ketones consisting of simple dehalogenation and reductive alkylation. Table VI, which lists reductive alkylations of α,α′-dihalo ketones, is divided into three sections: (1) formation of α-monoalkylated ketones, (2) production of α,α′-dialkyl ketones, and (3) reductive dimerization. Table VII, which includes the [3 + 4] cyclocoupling of α,α′-dihalo and more highly halogenated ketones, consists of the following six sections according to the structures of the C$_4$ receptors: (1) open-chain 1,3-dienes; (2) carbocycles such as cyclopentadiene, fulvenes, cyclohexadiene, etc.; (3) pyrrole derivatives; (4) furan derivatives; (5) thiophene; and (6) anthracene. Table VIII, consisting of five sections, is concerned with the [3 + 2] cyclocoupling of α,α′-dibromo ketones with: (1) simple olefins, (2) aryl-substituted olefins, (3) enamines, (4) carboxamides and a lactam, and (5) other C$_2$ substrates. Table IX lists other cyclocoupling reactions of α,α′-dibromo ketones. Table X covers miscellaneous reactions.

Within each table the substrates are arranged in order of increasing number of carbon atoms. Derivatives of alcohols, amines, ketones, and carboxylic acids are listed by the number of carbon atoms in the parent skeletons, respectively. For instance, the bromo ketones **143** and **144** are dealt with as C$_7$ and C$_8$ compounds, respectively. Substrates with the same number of carbon atoms are listed in order of increasing complexity of the structures, e.g., primary before secondary alkyl, alkyl before alkenyl, aliphatic before aromatic, five-

$$p\text{-}CH_3OC_6H_4COCH_2Br$$
144

membered before six-membered cyclic, monocyclic before bicyclic, hydro-carbon before heteroatom-containing, chloro before bromo compounds.

When there is more than one reference, the experimental data are taken from the first reference, and the remaining references are arranged in numerical order.

In all the tables the numbers in parentheses shown after the reactants indicate the stoichiometric ratio of reactants to the halides; no number means that an excess of reagents is used or that the stoichiometric relation is not critical for the reaction. The symbol e⁻ signifies electrochemical treatment with a mercury cathode. Yields are based on the reactant present in the lowest concentration. Unless otherwise noted, yields were either estimated by GC or obtained by product isolation; when both yields are given in the experiments, the chromatographically obtained ones are shown first, followed by the isolated ones. A dash (—) in the reaction conditions or yield column indicates that no information is given in the reference. In addition to the usual chemical symbols, the following abbreviations are used in the tables:

Ac	Acetyl
acac	Acetylacetonate
bbn	9-Bicyclo[3.3.1]nonyl
DME	1,2-Dimethoxyethane
DMF	N,N-Dimethylformamide
DMSO	Dimethyl sulfoxide
Ether	Diethyl ether
HMPA	Hexamethylphosphoramide
MVK	Methyl vinyl ketone
Na_2H_2EDTA	Ethylenediaminetetraacetic acid disodium salt
PC	Phthalocyanine
Py	Pyridine
THP	Tetrahydropyranyl
TMEDA	N,N,N',N'-Tetramethylethylenediamine
TMS	Trimethylsilyl
Ts	Tosyl, $p\text{-}CH_3C_6H_4SO_2$

TABLE I. INTRAMOLECULAR REACTIONS

	Halo Ketone	Reactants (equiv.)	Conditions	Products and Yields (%)	Refs.
C$_3$	(CH$_2$Cl)$_2$CO	K^a (2)	295–303°	CH$_2$=CH$_2$ (33.3)	17
		"	272–281°	" (16.2)	17
		"	257–265°	" (11.9)	17
C$_5$	(CH$_3$CHBr)$_2$CO	"	224–257°	CH$_3$CH=CHCH$_3$ (Z) (11.5); (E) (15.1)	17
	(CH$_3$)$_2$CBrCOCH$_2$Br	"	217–232°	(CH$_3$)$_2$C=CH$_2$ (22.9)	17
C$_7$	[(CH$_3$)$_2$CBr]$_2$CO	"	277–297°	i-C$_3$H$_7$C(CH$_3$)=CH$_2$, (CH$_3$)$_2$C=C(CH$_3$)$_2$ (total 11.6)	17
		Fe$_2$(CO)$_9$	DMF, 25°, 18 hr	CH$_2$=C(CH$_3$)COC$_3$H$_7$-i (80), (i-C$_3$H$_7$)$_2$CO (9)	28
	[2,7-dibromocycloheptanone structure]	K^a (2)	255–265°	[cyclohexene structure] (22)	17
C$_8$	CH$_2$=CH(CH$_2$)$_2$CBr(CH$_3$)COCH$_2$Br	Fe$_2$(CO)$_9$	C$_6$H$_6$, reflux	CH$_2$=CH(CH$_2$)$_2$COC(CH$_3$)=CH$_2$ (—)	32
C$_{11}$	(t-C$_4$H$_9$CHBr)$_2$CO	"	C$_6$H$_6$, 80°, 6 hr	[cyclobutanone structure with t-C$_4$H$_9$, CH$_3$, CH$_3$ substituents] cis/trans (5:1) (78)	28, 29
C$_{14}$	[2-bromo-2,6-di-t-butylcyclohexanone structure]			[bicyclic ketone structure] (I),	28

222

			Products (%)	Ref.
C₁₅ (C₆H₅CHBr)₂CO	Zn/Cu	C_6H_6, reflux, 48 hr	I (11), II (28), IV (40)	28
	Fe₂(CO)₉	C_6H_6, 25°, 23 hr	I (80), II (15), III (3)	28, 29
	Ni(CO)₄	C_6H_6, 24 hr	I (5), II (64)	28
	Fe₂(CO)₉	C_6H_6, 25°, 24 hr	(70)	28, 29
C₁₈ (I), R = H	Na/Hg	C_6H_6, 3 min	II, R = H (13.6)	31
	Ca	THF, − 70°, 23 min	II, R = H (26)	31
	Zn	Dioxane, 60°, 125 min	II, R = H (74)	31
	"	Dioxane, reflux, 25 min	II, R = H (70)	31

Structures:

(II): t-C₄H₉ substituted cyclohexenone with C₄H₉-t

(III): t-C₄H₉ substituted cyclohexenone with C₃H₇-i, CH₃

(IV): bicyclic ketone with CH₃, CH₃, C₄H₉-t, CH₃

Indanone with C₆H₅ (70)

(II): bicyclic cyclopentenone with C₆H₅, C₆H₅, R

(I), R = H: bicyclic dibromoketone with Br, Br, R, C₆H₅, C₆H₅

TABLE I. INTRAMOLECULAR REACTIONS (*Continued*)

Halo Ketone	Reactants (equiv.)	Conditions	Products and Yields (%)	Refs.
C_{18} (*Contd.*) I, R = H	$Fe_2(CO)_9$ "	C_6H_6, DMF, 25°, 24 hr	II, R = H (—) II, R = H (95)	28 28, 29
	Aq HI	Acetone, 10 min	II, R = H (89)	31
I, R = OCH_3	$Fe_2(CO)_9$	C_6H_6, 3.5 hr	II, R = OCH_3 (72)	30b
C_{19}	"	C_6H_6, 20 hr	(80)	30a
C_{21} (I)			(II)	
I, R = H	NaI	Acetone, reflux, 15 min	II, R = H (—)	137
I, R = Ac	"	Acetone, reflux, 4.5 hr	II, R = Ac (—)	"

^{a} Potassium vapor was used.

224

TABLE II. REDUCTION OF α,α'-DIHALO KETONES IN THE PRESENCE OF OXY OR AMINO NUCLEOPHILES

Halo Ketone	Reactants (equiv.)	Conditions	Products and Yields (%)	Refs.
C₃ ClCH₂COCH₂Cl	t-C₄H₉OK (1–2), (C₂H₅)₃B	THF, 20°, 3 hr	CH₃COCH₃ (55)	76
ClCH₂COCH₂Br	t-C₄H₉OK (1), (C₂H₅)₃B	"	CH₃COCH₃ (50)	76
BrCH₂COCH₂Br	t-C₄H₉OK (1–2), (C₂H₅)₃B	"	CH₃COCH₃ (41)	76
	2,6-(t-C₄H₉)₂C₆H₃OK (2), (C₂H₅)₃B	"	CH₃COCH₃ (42)	76
ICH₂COCH₂I	2e⁻, C₆H₅CO₂Na	C₆H₅CO₂H/DMF, 14°	C₆H₅CO₂CH₂OCH₃ (—)	20
	Zn/Cu	CH₃OH, −5 to 0°, 75 min	CH₃COCH₃ (100)	21
C₄ CH₃CHBrCOCH₂Br	Zn/Cu	CH₃OH, −35 to 0°, 75 min	C₂H₅COCH₃ (100)	21
C₅ (CH₃CHBr)₂CO	"	CH₃OH, −5 to 0°, 75 min	(C₂H₅)₂CO (38), CH₃CH(OCH₃)COC₂H₅ (62)	21
	2e⁻	CH₃OH, −10°	CH₃CH(OCH₃)COC₂H₅ (62), [structure: X—OH, CH₃, CH₃, CH₃ cyclopropane] (I) X = OCH₃ (30)	38
	2e⁻, CH₃OH (2)	CH₃CN, −10°	I, X = OCH₃ (50)	38
	2e⁻, (CH₃)₂NH (2)	CH₃CN, −20°	I, X = N(CH₃)₂ (70)	38
(CH₃)₃CBrCOCH₂Br	Zn/Cu	CH₃OH, −35 to 5°, 75 min	(CH₃)₂C(OR)COCH₃ (I), i-C₃H₇COCH₂OR (II), i-C₃H₇COCH₃ (III) R = CH₃ I/III (14:86) (—)	21
	Hg, AcONa	AcOH, 14 ± 2°, 72 hr	R = Ac I/II (72:28) (26)	27
	Hg, AcONa	AcOH/DMF, 14 ± 2°	R = Ac I/II (20:80) (53)	27
	Hg	t-C₄H₉CO₂H, 14 ± 2°	R = COC₄H₉-t I/II (87:13) (74)	27
	2e⁻, t-C₄H₉CO₂Na	AcOH/DMF, 14 ± 2°	R = Ac I/II (11:89) (43)	27
	2e⁻, t-C₄H₉CO₂Na	t-C₄H₉CO₂H/DMF, 14 ± 2°	II, R = COC₄H₉-t (65)	27
	2e⁻, CH₃OH (2)	CH₃CN, −10°	[structure: X—OH, CH₃, CH₃, CH₃ cyclopropane] (I) X = OCH₃ (50)	38

225

TABLE II. REDUCTION OF α,α'-DIHALO KETONES IN THE PRESENCE OF OXY OR AMINO NUCLEOPHILES (*Continued*)

Halo Ketone	Reactants (equiv.)	Conditions	Products and Yields (%)	Refs.
C₅ (CH₃)₂CBrCOCH₂Br (*Contd.*)	2e⁻, (CH₃)₂NH (20)	CH₃CN, −20°	I, X = N(CH₃)₂ (30)	38
	2e⁻, (CH₃)₂NH (2)	"	I, X = N(CH₃)₂ (70)	38
	2e⁻, (C₂H₅)₄NBr	DMF	[structure] CH₃, C=O, CH₃ (—)	37
(CH₃)₂CBrCOCH₂I	Zn/Cu	CH₃OH, −35 to −5°, 75 min	(CH₃)₂C(OCH₃)COCH₃ (I), i-C₃H₇COCH₃ (II) I/II (5–33:67–95) (—)	21
[dibromocyclopentanone structure]	Hg	AcOH	[OAc structure] (I) (36)	26
	"	AcOH/THF (1:9)	I (16)	26
	"	AcOH/C₆H₆ (1:9)	I (19)	26
C₆ (CH₃)₂CBrCOCHBrCH₃			(CH₃)₂C(OR)COC₂H₅ (I), i-C₃H₇COCH(OR)CH₃ (II), i-C₃H₇COC₂H₅ (III)	
	Zn/Cu	CH₃OH, 0°, 75 min	R = CH₃, I/II/III (76:24) (—)	21
	Hg	AcOH, 14 ± 2°, 72 hr	R = Ac I/II (80:20) (84)	27
	Hgᵃ	AcOH, 25°, 1–4 days	R = Ac I/II (55:45) (68)	23
	Hg, AcONa	AcOH/DMF, 25°, 1–4 days	R = Ac I/II (77:23) (61)	27
	Hg	t-C₄H₉CO₂H, 25°, 1–4 days	R = COC₄H₉-t I/II (98:2) (64)	27
	Fe₂(CO)₉	CH₃OH, 25°, 38 hr	R = CH₃ I/II (9:1) (45), III (<3)	28
	Fe₂(CO)₉, AcONa	DMF, 25°, 16 hr	R = Ac I/II (69:31) (46), III (3)	28
	2e⁻, AcONa	AcOH, 25°, 1–4 days	R = Ac I/II (87:13) (28), III (1)	27
	"	AcOH/DMF, 25°, 1–4 days	R = Ac I/II (33:67) (73)	27
	2e⁻, t-C₄H₉CO₂Na	t-C₄H₉CO₂H/DME, 1–4 days	R = COC₄H₉-t I/II (40:60) (71)	27

226

Substrate	Reagent	Conditions	Products (%)	Ref.
$C_2H_5CBr(CH_3)COCH_2Br$	Hg^a	AcOH	$C_2H_5C(OAc)(CH_3)COCH_3$ (I), $C_2H_5CH(CH_3)COCH_2OAc$ (II) I/II (70:30) (31), $C_2H_5CH(CH_3)COCH_3$ (4)	23
2,6-dibromocyclohexanone (structure)	Hg	"	(46), 2-acetoxycyclohexanone (structure), cyclohexanone (11)	26
C_7 $(C_2H_5CHBr)_2CO$	$2e^-$, NaOAc	AcOH, $25 \pm 1°$	$(n\text{-}C_3H_7)_2CO$ (90)	22
$(CH_3)_2CBrCOCHBrC_2H_5$			$(CH_3)_2C(OR)COC_3H_7\text{-}n$ (I), $i\text{-}C_3H_7COCH(OR)C_2H_5$ (II), $i\text{-}C_3H_7COC_3H_7\text{-}n$ (III)	
	Hg	AcOH, $14 \pm 2°$, 72 hr	R = Ac I/II (85:15) (90), III (4)	27
	Hg, AcONa	AcOH/DMF, $14 \pm 2°$, 72 hr	R = Ac I/II (85:15) (82), III (2)	27
	Hg	$t\text{-}C_4H_9CO_2H$, $14 \pm 2°$, 72 hr	R = $COC_4H_9\text{-}t$ I/II (99:1) (72), III (2)	27
	$2e^-$, AcONa	AcOH, $14 \pm 2°$, 72 hr	R = Ac I/II (86:14) (72), III (23)	27
	"	AcOH/DMF, $14 \pm 2°$, 72 hr	R = Ac I/II (50:50) (74), III (8)	27
	$2e^-$, $t\text{-}C_4H_9CO_2Na$	$t\text{-}C_4H_9CO_2H$/DMF, $14 \pm 2°$, 72 hr	R = $COC_4H_9\text{-}t$ I/II (30:70) (66), III (4)	27
$[(CH_3)_2CBr]_2CO$	Zn/Cu	CH_3OH, -5 to $0°$, 75 min	$i\text{-}C_3H_7COC(OR)(CH_3)_2$ (I), $(i\text{-}C_3H_7)_2CO$ (II) I, R = CH_3 (72), II (28)	21
	Hg^a	CH_3OH, $25°$, 1–4 days	I, R = CH_3 (71)	23
	Hg	AcOH, $14 \pm 2°$, 72 hr	I, R = Ac (73), II (1)	27
	Hg^a	AcOH, $25°$, 1–4 days	I, R = Ac (71)	23
	"	AcOD, $25°$, 1–4 days	$(CH_3)_2CDCOC(OAc)(CH_3)_2$ (60)	23
	"	$C_2H_5CO_2H$, $25°$, 1–4 days	I, R = C_2H_5 (80)	23
	Hg	$t\text{-}C_4H_9CO_2H$, $14 \pm 2°$, 72 hr	I, R = $COC_4H_9\text{-}t$ (62), II (1)	27
	Hg^a	$t\text{-}C_4H_9CO_2H$, $25°$, 1–4 days	I, R = $COC_4H_9\text{-}t$ (75)	23
	$Fe_2(CO)_9$	H_2O/DMF (5:95), 24 hr	II (25), $(CH_3)_2CBrCOC_3H_7\text{-}i$ (15)	28

TABLE II. REDUCTION OF α,α′-DIHALO KETONES IN THE PRESENCE OF OXY OR AMINO NUCLEOPHILES (*Continued*)

Halo Ketone	Reactants (equiv.)	Conditions	Products and Yields (%)	Refs.
C_7 $[(CH_3)_2CBr]_2CO$ (*Contd.*)	$Fe_2(CO)_9$, AcONa	DMF, 30°, 19 hr	I, R = Ac (60), II (20)	28, 29
	$2e^-$, LiCl (1)	H_2O/DMF (5:95), −8 to 40°	I, R = H (—)ᵇ	20
	$(n\text{-}C_4H_9)_4NBF_4$ (0.2)	C_2H_5OH/CH_3CN (1:99), 14°	I, R = H (—)ᵇ	20
	$2e^-$, $(n\text{-}C_4H_9)_4NBF_4$ (0.2), AcONa, AcOHᶜ	DMF, −32°	R = Ac I/II (17:1) (—)ᵇ	20
	$2e^-$, AcONa	AcOH/DMF, 14 ± 2°, 72 hr	I, R = Ac (61), II (2)	27, 20
	"	AcOH, 25 ± 1°	R = Ac I/II (91.9:1.2) (75)ᵈ	22
	$2e^-$ ᵉ	"	R = Ac I/II (94:1) (81)ᵈ	22
	$2e^-$, $t\text{-}C_4H_9CO_2Na$	$t\text{-}C_4H_9CO_2H$/DMF, 14 ± 2°, 72 hr	I, R = Ac (57)	27
	$2e^-$	CH_3OH	$\begin{array}{c}X \;\text{—OH}\\ CH_3 \diagup \diagdown CH_3\\ H_3C \quad CH_3\end{array}$ CH$_3$ (I) X = OCH_3 (90), $i\text{-}C_3H_7C(CH_3)_2CO_2CH_3$ (4), $i\text{-}C_3H_7COC(OCH_3)(CH_3)_2$ (6)	37
	"	CH_3OH, <0°	I, X = OCH_3 (100)	37, 38
	$2e^-$, CH_3OH (2)	CH_3CN, −5°	I, X = OCH_3 (85)	38
	$2e^-$, $(CH_3)_2NH$ (2)	"	I, X = $N(CH_3)_2$ (80)	38
	$2e^-$, CH_3NH_2 (2)	CH_3CN, −20°	I, X = $NHCH_3$ (60)	38
	$2e^-$, $C_6H_5NH_2$ (2)	CH_3CN, −5°	I, X = NHC_6H_5 (80)	38
[2,2-dibromo-substituted cyclopentanone structure]	$2e^-$, AcONa	AcOH, 25 ± 1°	[cyclopentanone with OAc] (2.1), [dimethyl cyclopentanone] (32)	22

228

Substrate	Reagent	Conditions	Product(s) and Yield(s) (%)	Refs.
	Hg[a]	AcOH, 25°, 1–4 days	(I), (II),	23
	t-C₄H₉OK (2), (C₂H₅)₃B	THF	I/cis-II/trans-II (44 : 32 : 24) (66), 2-methylcyclohexanone (III) (1) III (50)	76
	Hg " "	AcOH/C₆H₆ (1 : 9) AcOH/THF AcOH	(I), cycloheptanone (II) I (14), II (35) I (13), II (28) I (16), II (32)	26 26 26
	Zn/Cu, NH₄Cl "	CH₃OH, 15 min	(100) (100)	75, 104
C₈ (CH₃)₂CBrCOCHBrC₃H₇-n	Hg[a] "	AcOH, 25°, 1–4 days t-C₄H₉CO₂H, 25°, 1–4 days	(CH₃)₂C(OR)COC₄H₉-n (I), i-C₃H₇COCH(OAc)C₃H₇-n (II) R = Ac I/II (70 : 30) (75) R = COC₄H₉-t I/II (96 : 4) (74)	74, 45, 110
(CH₃)₂CBrCOCHBrC₃H₇-i			(CH₃)₂C(OR)COCH₂C₃H₇-i (I), i-C₃H₇COCH(OR)C₃H₇-i (II), i-C₃H₇COCH₂C₃H₇-i (III)	23 23

TABLE II. REDUCTION OF α,α'-DIHALO KETONES IN THE PRESENCE OF OXY OR AMINO NUCLEOPHILES (*Continued*)

Halo Ketone	Reactants (equiv.)	Conditions	Products and Yields (%)	Refs.
C_8 $(CH_3)_2CBrCOCHBrC_3H_7$-i (*Contd.*)	Hg	AcOH, $14 \pm 2°$, 72 hr	R = Ac I/II (91:9) (89), III (2)	27
	Hg^a	AcOH, $25°$, 1–4 days	R = Ac I/II (93:7) (60)	23
	Hg	t-$C_4H_9CO_2H$, $14 \pm 2°$, 72 hr	R = COC_4H_9-t I/II (99:1) (68), III (1)	27
	Hg^a	$(C_2H_5)_3CCO_2H$, $25°$, 1–4 days	I, R = $COC(C_2H_5)_3$ (76)	23
	$2e^-$, AcONa	AcOH, $25 \pm 1°$	R = Ac I/II/III (65:12:15) (73)	22
	"	AcOH/DMF, $14 \pm 2°$ 72 hr	R = Ac I/II (98:2) (74), III (12)	27
	$2e^-$, t-$C_4H_9CO_2Na$	t-$C_4H_9CO_2H$/DMF, $14 \pm 2°$, 72 hr	R = COC_4H_9-t I/II (99:1) (50), III (6)	27
$(CH_3)_2CBrCOCBr(CH_3)C_2H_5{}^f$	$2e^-$, AcONa	AcOH	$(CH_3)_2C(OAc)COCH(CH_3)C_2H_5$ (I), i-$C_3H_7COC(OAc)(CH_3)C_2H_5$ (II), i-$C_3H_7COCH(CH_3)C_2H_5$ (III) I/II/III (42:56:1) $(85)^{d,j}$	22
	"	AcOH, $25 \pm 1°$	OAc (I), 2,6-dimethylcyclohexanone (II) I/II (59:2) (68)	22
	Hg	AcOH	(2), cyclooctanone (61)	26
	Zn/Cu, NH_4Cl	CH_3OH/ether, 18 hr	(100)	97, 74, 96

230

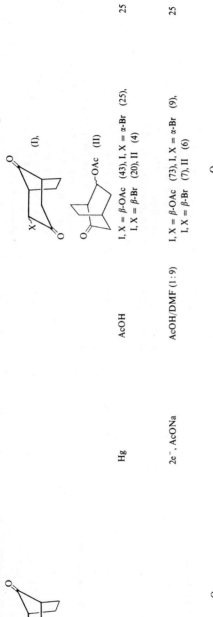

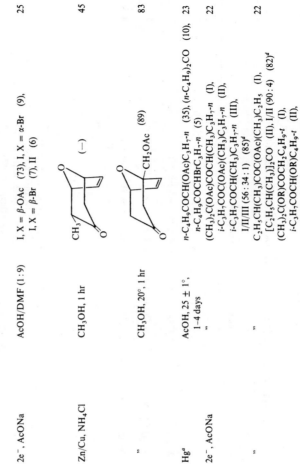

C_9 $(n\text{-}C_3H_7CHBr)_2CO$ Hg^a AcOH, 25 ± 1°, 1–4 days $n\text{-}C_4H_9COCH(OAc)C_3H_7\text{-}n$ (35), $(n\text{-}C_4H_9)_2CO$ (10),   23
$n\text{-}C_4H_9COCHBrC_3H_7\text{-}n$ (5)

$(CH_3)_2CBrCOCBr(CH_3)C_3H_7\text{-}n$ $2e^-$, AcONa   " $(CH_3)_2C(OAc)COCH(CH_3)C_3H_7\text{-}n$ (I),   22
$i\text{-}C_3H_7\text{-}COC(OAc)(CH_3)C_3H_7\text{-}n$ (II),
$i\text{-}C_3H_7\text{-}COCH(CH_3)C_3H_7\text{-}n$ (III),
I/II/III (56 : 34 : 1) (85)^d

$[C_2H_5CBr(CH_3)]_2CO$ "   " $C_2H_5CH(CH_3)COC(OAc)(CH_3)C_2H_5$ (I),   22
$[C_2H_5CH(CH_3)]_2CO$ (II), I/II (90 : 4) (82)^d

$(CH_3)_2CBrCOCHBrC_4H_9\text{-}t$ "   " $(CH_3)_2C(OR)COCH_2C_4H_9\text{-}t$ (I),
$i\text{-}C_3H_7COCH(OR)C_4H_9\text{-}t$ (II),
$i\text{-}C_3H_7COCH_2C_4H_9\text{-}t$ (III)

Hg   AcOH   I, X = β-OAc (43), I, X = α-Br (25),   25
I, X = β-Br (20), II (4)

$2e^-$, AcONa   AcOH/DMF (1:9)   I, X = β-OAc (73), I, X = α-Br (9),   25
I, X = β-Br (7), II (6)

Zn/Cu, NH$_4$Cl   CH$_3$OH, 1 hr   (—)   45

"   CH$_3$OH, 20°, 1 hr   CH$_2$OAc (89)   83

231

TABLE II. REDUCTION OF α,α'-DIHALO KETONES IN THE PRESENCE OF OXY OR AMINO NUCLEOPHILES (*Continued*)

Halo Ketone	Reactants (equiv.)	Conditions	Products and Yields (%)	Refs.
C₉ (CH₃)₂CBrCOCHBrC₄H₉-t (*Contd.*)	Hg "	AcOH, 14 ± 2°, 72 hr t-C₄H₉CO₂H, 14 ± 2°, 72 hr	R = Ac I/II (95:5) (58), III (1) R = COC₄H₉-t I/II (99:1) (50), III (2)	27 27
	2e⁻, AcONa	AcOH/DMF, 14 ± 2°, 72 hr	I, R = Ac (48), III (5)	27
	2e⁻, t-C₄H₉CO₂Na	t-C₄H₉CO₂H/DMF, 14°, 72 hr	I, R = COC₄H₉-t (67), III (7)	27
C₆H₅CHBrCOCH₂Br	2e⁻, AcONa	AcOH, 25°	C₆H₅CH₂COCH₃ (100, 85)	24, 22
[cyclononane ring, Br—CH, O=C, CH—Br]	Hg	AcOH	Cyclononanone (44), [cyclononanone, 2-Br, O] (22)	26
C₁₀ (CH₃)₂CBrCOCHBrCH₂C₄H₉-t	Hg "	AcOH, 14 ± 2°, 72 hr t-C₄H₉CO₂H, 14 ± 2°, 72 hr	(CH₃)₂C(OR)CO(CH₂)₂C₄H₉-t (I), i-C₃H₇CH(OR)CH₂C₄H₉-t (II), i-C₃H₇CO(CH₂)₂C₄H₉-t (III) R = Ac I/II (95:5) (85) R = COC₄H₉-t I/II (99:1) (85)	27 27
	2e⁻, AcONa 2e⁻, t-C₄H₉CO₂Na	AcOH, 14 ± 2°, 72 hr t-C₄H₉CO₂H, 14 ± 2°, 72 hr	R = Ac I/II (84:6) (37), III (13) R = COC₄H₉-t I/II (40:60) (61), III (20)	27 27
C₂H₅CBr(CH₃)COCBr(CH₃)C₃H₇-n	2e⁻, AcONa	AcOH	C₂H₅CH(OAc)COCH(CH₃)C₃H₇-n (I), C₂H₅CH(CH₃)COC(OAc)(CH₃)C₃H₇-n (II), C₂H₅CH(CH₃)COCH(CH₃)C₃H₇-n (III) I/II/III (56:39:0.2) (86)ᵈ	22
C₆H₅CHBrCOCHBrCH₃			C₆H₅CH(OR)COC₂H₅ (I), C₆H₅CH₂COCH(OR)CH₃ (II), C₆H₅CH₂COC₂H₅ (III)	

I

I (α- and β-Br)

I (β-Br)

I (α-Br)

Zn/Cu	CH$_3$OH, −5 to 0°, 75 min	R = CH$_3$ II/III (28:72) (—)	21
Hg, AcONa	AcOH, 25°	R = Ac I (7), II (59), III (16)	24
Hg, AcONa	"	R = Ac I (12), II (66), III (16)	24
2e⁻, AcONa	"	R = Ac I (8), II (21), III (66)	24
		4-t-Butylcyclohexanone (II),	

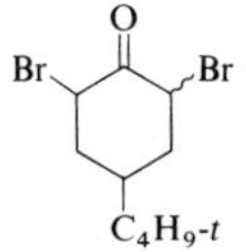

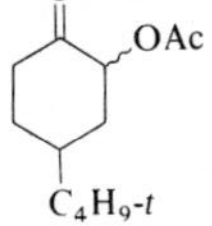

(III)

Hg	AcOH	II (65), III (24)	25
2e⁻, AcONa	AcOH, 25 ± 1°	II/III (1:99) (90)	25
CH$_3$OK (2), (n-C$_4$H$_9$)$_3$B	THF, 20°	II (20)	76
i-C$_3$H$_7$ONa (2), (n-C$_4$H$_9$)$_3$B	"	II (52)	76
i-C$_3$H$_7$OK (2), (n-C$_4$H$_9$)$_3$B	"	II (60)	76
t-C$_4$H$_9$OK (1), (n-C$_4$H$_9$)$_3$B	"	II (20)	76
(C$_2$H$_5$)$_3$COK (2) C$_2$H$_5$B-bbn	THF	II (70)	76
2,4,6-(t-C$_4$H$_9$)$_3$C$_6$H$_2$OK (2), (n-C$_4$H$_9$)$_3$B	THF, 20°	II (40)	76
2e⁻, AcONa	AcOH, 25 ± 1°	II/III (1:99) (90)	22
t-C$_4$H$_9$OK (2), (C$_2$H$_5$)$_3$B	THF	II (10)	76

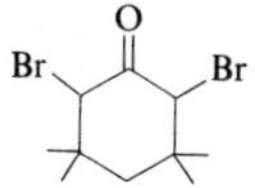

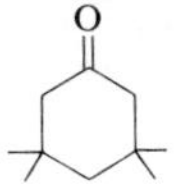

(I),

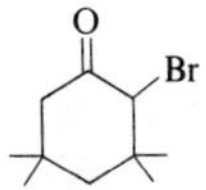

(II)

Hg^a	AcOH	I (76), II (5)	25
2e⁻, AcONa	AcOH/DMF (1:9)	I (80), II (1)	25

233

Halo Ketone	Reactants (equiv.)	Conditions	Products and Yields (%)	Refs.
C$_{10}$ (*Contd.*)	Hg	AcOH	(15), cyclodecanone (58)	26
	2e$^-$, CH$_3$OH (2)	CH$_3$CN, $-10°$	(40), (I), (II), (III), (IV), (V)	38

234

C_{11} Reactant	Reagent	Conditions	Product(s) (% Yield)	Refs.
$C_6H_5CBr(C_2H_5)COCH_2Br$	Hg	AcOH	I (5), II (24), III (14), IV (25), V (14) $C_6H_5C(OAc)(C_2H_5)COCH_3$ (I), $C_6H_5CH(C_2H_5)COCH_2OAc$ (II), $C_6H_5CH(C_2H_5)COCH_3$ (III)	25
	$2e^-$, AcONa	AcOH/DMF (1:9)	I (7.5), II (37), III (22), IV (17), V (11.5)	25
$C_6H_5CHBrCOCBr(CH_3)_2$	Hg, AcONa	AcOH, 25°	I (29), II (8), III (20) $C_6H_5CH(OR)COC_3H_7\text{-}i$ (I), $C_6H_5CH_2COC(OR)(CH_3)_2$ (II), $C_6H_5CH_2COC_3H_7\text{-}i$ (III)	24
	$2e^-$, AcONa	"	I (43), II (11), III (38)	24
	Zn/Cu	CH_3OH, −5°, 75 min	R = CH_3, II/III (84:16) (—)	21
	Hg, AcONa	AcOH, 25°	R = Ac I (6), II (88)	24
	$2e^-$, AcONa	"	R = Ac I (3), II (88), III (8)	24
$C_6H_5CBr(CH_3)COCHBrCH_3$	Hg, AcONa	"	I (62), II (17), III (12) $C_6H_5C(OAc)(CH_3)COC_2H_5$ (I), $C_6H_5CH(CH_3)COCH(OAc)CH_3$ (II), $C_6H_5CH(CH_3)COC_2H_5$ (III)	24
	$2e^-$, AcONa	"	I (64), II (15), III (14)	24
[macrocyclic ketone with two Br substituents]	Hg	AcOH/THF (1:9)	[OAc macrocyclic ketone] (I) (21)	26
	"	AcOH/C_6H_6 (1:9)	cycloundecanone (II) (9), I (28), II (7)	26
C_{12} [cyclohexanone with i-C_3H_7 and Br (2 positions)]	$Fe_2(CO)_9$	CH_3OH, 25°, 24 hr	[cyclohexanone, $C_3H_7\text{-}i$, OCH_3, i-C_3H_7] (I) (1), cis-I/trans-I (2:1) (61), [cyclohexenone, i-C_3H_7, $C_3H_7\text{-}i$] (21)	28

235

TABLE II. REDUCTION OF α,α'-DIHALO KETONES IN THE PRESENCE OF OXY OR AMINO NUCLEOPHILES (*Continued*)

Halo Ketone	Reactants (equiv.)	Conditions	Products and Yields (%)	Refs.
C_{12} (*Contd.*)	Hg	AcOH/THF (1:9)	(I) (9), cyclododecanone (II) (8)	26
	"	AcOH/C_6H_6 (1:9)	I (8), II (6)	26
C_{15} ($C_6H_5CHBr)_2CO$	Zn/Cu	CH_3OH, −5 to 0°, 75 min	$C_6H_5CH(OR)COCH_2C_6H_5$ (I) I, R = CH_3 (—)	21
	Hg, AcONa	AcOH, 25°	I, R = Ac (100)	24
	NaI	CH_3OH, reflux, 1 hr	I, R = CH_3 (57)	138
		Aq acetone, reflux, 50 min	I, R = H (93)	138
($C_6H_5)_2CBrCOCH_2Br$	Zn/Cu	CH_3OH, −40 to −35°, 210 min	$(C_6H_5)_2C(OCH_3)COCH_3$ (I), $(C_6H_5)_2CHCH_2CO_2CH_3$ (II) I (16), II (42)	39
	"	CH_3OH, 0°, 210 min	I (35), II (65)	39
	"	CH_3OH, 25°, 210 min	I (30), II (45)	39
C_{18}	Aq HI	1 min	(73)	30b

C$_{21}$

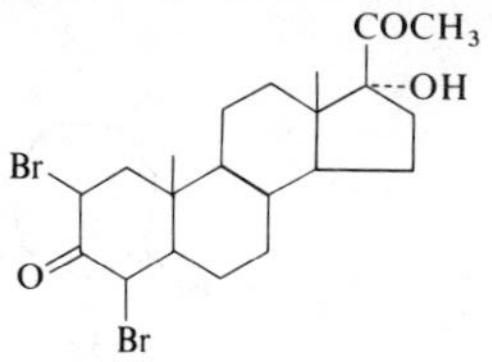

COCH$_3$

---OH

Br

O

Br

NaI — Acetone, reflux, 2 hr — (86) — 139

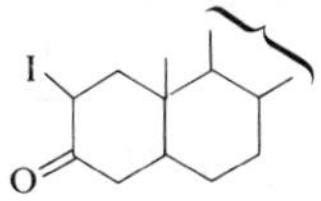

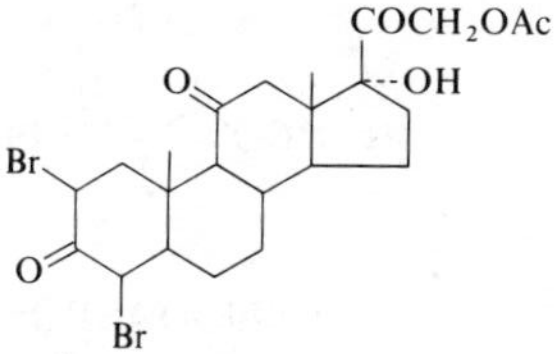

COCH$_2$OAc

---OH

O

Br

Br

CrCl$_2$, HCl — CH$_3$OH/acetone, $-45°$, 8 min — (21) — 137

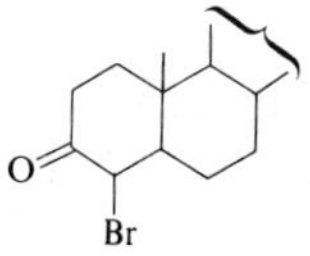

C$_{27}$

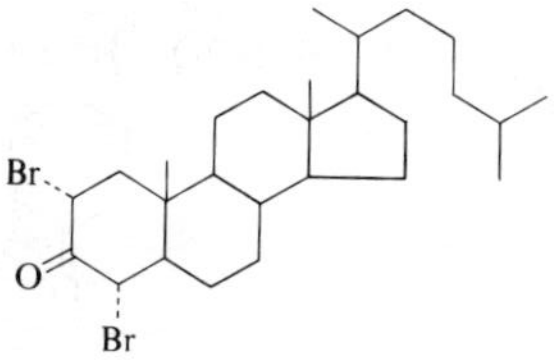

Br

O

Br

Cr(OAc)$_2$ — AcOH/CHCl$_3$, 30 min — (20) — 137, 140

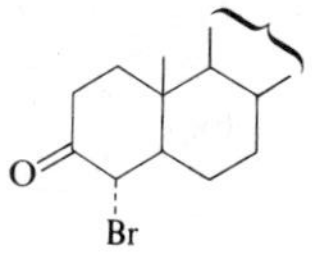

[a] Ultrasonically dispersed mercury was used.

[b] An α,β-unsaturated ketone was formed as a byproduct.

[c] Sodium acetate in acetic acid was added after the electrolysis.

[d] A 1,4-diketone was also formed.

[e] Constant-current electrolysis was used.

[f] The starting material was a 55 : 45 mixture of the dibromide and the corresponding monobromide; the yields were 55% of the acetoxy ketone and 45% of the parent ketone.

TABLE III. REDUCTIVE DEHALOGENATION OF α-MONOHALO KETONES TO THE PARENT KETONES

	Halo Ketone	Reactants (equiv.)	Conditions	Products and Yields (%)	Refs.
C_5	2-chlorocyclopentanone	LiI, $BF_3O(C_2H_5)_2$	—	cyclopentanone (100, 97)	56
C_6	ethyl 3-bromo-2-oxocyclopentanecarboxylate	$Fe(CO)_5$ (1.7)	Xylene, reflux, 6 hr	$CO_2C_2H_5$ (73)	50
	2-bromocyclohexanone	LiI, $BF_3O(C_2H_5)_2$	—	cyclohexanone (100)	56
C_7	$n\text{-}C_3H_7COCHBrC_2H_5$	VCl_2	Aq THF, reflux, 2–3 min	$(n\text{-}C_3H_7)_2CO$ (88)	55
	$i\text{-}C_3H_7COCBr(CH_3)_2$	$2e^-$, $LiCl$	Aq DMF, −5 to 25°	$(i\text{-}C_3H_7)_2CO$ (—)	20
	ethyl 3-bromo-2-oxocyclohexanecarboxylate	$Fe(CO)_5$ (2)	Xylene, reflux, 6 hr	$CO_2C_2H_5$ (68)	50
	2-bromocycloheptanone	$TiCl_3$	Aq CH_3CN, reflux, 18 hr	(I) (86)	54
	2-bromocycloheptanone	VCl_2	Aq THF, reflux, 2–3 min	I (96)	55

Substrate	Reagent	Conditions	Product (Yield)	Refs.
structure with OAc, Cl, H	CrCl₂	Acetone, 15 min	(>80)	141
bicyclic Cl, H ketone	TiCl₃	Aq CH₃CN, reflux, 18 hr	(80)	54
C_8 t-$C_4H_9COCBr(CH_3)_2$	$(t$-$C_4H_9)_2CuLi$	1. Ether, $-78°$; 2. H_2O	t-$C_4H_9COC_3H_7$-i (51)	142
$C_6H_5COCH_2F$	$2e^-$	C_2H_5OH, 10 min	$C_6H_5COCH_3$ (—)	13
$C_6H_5COCH_2Cl$	LiI, $BF_3O(C_2H_5)_2$	—	$C_6H_5COCH_3$ (100)	56
$RCOCH_2Br$ (I)			$RCOCH_3$ (II)	
I, R = C_6H_5	Fe(II)–polyPC (1), C_6H_5SH	C_6H_6, 80°, 24 hr	II, R = C_6H_5 (16)	11, 42
	Zn/Cu, NaI, collidine	DMSO, 90°, 30 min	II, R = C_6H_5 (22)ᵃ	15
	Fe(CO)₅ (2)	Xylene, reflux, 4 hr	II, R = C_6H_5 (57)	50
	$Co_2(CO)_8$ (0.1), $C_6H_5CH_2N(C_2H_5)_3Cl$ (0.5), NaOH	H_2O/C_6H_6, 2 hr	II, R = C_6H_5 (23)ᵃ	51
	$CuC≡CC_6H_5$	$(CH_2OH)_2$, 140°, 16 hr	II, R = C_6H_5 (47)	143
	TiCl₃	Aq CH₃CN, reflux, 18 hr	II, R = C_6H_5 (100)	54
I, R = p-ClC_6H_4	VCl₂	Aq THF	II, R = C_6H_5 (92)	55
I, R = p-BrC_6H_4	Mo(CO)₆, Al_2O_3	THF, 18 hr	II, R = p-ClC_6H_4 (83)	12
	Mo(CO)₆	DME, 85 to 90°, 48 hr	II, R = p-BrC_6H_4 (25)ᵇ	43
	Mo(CO)₆, Al_2O_3	THF, 18 hr	II, R = p-BrC_6H_4 (71)	12
	Zn/Cu, NaI, collidine	DMSO, 70°, 30 min	II, R = p-BrC_6H_4 (22)ᵃ	15

239

Halo Ketone	Reactants (equiv.)	Conditions	Products and Yields (%)	Refs.
C_8 I, R = p-BrC$_6$H$_4$ (*Contd.*)	Co$_2$(CO)$_8$ (0.1), C$_6$H$_5$CH$_2$N(C$_2$H$_5$)$_3$Cl (0.5), NaOH	H$_2$O/C$_6$H$_6$, 2 hr	II, R = p-BrC$_6$H$_4$ (69)[a]	51
	TiCl$_3$	Aq CH$_3$CN, reflux, 18 hr	II, R = p-BrC$_6$H$_4$ (98)	54
	VCl$_2$	Aq THF, 2–3 min	II, R = p-BrC$_6$H$_4$ (98)	55
	LiI, BF$_3$O(C$_2$H$_5$)$_2$	—	II, R = p-BrC$_6$H$_4$ (100)	56
I, R = p-CH$_3$OC$_6$H$_4$	Mo(CO)$_6$	DME, 85 to 90°, 48 hr	II, R = p-CH$_3$OC$_6$H$_4$ (46)[b]	43
	Mo(CO)$_6$, Al$_2$O$_3$	—	II, R = p-CH$_3$OC$_6$H$_4$ (60)	12
	Zn/Cu, NaI, collidine	DMSO, 170°, 60 min	II, R = p-CH$_3$OC$_6$H$_4$ (31)[a]	15
	Co$_2$(CO)$_8$ (0.1), C$_6$H$_5$CH$_2$N(C$_2$H$_5$)$_3$Cl (0.5), NaOH	H$_2$O/C$_6$H$_6$	II, R = p-CH$_3$OC$_6$H$_4$ (25)[a]	51
I, R = p-O$_2$NC$_6$H$_4$	Mo(CO)$_6$	DME, 85 to 90°, 48 hr	II, R = p-O$_2$NC$_6$H$_4$ (20)[a]	43
	Fe(CO)$_5$ (1.9)	Xylene, reflux, 6 hr	(54)	50
	CrCl$_2$	Acetone	(>72)	141

240

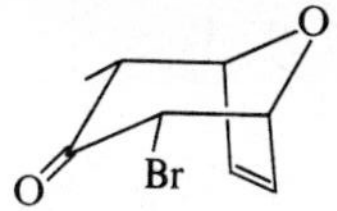

C₉

Reactant	Reagents	Conditions	Product(s) (%)	Refs.
(bicyclic bromo diketone structure)	Zn/Cu, NH₄Cl	CH₃OH, 15 min	(bicyclic diketone structure) (100)	45, 74
R¹COCBr(R²)CH₃ (I) I, R¹ = t-C₄H₉, R² = C₂H₅	(t-C₄H₉)₂CuLi	1. Ether, −78° 2. H₂O	R¹COCH(R²)CH₃ (II) II, R¹ = t-C₄H₉, R² = C₂H₅ (75)	142
I, R¹ = C(CH₃)₂C₂H₅, R² = CH₃	″	1. Ether, −78° 2. H₂O	II, R¹ = C(CH₃)₂C₂H₅, R² = CH₃ (56)	142
I, R¹ = C₆H₅, R² = H	Fe(II)–polyPC (1), C₆H₅SH	Aq CH₃OH, 80°, 30 min	II, R¹ = C₆H₅, R² = H (58)	11
	″	C₆H₆, 80°, 96 hr	II, R¹ = C₆H₅, R² = H (56)	11
(indanone structure) (I), X = H	Fe(CO)₅ (2)	Xylene, reflux, 4 hr	(indanone product) (II), R = X = H (55)	50
	1. Fe(CO)₅ (2) 2. D₂O	″	II, R = X = H (—)	50
	Fe(CO)₅ (2)	Xylene-d₈, reflux, 5 hr	II, R = D, X = H (57)	50
I, X = Br	″	Xylene, reflux, 4 hr	II, R = H, X = Br (45)	50
(bicyclic bromo ketone structure)	Zn, NH₄Cl	CH₃OH	(bicyclic ketone structure) (—)	47
(methyl-bicyclic bromo diketone structure)	Zn/Cu, NH₄Cl	CH₃OH, 1 hr	(methyl-bicyclic diketone structure) (>90)	45, 74

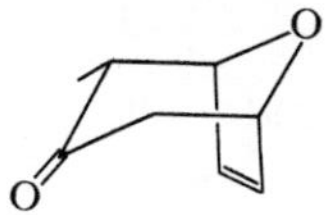

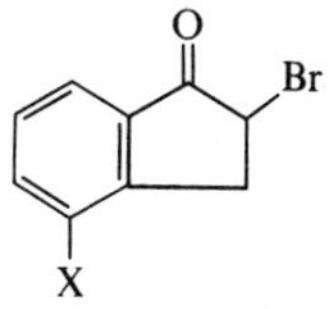

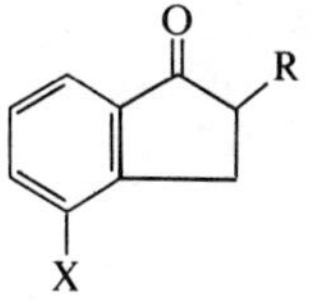

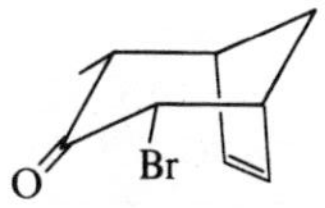

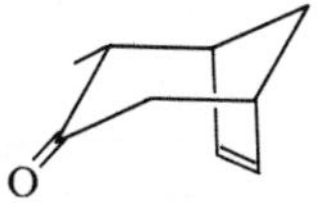

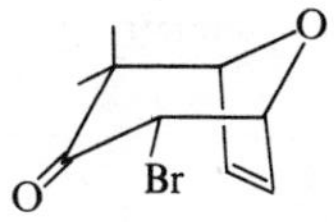

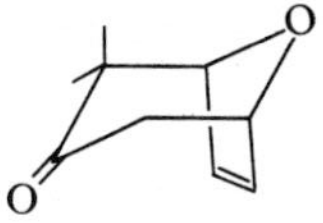

241

TABLE III. REDUCTIVE DEHALOGENATION OF α-MONOHALO KETONES TO THE PARENT KETONES (*Continued*)

Halo Ketone	Reactants (equiv.)	Conditions	Products and Yields (%)	Refs.
C_{10} t-$C_4H_9COCBr(C_2H_5)_2$	$(t$-$C_4H_9)_2CuLi$	Ether, $-78°$	t-$C_4H_9COCH(C_2H_5)_2$ (78)	142
$C_2H_5C(CH_3)_2COCBr(CH_3)C_2H_5$	Li	1. Ether / 2. H_2O	$C_2H_5C(CH_3)_2COCH(CH_3)C_2H_5$ (97)	40
"	"	1. Ether / 2. D_2O	$C_2H_5C(CH_3)_2COCD(CH_3)C_2H_5$ (50)	40
(I) [decalone with Br]			(II), (III) [decalone structures]	
cis-I	Zn	AcOH, 4.5 hr	II/III (37 : 63) (—)	46
	Zn, collidine·HCl	Collidine, 23 hr	II/III (77 : 23) (—)	46
trans-I	Zn	AcOH, 5.5 hr	II/III (39 : 61) (—)	46
	Zn, collidine·HCl	Collidine, 39 hr	II/III (74 : 26) (—)	46
	"	CH_3OH, 19 hr	II/III (59 : 41) (—)	46
	"	CH_3CN, 8 hr	II/III (52 : 48) (—)	46
[Br-decalone structure]	Zn	AcOH, 4 hr	[decalone structure] (—)	46
[Br-decalone structure]	"	AcOH, 3.5 hr	[decalone structure] (—)	46

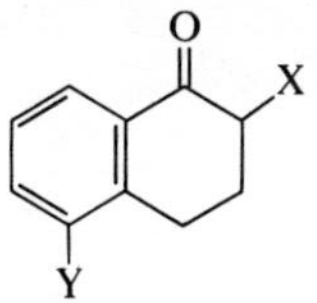

Structure (I):

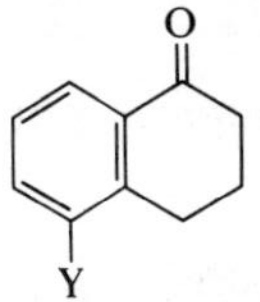

I, X = Br, Y = H
I, X = Br, Y = OCH$_3$
I, X = I, Y = OCH$_3$

Structure (I): (bicyclic with R, Br)

I, R = H

I, R = D

Structure (II):

II, Y = H (56)
II, Y = OCH$_3$ (100, 98)
II, Y = OCH$_3$ (100)

Structure (with R^1, R^2):

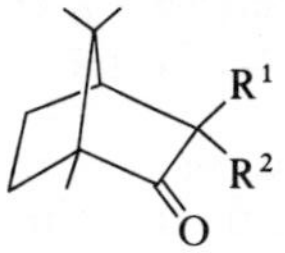

(II) R^1 = R^2 = H,
(III) R^1 = H, R^2 = D,
(IV) R^1 = D, R^2 = H,
(V) R^1 = R^2 = D

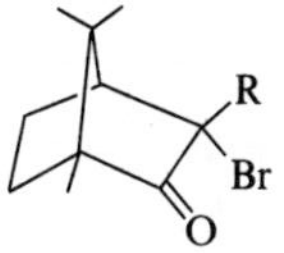

Reagent	Conditions	Product	Ref
Fe(CO)$_5$ (2)	Xylene, reflux, 4 hr	II, Y = H (56)	50
LiI, BF$_3$O(C$_2$H$_5$)$_2$	Ether	II, Y = OCH$_3$ (100, 98)	56
"	"	II, Y = OCH$_3$ (100)	56
Zn	AcOD, 90°, 14.5 hr	II/IV (12:88) (75)	144
Fe$_2$(CO)$_9$	D$_2$O/DMF (5:95), 60°, 20 min	II/IV (13:87) (80)	28, 29
TiCl$_3$	Aq CH$_3$CN, reflux, 18 hr	II (84)	54
(C$_2$H$_5$)$_2$Zn	C$_6$H$_6$, reflux, 48 hr	II 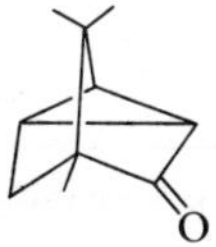 (—), (—)	22
LiI, BF$_3$O(C$_2$H$_5$)$_2$	—	II (95)	56
Zn	AcOH, 90°, 14.5 hr	II/III/V (8:91:0.5) (89)	144

243

TABLE III. Reductive Dehalogenation of α-Monohalo Ketones to the Parent Ketones (*Continued*)

Halo Ketone	Reactants (equiv.)	Conditions	Products and Yields (%)	Refs.
C$_{10}$ (*Contd.*) BrCH$_2$— [structure with H, Br, C=O]	Zn, HBr(g)	CH$_2$Cl$_2$, 2.5–4 hr	BrCH$_2$— [structure with C=O] (75)	49
C$_{11}$ t-C$_4$H$_9$C(CH$_3$)$_2$COCBr(CH$_3$)$_2$	(t-C$_4$H$_9$)$_2$CuLi	Ether, −78°	t-C$_4$H$_9$C(CH$_3$)$_2$COC$_3$H$_7$-i (61)	142
[bicyclic structure with Br, OCH$_3$, C=O]	LiI, BF$_3$O(C$_2$H$_5$)$_2$	—	[bicyclic structure with OCH$_3$, C=O] (100)	56
C$_{12}$ RCOCH$_2$Br (I) I, R = 1-adamantyl	Mo(CO)$_6$	DME, 85 to 90°, 48 hr	RCOCH$_3$ (II) II, R = 1-adamantyl (14)[b]	43, 12
	Mo(CO)$_6$, Al$_2$O$_3$	—	II, R = 1-adamantyl (88)	12
	Co$_2$(CO)$_8$ (0.1), C$_6$H$_5$CH$_2$N(C$_2$H$_5$)$_3$Cl (0.5), NaOH	H$_2$O/C$_6$H$_6$, 2 hr	II, R = 1-adamantyl (60)	51
I, R = 2-naphthyl	Mo(CO)$_6$, Al$_2$O$_3$	—	II, R = 2-naphthyl (80)	12
	Co$_2$(CO)$_8$ (0.1), C$_6$H$_5$CH$_2$N(C$_2$H$_5$)$_3$Cl (0.5), NaOH	H$_2$O/C$_6$H$_6$	II, R = 2-naphthyl (64)	51
	Co$_2$(CO)$_8$ (1), C$_6$H$_5$CH$_2$N(C$_2$H$_5$)$_3$Cl (0.5), NaOH	H$_2$O/C$_6$H$_6$	II, R = 2-naphthyl (97)	51

Substrate	Reagent	Conditions	Product(s) and Yield(s) (%)	Refs.
C_{13} 2-chloro-2-methyl-3-phenylcyclopentanone	Zn	AcOH	3-phenylcyclopentanone (—)	73
1-bromocyclododecanone	$TiCl_3$	Aq CH_3CN, reflux, 18 hr	(I) (90)	54
	$(CH_3)_2CuLi$	1. Ether 2. H^+	I (95)	60
	"	1. Ether 2. D_2O	I-d^c (88)	60
	LiI, $BF_3O(C_2H_5)_2$	—	I (100)	56
C_{14} $p\text{-}C_6H_5C_6H_4COCH_2X$ (I), X = Cl	$Mo(CO)_6$	DME, 85–90°, 48 hr	$COCH_3$/CH_3 pentamethyl arene (I) (51)	43
	$(C_6H_5)_3PMo(CO)_5$	"	I (76)	43
	$W(CO)_6$	DME, 85–90°, 8 days	I (12)	43
	$Fe(II)\text{-}PC$ (1), C_6H_5SH	C_6H_6, 80°, 48 hr	$p\text{-}C_6H_5C_6H_4COCH_3$ (II) II (83)	42
	$Fe(II)\text{-}polyPC$ (0.1), C_6H_5SH	Aq CH_3OH, 80°, 2 hr	II (67)	42
	$Fe(II)\text{-}polyPC$ (1), C_6H_5SH	Aq CH_3OH, 80°, 30 min	II (92)	42
	"	C_6H_6, 80°, 48 hr	II (75)	42

TABLE III. REDUCTIVE DEHALOGENATION OF α-MONOHALO KETONES TO THE PARENT KETONES (*Continued*)

Halo Ketone	Reactants (equiv.)	Conditions	Products and Yields (%)	Refs.
C$_{14}$ I, X = Cl (*Contd.*)	Fe(acac)$_3$ (1), C$_6$H$_5$SH	C$_6$H$_6$, 80°, 48 hr	II (72)	42
	Fe(OAc)$_3$ (1), C$_6$H$_5$SH	"	II (31)	42
	FeCl$_3$ (1), C$_6$H$_5$SH	"	II (11)	42
	FeCl$_3$ (1), py(2), C$_6$H$_5$SH	"	II (36)	42
	Co(III)–PC (1), C$_6$H$_5$SH	"	II (64)	42
I, X = Br	Fe(II)–PC (1), C$_6$H$_5$SH	Aq CH$_3$OH, 25°, 5 min	II (31)	11
	Fe(II)–PC (1), C$_6$H$_5$SH	Aq CH$_3$OH, 80°, 5 min	II (59)	11
	Mo(CO)$_6$	DME, 85–90°, 48 hr	II (51)^b	43
	Mo(CO)$_6$, Al$_2$O$_3$	THF, 18 hr	II (73)	12
	Co$_2$(CO)$_8$ (0.1), C$_6$H$_5$CH$_2$N(C$_2$H$_5$)$_3$Cl (0.5), NaOH	H$_2$O/C$_6$H$_6$, 2 hr	II (58)	51
	Co$_2$(CO)$_8$ (1), C$_6$H$_5$CH$_2$N(C$_2$H$_5$)$_3$Cl (0.5), NaOH	"	II (64)	51
	VCl$_2$	Aq THF, reflux, 2–3 min	II (96)	55
	Fe(II)–PC (1), C$_6$H$_5$SH	C$_6$H$_6$, 80°, 24 hr	II (37)	11
	Fe(II)–polyPC (0.1), C$_6$H$_5$SH	Aq CH$_3$OH, 25°, 2 hr	II (67)	11
	Fe(II)–polyPC (1), C$_6$H$_6$SH	Aq CH$_3$OH, 25°, 30 min	II (47)	11
	"	Aq CH$_3$OH, 80°, 5 min	II (87)	11
	Fe(II)–polyPC (0.1), C$_6$H$_5$SH	C$_6$H$_6$, 80°, 120 hr	II (34)	11

Substrate	Reagent	Conditions	Product (Yield)	Refs
C_{18} [OSi(CH$_3$)$_2$C$_4$H$_9$-t, Cl cyclobutanone]	Fe(II)–polyPC (1), C_6H_5SH	C_6H_6, 80°, 12 hr DME	II (67)	11
	1. $Fe(CO)_5$ 2. D_2O		p-$C_6H_5C_6H_4COCH_2D$ (—)	14
	Zn	AcOH	[OSi(CH$_3$)$_2$C$_4$H$_9$-t cyclobutanone] (—)	145
C_{19} [Br, C$_6$H$_5$, C$_6$H$_5$ cyclopropanone]	HI	Aq acetone, 10 min	[structure] (81)	31
	Zn	C_2H_5OH, 3 hr	[C$_6$H$_5$ structure] (99)	44
C_{19} [OAc, Br steroid]	CrCl$_2$	Acetone, 10–30 min	[steroid] (I) (63)	146
[I steroid]	Zn	C_2H_5OH/dioxane, reflux, 7 hr	[steroid] (56)	146

247

TABLE III. REDUCTIVE DEHALOGENATION OF α-MONOHALO KETONES TO THE PARENT KETONES (*Continued*)

Halo Ketone	Reactants (equiv.)	Conditions	Products and Yields (%)	Refs.
C_{21}				
[steroid structure with COCH₃, ---OH, I, O]	$CrCl_2$	Acetone, 5 min	[structure] (83)	139
[steroid structure with COCH₂OAc, ---OH, Br, O] (I)	NaI	Acetone, reflux, 1 hr	[structure] (—)	137
2- and 4-Br-I				
[steroid structure with COCH₂OAc, ---Br, I, X---, O] (I) I, X = H I, X = OAc	$CrCl_2$ "	Acetone, 10 min "	[structure] (II) II, X = H (ca. 15) II, X = OAc (ca. 60)	147 147

248

	Reagents	Conditions	Products	Ref.
C_{27}	"	Acetone, 30 min	(I) (85)	41
	Zn	AcOH, reflux, 10 hr	I (66)	41
	$(CH_3)_2CuLi$ (2)	Ether, 0°	I (80), (12)	52
	LiI, $BF_3O(C_2H_5)_2$	Ether/C_6H_6	I (100)	56
	$(CH_3)_2CuLi$ (2)	Ether, 0°, 2 min	(95)	48

[a] A moderate amount of $(RCOCH_2)_2$ was also obtained.

[b] The ketone $RCOCH=C(R)CH_3$ was also obtained.

[c] The product was 2-monodeuteriocyclododecanone.

249

TABLE IV. REDUCTIVE ALKYLATION OF α-MONOHALO KETONES

A. With Alkyl Halides, Alkyl Sulfonates, etc.

Halo Ketone	Reactants (equiv.)	Conditions	Products and Yields (%)	Refs.
C_5 $CH_3COCBr(CH_3)_2$	Zn, $R^1CH(SR^2)Cl$ (I)		$CH_3COC(CH_3)_2CHR^1(SR^2)$ (II)	
	I, $R^1 = H$, $R^2 = n\text{-}C_3H_7$	$AcOC_2H_5$, 1 hr	II, $R^1 = H$, $R^2 = n\text{-}C_3H_7$ (73)	59
	I, $R^1 = CH_3$, $R^2 = CH_2CH{=}CH_2$	"	II, $R^1 = CH_3$, $R^2 = CH_2CH{=}CH_2$ (51)	59
C_7 $n\text{-}C_3H_7COCHBrC_2H_5$	Zn, $R^1CH(SR^2)Cl$ (I)		$n\text{-}C_3H_7COCH(C_2H_5)CHR^1(SR^2)$ (II)	
	I, $R^1 = H$, $R^2 = n\text{-}C_4H_9$	"	II, $R^1 = H$, $R^2 = n\text{-}C_4H_9$ (77)	59
	I, $R^1 = H$, $R^2 = n\text{-}C_3H_7$	"	II, $R^1 = H$, $R^2 = n\text{-}C_3H_7$ (67)	59
	I, $R^1 = CH_3$, $R^2 = C_2H_5$	"	II, $R^1 = CH_3$, $R^2 = C_2H_5$ (82)	59
	I, $R^1 = CH_3$, $R^2 = CH_2CH{=}CH_2$	"	II, $R^1 = CH_3$, $R^2 = CH_2CH{=}CH_2$ (62)	59
	I, $R^1 = CH_3$, $R^2 = n\text{-}C_4H_9$	"	II, $R^1 = CH_3$, $R^2 = n\text{-}C_4H_9$ (61)	59
$i\text{-}C_3H_7COCBr(CH_3)_2$	Li, RI (I)		$i\text{-}C_3H_7COCR(CH_3)_2$ (II)	
	I, $R = CH_3$	Ether	II, $R = CH_3$ (94)	40
	I, $R = C_2H_5$	HMPA	II, $R = C_2H_5$ (50)	40
	I, $R = i\text{-}C_3H_7$	"	II, $R = i\text{-}C_3H_7$ (20)	40

(I)

(II),

(III)

250

Substrate	Reagent	Conditions	Product(s) (% Yield)	Refs.
I (0.016M)	Zn, CH₃I	DMSO/C₆H₆ (1:10), 8 hr	II (43), III (2)	57
I (0.060M)	"	"	II (23), III (—)	57
I (0.28M)	"	"	II (6), III (58), cycloheptanone (8)	57
C₈ t-C₄H₉COCBr(CH₃)₂	Li, RI (I) I, R = CH₃	Ether	t-C₄H₉COCR(CH₃)₂ (II) II, R = CH₃ (93)	40
	I, R = C₂H₅	HMPA	II, R = C₂H₅ (15)	40
	1. (t-C₄H₉)₂CuLi 2. RI (I) I, R = CH₃	Ether, −78°	t-C₄H₉COCR(CH₃)₂ (II) II, R = CH₃ (51)	142
	I, R = C₂H₅	"	II, R = C₂H₅ (42)	142
C₆H₅COCH₂Br	CuC≡CR (I) I, R = n-C₃H₇	Neat, 140°, 5 min	II, R = n-C₃H₇ (29)	143
	I, R = C₆H₅	C₆H₅NO₂, 240°, 5 min	II, R = C₆H₅ (54)	143

(II) (furan structure, C₆H₅ … R)

Substrate	Reagent	Conditions	Product(s) (% Yield)	Refs.
(bromo dimethylcyclohexanedione)	CuC≡CC₆H₅	—	(43)	143

Substrate	Reagent	Conditions	Product(s) (% Yield)	Refs.
(bromocyclooctanone)	Zn, CH₃I	DMSO/C₆H₆ (1:10), 8 hr	(66), cyclooctanone (7)	57
C₉ n-C₄H₉COCHBrC₃H₇-n	Zn, R¹CH(SR²)Cl (I) I, R¹ = H, R² = n-C₃H₇	AcOC₂H₅, 1 hr	n-C₄H₉COCH(C₃H₇-n)CHR¹(SR²) (II) II, R¹ = H, R² = n-C₃H₇ (58)	59
	I, R¹ = H, R² = n-C₄H₉	"	II, R¹ = H, R² = n-C₄H₉ (59)	59
	I, R¹ = CH₃, R² = C₂H₅	"	II, R¹ = CH₃, R² = C₂H₅ (55)	59

251

TABLE IV. REDUCTIVE ALKYLATION OF α-MONOHALO KETONES (*Continued*)

A. With Alkyl Halides, Alkyl Sulfonates, etc. (*Continued*)

Halo Ketone	Reactants (equiv.)	Conditions	Products and Yields (%)	Refs.
C$_9$ n-C$_4$H$_9$COCHBrC$_3$H$_{7}$-n (*Contd.*)	I, R^1 = CH$_3$, R^2 = CH$_2$CH=CH$_2$	AcOC$_2$H$_5$, 1 hr	II, R^1 = CH$_3$, R^2 = CH$_2$CH=CH$_2$ (56)	59
	I, R^1 = CH$_3$, R^2 = n-C$_4$H$_9$	"	II, R^1 = CH$_3$, R^2 = n-C$_4$H$_9$ (53)	59
	I, R^1 = CH$_3$, R^2 = C$_6$H$_5$	"	II, R^1 = CH$_3$, R^2 = C$_6$H$_5$ (51)	59
C$_2$H$_5$C(CH$_3$)$_2$COCBr(CH$_3$)$_2$	1. (t-C$_4$H$_9$)$_2$CuLi 2. CH$_3$I	Ether, −78°	C$_2$H$_5$C(CH$_3$)$_2$COC$_4$H$_9$-t (51)	142
C$_{10}$ C$_2$H$_5$C(CH$_3$)$_2$COCBr(CH$_3$)C$_2$H$_5$	Li, CH$_3$I	Ether	[C$_2$H$_5$C(CH$_3$)$_2$]$_2$CO (45)	40

(I) structure with Br, O, X; I, X = H; I, X = OCH$_3$

(II) structure with R, O, X

Halo Ketone	Reactants (equiv.)	Conditions	Products and Yields (%)	Refs.
I, X = H	Zn, CH$_3$I (2)	DMSO/C$_6$H$_6$ (1:10), 8 hr	II, R = CH$_3$, X = H (85)	57, 58
I, X = OCH$_3$	Zn, CH$_3$I (2)	DMSO/C$_6$H$_6$ (1:10)	II, R = CH$_3$, X = OCH$_3$ (75)	58
	Zn, C$_2$H$_5$I	DMSO	II, R = C$_2$H$_5$, X = OCH$_3$ (15)	58
	"	DMSO/C$_6$H$_6$ (1:10)	II, R = C$_2$H$_5$, X = OCH$_3$ (<5)	58
	"	HMPA/C$_6$H$_6$ (2:3)	II, R = C$_2$H$_5$, X = OCH$_3$ (55)	58
	Zn, CH$_2$=CHCH$_2$Br (1)	DMSO/C$_6$H$_6$	II, R = CH$_2$CH=CH$_2$, X = OCH$_3$ (75)	57
	"	"	II, R = CH$_2$CH=CH$_2$, X = OCH$_3$ (60),	58

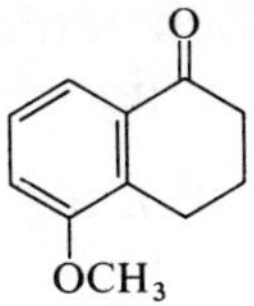

C$_{11}$ n-C$_5$H$_{11}$COCHBrC$_4$H$_9$-n

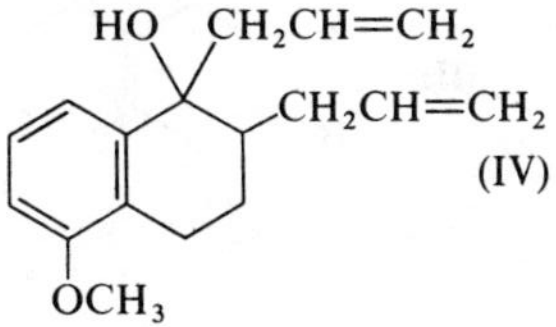

Zn, CH$_2$=CHCH$_2$Br (2)

Zn, C$_6$H$_5$CH$_2$Br

(III) (20),

HO CH$_2$CH=CH$_2$

  CH$_2$CH=CH$_2$

(IV) (20)

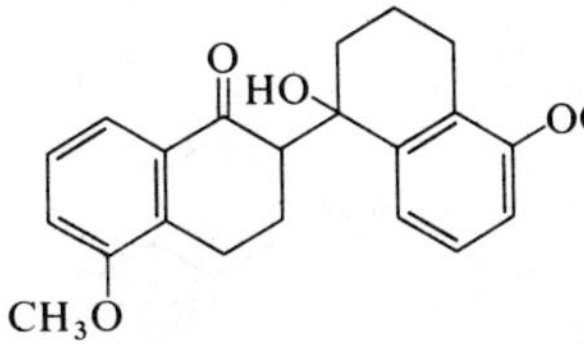

IV (—)

II, R = CH$_2$C$_6$H$_5$, X = OCH$_3$ (20), 58

III (20),

57, 58

58

Zn, R^{1}CH(SR^2)Cl (I)

I, R^1 = H, R^2 = n-C$_4$H$_9$
I, R^1 = CH$_3$, R^2 = C$_2$H$_5$

AcOC$_2$H$_5$, 1 hr
"

n-C$_5$H$_{11}$COCH(C$_4$H$_9$-n)CHR^1(SR^2) (II)

II, R^1 = H, R^2 = n-C$_4$H$_9$ (57)
II, R^1 = CH$_3$, R^2 = n-C$_4$H$_9$ (51)

59
59

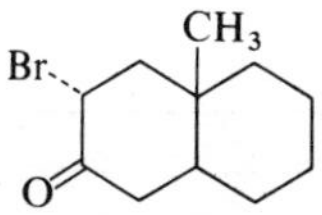

Zn, CH$_3$I

DMSO/C$_6$H$_6$, 5 hr

(25)

58

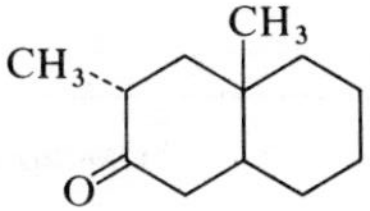

253

TABLE IV. REDUCTIVE ALKYLATION OF α-MONOHALO KETONES (*Continued*)

A. *With Alkyl Halides, Alkyl Sulfonates, etc.* (*Continued*)

Halo Ketone	Reactants (equiv.)	Conditions	Products and Yields (%)	Refs.
C$_{11}$ (*Contd.*)	Zn, CH$_3$I	DMSO/C$_6$H$_6$, 5 hr	(20) (II), (III)	58
I, X = H	"	DMSO/C$_6$H$_6$ (1:10), 8 hr	X = H, II (57), III (43)	57
I, X = OCH$_3$	"	—	X = OCH$_3$, II (60), III (20)	58
C$_{12}$ (C$_2$H$_5$)$_3$CCOCBr(CH$_3$)C$_2$H$_5$	Li, C$_2$H$_5$I	HMPA	(C$_2$H$_5$)$_3$CCOC(C$_2$H$_5$)$_2$CH$_3$ (14)	40

C13 n-C$_6$H$_{13}$COCHBrC$_5$H$_{11}$-n (I)

Starting material	Reagents	Conditions	Products	Ref.
(cyclododecanone, Br)	Zn, CH$_3$I	DMSO/C$_6$H$_6$ (1:10), 8 hr	(I) R = CH$_3$ (99)	57
	Zn, C$_2$H$_5$OTs	14 days	I, R = C$_2$H$_5$ (13)	57
	Zn, C$_2$H$_5$I	HMPA/C$_6$H$_6$ (1:10)	I, R = C$_2$H$_5$ (35)	58
	(CH$_3$)$_2$CuLi, CH$_3$I	Ether	I, R = CH$_3$ (67)	60
I (0.013M)	Zn, CH$_2$I	DMSO/C$_6$H$_6$ (1:10), 8 hr	n-C$_6$H$_{13}$COCH(CH$_3$)C$_5$H$_{11}$-n (II), (III) II (67), III (<2)	57
I (0.051M)	"	"	II (61), III (2)	57
I (0.18M)	"	"	II (33), III (30)	57
C19 (steroid, Br, AcO)	1. CH$_3$MgI 2. CH$_3$I 3. H$_2$SO$_4$ 4. Ac$_2$O	1. THF, 20°, 15 min 2. Reflux, 45 min THF, reflux, 18 hr CH$_3$OH, reflux Py	(45), (50)	148

TABLE IV. REDUCTIVE ALKYLATION OF α-MONOHALO KETONES (*Continued*)

A. *With Alkyl Halides, Alkyl Sulfonates, etc.* (*Continued*)

Halo Ketone	Reactants (equiv.)	Conditions	Products and Yields (%)	Refs.
C_{21}	1. CH_3MgI 2. CH_3I	Ether 1. THF/ether, 15 min 2. Reflux, 1 hr	(30), (–)	149
C_{27}	Zn, CH_3I	$DMSO/C_6H_6$ (1:10), 8 hr	(40), (40)	57

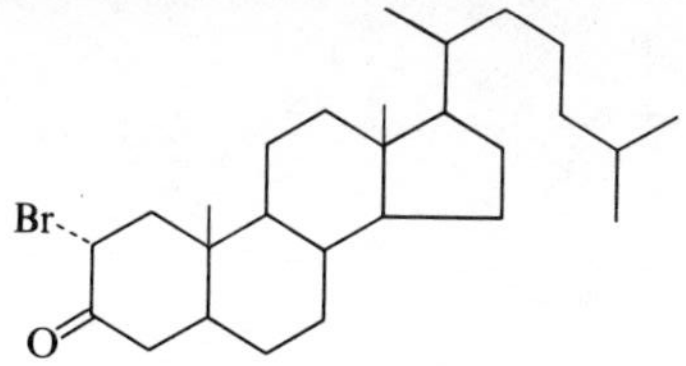

52

(12,

(80)

(45),   148

(50)

(20)   148

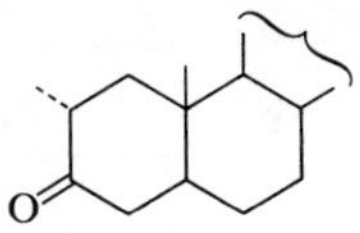

Ether, 0°

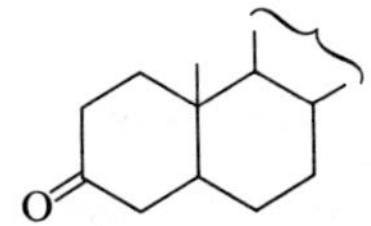

THF
,,
CH$_3$OH
Py

Ether
1. THF, 20°, 30 min
2. Reflux, 1 hr

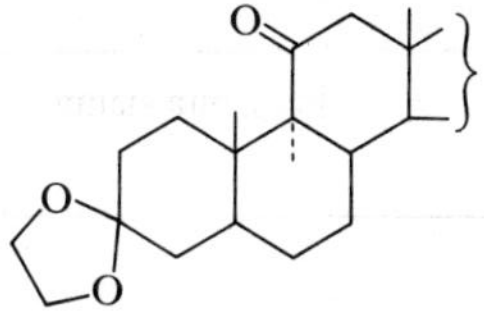

(CH$_3$)$_2$CuLi (2)

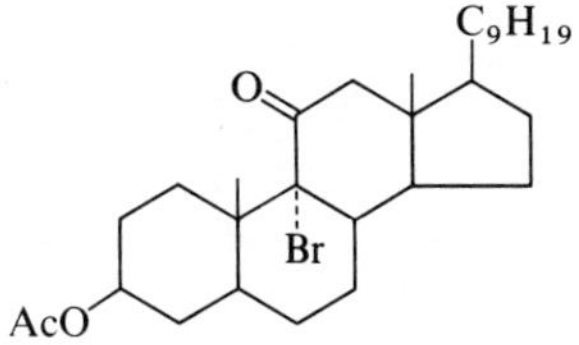

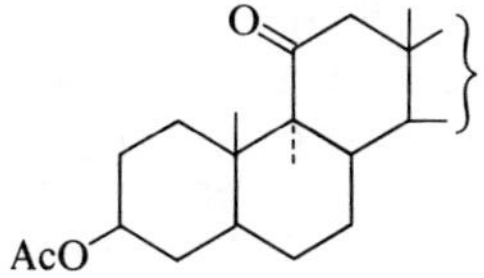

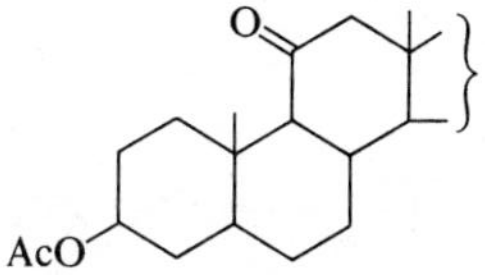

1. CH$_3$MgI
2. CH$_3$I
3. H$_2$SO$_4$
4. Ac$_2$O

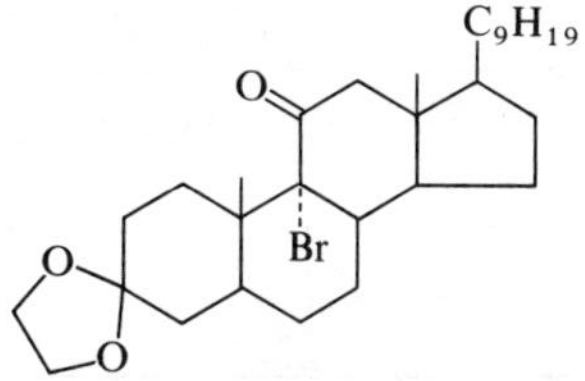

1. CH$_3$MgI
2. CH$_3$I

C$_9$H$_{19}$

C$_9$H$_{19}$

AcO

C$_{28}$

TABLE IV. REDUCTIVE ALKYLATION OF α-MONOHALO KETONES (*Continued*)

B. With Olefins or Acetylenes

Halo Ketone	Reactants (equiv.)	Conditions	Products and Yields (%)	Refs.
C$_8$ C$_6$H$_5$COCH$_2$Br	Zn/Cu, (I) (0.2),		C$_6$H$_5$COCH$_2$CH$_2$⟨(CH$_2$)$_n$⟩ (II),	
	NaI, NaHCO$_3$		C$_6$H$_5$COCH$_2$CH=⟨(CH$_2$)$_n$⟩ (III)	
	I, n = 1	DMSO, 60°, 1 hr	n = 1 II (43), III (5)	15
	I, n = 2	DMSO, 80°, 1 hr	II, n = 2 (54)	15
	I, n = 3	DMSO, 60°, 1 hr	n = 3 II (50), III (6)	15
	Zn/Cu, C$_6$H$_5$CH=CH$_2$ (0.5–3), NaI, collidine	DMSO, 90°, 30 min	(C$_6$H$_5$COCH$_2$CH$_2$CHC$_6$H$_5$)$_2$ (10), (C$_6$H$_5$COCH)$_2$)$_2$ (7)	15
	Zn/Cu, (C$_6$H$_5$)$_2$C=CH$_2$, NaI, NaHCO$_3$	DMSO, 140°, 1 hr	C$_6$H$_5$CO(CH$_2$)$_2$CH(C$_6$H$_5$)$_2$ (56)	15
	Zn/Cu, (p-CH$_3$OC$_6$H$_4$)$_2$C=CH$_2$, NaI, NaHCO$_3$	"	C$_6$H$_5$CO(CH$_2$)$_2$CH(C$_6$H$_4$OCH$_3$-p)$_2$ (54)	15
	Zn/Cu,	DMSO, 60°, 1 hr	(23),	15

(23)

NaI, NaHCO₃

Zn/Cu,

$CH_2=CRCR=CH_2$ (I) (0.2) —

$(C_6H_5COCH_2CH_2CR=CRCH_2)_2$ (II),

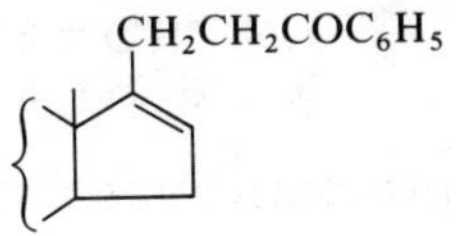

$C_6H_5CO(CH_2)_2CR(CR=CH_2)CH_2$—$CR=CR$ (III)

61

61

DMSO, 60°

I, R = CH₃

cis,cis-II/*cis,trans*-II/*trans,trans*-II (15:47:27) (55), III (7)

I, R = C₆H₅ "

cis,cis-II/*cis,trans*-II/*trans,trans*-II (28:60:12) (73)

Zn/Cu,

$CH_2=C(CH_3)C≡CR$ (I) (0.2) —

$[C_6H_5COCH_2CH_2C(CH_3)=C=CR]_2$ (II),

$C_6H_5CO(CH_2)_2C(CH_3)(C≡CR)CR=C$ (III),

$C_6H_5(CH_2)_2C(CH_3)C=C$

$[C_6H_5COCH_2CH_2C(CH_3)C≡CR]_2$ (IV)

DMSO

I, R = C₂H₅   II/III/IV (total 78)   61

I, R = C₆H₅   II/III/IV (total 91)   61

Zn/Cu,

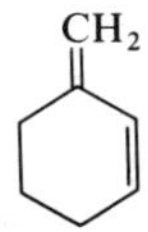

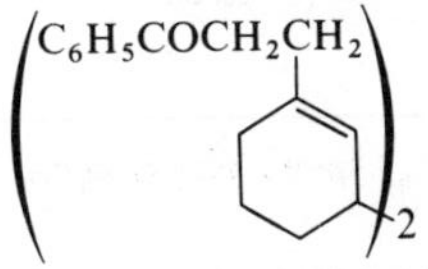

R = C₂H₅ II/III/IV (total 78)

R = C₆H₅ II/III/IV (total 91)

$C_6H_5COCH_2CH_2$ (62)

61

TABLE IV. REDUCTIVE ALKYLATION OF α-MONOHALO KETONES (*Continued*)

C. *With Aldehydes*

Halo Ketone	Reactants (equiv.)	Conditions	Products and Yields (%)	Refs.
C$_6$ 2-bromocyclohexanone (structure)	Zn, RCHO (I)		(structure) CHR (II)	
	I, R = CH$_3$	DMSO/C$_6$H$_6$ (1:10), 24 hr	II, R = CH$_3$ (57)[a]	57
	I, R = C$_2$H$_5$	DMSO/C$_6$H$_6$ (1:10)	II, R = C$_2$H$_5$ (67)[a]	57
	Zn (1.5), RCHO (I) (1.1), (C$_2$H$_5$)$_2$AlCl (1.1), CuBr (0.05)		(structure) CH(OH)R (II)	
	I, R = i-C$_3$H$_7$	THF/hexane, −20°	II, R = i-C$_3$H$_7$ (93)[b]	63
	I, R = C$_6$H$_5$	"	II, R = C$_6$H$_5$ (97)[c]	63
C$_7$ n-C$_3$H$_7$COCHBrC$_2$H$_5$	Zn, n-C$_3$H$_7$CHO	—	n-C$_3$H$_7$COCH(C$_2$H$_5$)CHOHC$_3$H$_7$-n (49)	62
C$_8$ (structure, 2-bromo-2-methylcyclohexanone)	Zn/Ag (1.5), C$_6$H$_5$CHO (1.1), (C$_2$H$_5$)$_2$AlCl (1.1), CuBr (0.05)	THF/hexane, −20°, 55 min	(structure) CHOHC$_6$H$_5$ (100)[d]	63
C$_8$ C$_6$H$_5$COCH$_2$Br	Zn (1.5), RCHO (I) (1.1), (C$_2$H$_5$)$_2$AlCl (1.1), CuBr (0.05)		C$_6$H$_5$COCH$_2$CHOHR (II)	
	I, R = i-C$_3$H$_7$	THF/hexane, −20°	II, R = i-C$_3$H$_7$ (92)	63
	I, R = C$_6$H$_5$	"	II, R = C$_6$H$_5$ (95)	63

Substrate	Reagents	Conditions	Product	Ref.
(2-bromocyclooctanone)	I, R = CH=CHC$_6$H$_5$	"	II, R = CH=CHC$_6$H$_5$ (92)	63
2-bromocyclooctanone	Zn, CH$_3$CHO	DMSO/C$_6$H$_6$ (1:10), 24 hr	[2-ethylidenecyclooctanone, =CHCH$_3$] (57)^a	57
C$_9$ n-C$_4$H$_9$COCHBrC$_3$H$_7$-n	Zn, RCHO (I) I, R = CH$_3$ I, R = n-C$_3$H$_7$	DMSO/C$_6$H$_6$ (1:10)	n-C$_4$H$_9$COCH(C$_3$H$_7$-n)CHOHR (II) II, R = CH$_3$ (63) II, R = n-C$_3$H$_7$ (58)	62 62
C$_{10}$ (2-bromo-5-methoxy-1-tetralone)	Zn, RCHO (I) I, R = C$_2$H$_5$	DMSO/C$_6$H$_6$ (1:10)	[5-methoxy, =CHC$_2$H$_5$] II (95)^a	57
"	"	DME, 3 hr	II + [—CHOHC$_2$H$_5$, OCH$_3$ structure]	58
"	"	DME, 18 hr	(total 95) II (70)^a	58
C$_{12}$ (2-bromocyclododecanone)	Zn, RCHO (I) I, R = CH$_3$ I, R = C$_2$H$_5$	DMSO/C$_6$H$_6$ (1:10) "	[cyclododecanone, =CHR] II, R = CH$_3$ (70)^a II, R = C$_2$H$_5$ (85)^a	57 57

TABLE IV. REDUCTIVE ALKYLATION OF α-MONOHALO KETONES (*Continued*)

C. With Aldehydes (*Continued*)

Halo Ketone	Reactants (equiv.)	Conditions	Products and Yields (%)	Refs.
C_{27} 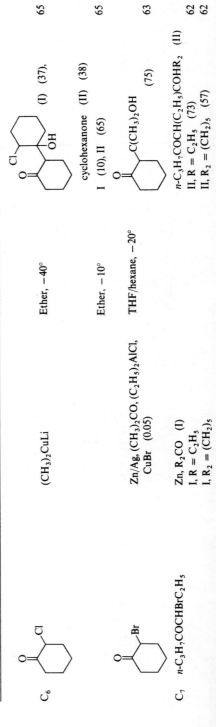	Zn, CH₃CHO	DMSO/C₆H₆ (1:10)	(97)	57, 58

D. With Ketones

C_6	(CH₃)₂CuLi	Ether, −40°	(I) (37),	65
			cyclohexanone (II) (38)	65
		Ether, −10°	I (10), II (65)	65
	Zn/Ag, (CH₃)₂CO, (C₂H₅)₂AlCl, CuBr (0.05)	THF/hexane, −20°	(75)	63
C_7 n-C₃H₇COCHBrC₂H₅	Zn, R₂CO (I) I, R = C₂H₅ I, R₂ = (CH₂)₅		n-C₃H₇COCH(C₂H₅)COHR₂ (II) II, R = C₂H₅ (73) II, R₂ = (CH₂)₅ (57)	62 62

262

	Substrate	Reagents	Conditions	Product (%)	Ref.
	[2-bromocycloheptanone]	Zn, $(CH_3)_2CO$	$DMSO/C_6H_6$ (1:10)	[cycloheptanone with $=C(CH_3)_2$ substituent] (16)[a]	57
C_8	$C_6H_5COCH_2Br$	Zn (1.5), cyclohexanone (1.1), $(C_2H_5)_2AlCl$ (1.1), $CuBr$ (0.05)	THF/hexane, $-20°$	$C_6H_5COCH_2$—[1-hydroxycyclohexyl] OH (83)	63
C_9	$n\text{-}C_4H_9COCHBrC_3H_7\text{-}n$	Zn, [cyclopropyl]—$COCH_3$	—	$n\text{-}C_4H_9COCH(C_3H_7\text{-}n)COHCH_3$ [cyclopropyl] (70)	62
C_{10}	[bicyclic H / Br structure]	Zn (1.5), $(CH_3)_2CO$ (1.1), $(C_2H_5)_2AlCl$ (1.1), $CuBr$ (0.05)	THF/hexane	[bicyclic structure with $C(CH_3)_2OH$ and H] (79)	63
C_{11}	$n\text{-}C_5H_{11}COCHBrC_4H_9\text{-}n$	Zn, $C_6H_5COCH_3$	THF/hexane	$n\text{-}C_5H_{11}COCH(C_4H_9\text{-}n)C(OH)$- $(CH_3)C_6H_5$ (53)	62

E. With Acyl Chlorides or Acid Anhydrides

	Substrate	Reagents	Conditions	Product (%)	Ref.
C_3	CH_3COCH_2Cl	$Na_2Fe(CO)_4$, $RCOCl$ (1), Py I, $R = CH_2CH(CH_3)_2$ I, $R = p\text{-}CH_3OC_6H_4$	THF, $25°$, 10 min "	$CH_3C(OCOR)\!=\!CH_2$ (II) II, $R = CH_2CH(CH_3)_2$ (97) II, $R = p\text{-}CH_3OC_6H_4$ (71)	66 66
C_5	$C_2H_5COCHBrCH_3$	Zn, $RCOCl$ (I) I, $R = CH_3$ I, $R = n\text{-}C_3H_7$ I, $R = C_6H_5$	$AcOC_2H_5$ " "	$C_2H_5COCH(COR)CH_3$ (II) II, $R = CH_3$ (46) II, $R = n\text{-}C_3H_7$ (49) II, $R = C_6H_5$ (51)	64 64 64

TABLE IV. REDUCTIVE ALKYLATION OF α-MONOHALO KETONES (*Continued*)

E. *With Acyl Chlorides or Acid Anhydrides* (*Continued*)

Halo Ketone	Reactants (equiv.)	Conditions	Products and Yields (%)	Refs.
C_6 (2-chlorocyclohexanone)	$(CH_3)_2CuLi$, CH_3COCl	Ether, $-10°$	(cyclohexanone with COCH$_3$) (50), cyclohexanone (50)	65
C_8 $C_6H_5COCH_2Cl$	$Na_2Fe(CO)_4$, $RCOCl$ (I) (1), Py	THF, $25°$, 10 min	$C_6H_5C(OCOR)=CH_2$ (II)	66
	I, R = CH_3	"	II, R = CH_3 (88)	66
	I, R = i-C_3H_7	"	II, R = i-C_3H_7 (79)	66
	I, R = n-C_4H_9	"	II, R = n-C_4H_9 (74)	66
	I, R = t-C_4H_9	"	II, R = t-C_4H_9 (87)	66
	I, R = C_6H_5	"	II, R = C_6H_5 (78)	66
C_9 n-$C_4H_9COCHBrC_3H_7$-n	Zn, $RCOCl$ (I) (1.5)	$AcOC_2H_5$	n-$C_4H_9COCH(COR)C_3H_7$-n (II)	64
	I, R = CH_3	"	II, R = CH_3 (56)	64
	I, R = n-C_3H_7	"	II, R = n-C_3H_7 (53)	64
	I, R = n-C_4H_9	"	II, R = n-C_4H_9 (48)	64
	I, R = C_6H_5	"	II, R = C_6H_5 (55)	64
C_{10} (2-bromo-5-methoxy-1-tetralone)			(1-acetoxy-5-methoxy-3,4-dihydronaphthalene) (I),	

	Reactant	Reagent	Solvent	Product	Refs.
		Zn, CH$_3$COCl	DME	*(2-acetyl-5-methoxy-1-tetralone)* (II) I (60)	58
		Zn, Ac$_2$O	"	I (30), II (30),	58
				(5-methoxy-1-tetralone) (40)	58
		Zn, RCOCl (I) (1.5)	AcOC$_2$H$_5$	n-C$_5$H$_{11}$COCH(COR)C$_4$H$_9$-n (II)	
		I, R = CH$_3$		II, R = CH$_3$ (52)	64
		I, R = n-C$_3$H$_7$	"	II, R = n-C$_3$H$_7$ (50)	64
		I, R = n-C$_4$H$_9$	"	II, R = n-C$_4$H$_9$ (45)	64
		I, R = C$_6$H$_5$	"	II, R = C$_6$H$_5$ (47)	64
C$_{11}$	n-C$_5$H$_{11}$COCHBrC$_4$H$_9$-n (I)				

F. With α-Bromo Ketones (Dimerization)

	Reactant	Reagent	Solvent	Product	Refs.
C$_4$	R^{1}COCHBrR^2 (I)	Ni(CO)$_4$ (0.6)	DMF, 30°, 5 hr	$R^1COCHR^2C(\overset{O}{\triangle})CHR^2$ (II), R^1	
	I, R^1 = R^2 = CH$_3$			II, R^1 = R^2 = CH$_3$ (70)	67
	I, R^1 = C$_2$H$_5$, R^2 = H	"		II, R^1 = C$_2$H$_5$, R^2 = H (52)	67
C$_5$	C$_2$H$_5$COCHBrCH$_3$	Mg	Ether	*(tetrasubstituted furan)* (I) (—)	150

TABLE IV. REDUCTIVE ALKYLATION OF α-MONOHALO KETONES (*Continued*)

F. With α-Bromo Ketones (Dimerization) (*Continued*)

Halo Ketone	Reactants (equiv.)	Conditions	Products and Yields (%)	Refs.
C5 C2H5COCHBrCH3 (*Contd.*)	Zn	—	I (47), (C2H5)2CO (1)	58
	"	—	I (98)	57
	Ni(CO)4 (0.6)	DMF, 30°, 5 hr	$C_2H_5COCH(CH_3)C\overset{O}{\diagup\diagdown}CHCH_3$ with C_2H_5 (84)	67
2-bromocyclopentanone	Zn	DMSO/C6H6 (1:10)	(dimer) (18), cyclopentanone (3)	58
C6 t-C4H9COCH2Br	Ni(CO)4 (0.6)	DMF, 30°, 5 hr	$t\text{-}C_4H_9COCH_2C\overset{O}{\diagup\diagdown}CH_2$ with $C_4H_9\text{-}t$ (61)	67
2-bromocyclohexanone	Zn	DMSO/C6H6 (1:10, 8 hr)	(dibenzofuran-type dimer) (62), cyclohexanone (6)	57

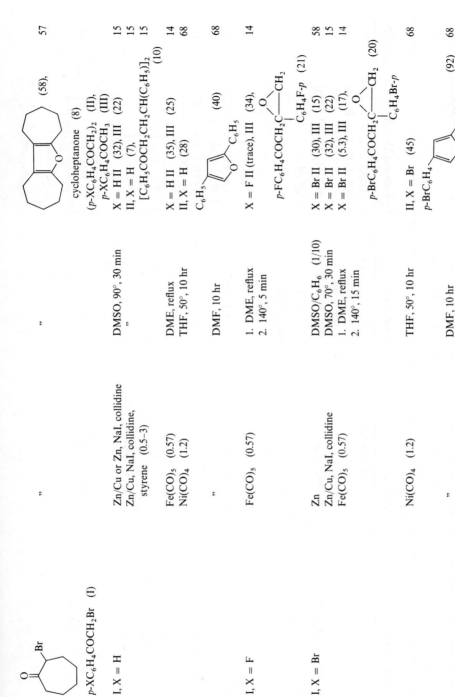

	Reagent	Conditions	Product(s) (%)	Refs.
C$_7$ (2-bromocycloheptanone structure)	"	"	cycloheptanone fused product (58),	57
C$_8$ p-XC$_6$H$_4$COCH$_2$Br (I)			cycloheptanone (8)	
			(p-XC$_6$H$_4$COCH$_2$)$_2$ (II), p-XC$_6$H$_4$COCH$_3$ (III)	
I, X = H	Zn/Cu or Zn, NaI, collidine	DMSO, 90°, 30 min	X = H II (32), III (22)	15
	Zn/Cu, NaI, collidine, styrene (0.5–3)	"	II, X = H (7), [C$_6$H$_5$COCH$_2$CH$_2$CH(C$_6$H$_5$)]$_2$ (10)	15
	Fe(CO)$_5$ (0.57)	DME, reflux	X = H II (35), III (25)	14
	Ni(CO)$_4$ (1.2)	THF, 50°, 10 hr	II, X = H (28)	68
	"	DMF, 10 hr	furan (C$_6$H$_5$-substituted) (40)	68
I, X = F	Fe(CO)$_5$ (0.57)	1. DME, reflux 2. 140°, 5 min	X = F II (trace), III (34), p-FC$_6$H$_4$COCH$_2$C(O epoxide)CH$_2$ C$_6$H$_4$F-p (21)	14
I, X = Br	Zn	DMSO/C$_6$H$_6$ (1/10)	X = Br II (30), III (15)	58
	Zn/Cu, NaI, collidine	DMSO, 70°, 30 min	X = Br II (32), III (22)	15
	Fe(CO)$_5$ (0.57)	1. DME, reflux 2. 140°, 15 min	X = Br II (5.3), III (17), p-BrC$_6$H$_4$COCH$_2$C(O epoxide)CH$_2$ C$_6$H$_4$Br-p (20)	14
	Ni(CO)$_4$ (1.2)	THF, 50°, 10 hr	II, X = Br (45)	68
	"	DMF, 10 hr	furan (p-BrC$_6$H$_4$, C$_6$H$_4$Br-p) (92)	68

TABLE IV. REDUCTIVE ALKYLATION OF α-MONOHALO KETONES (Continued)

F. With α-Bromo Ketones (Dimerization) (Continued)

Halo Ketone	Reactants (equiv.)	Conditions	Products and Yields (%)	Refs.
C$_8$ I, X = OCH$_3$ (Contd.)	Zn/Cu, NaI, collidine	DMSO, 170°, 60 min	X = OCH$_3$ II (25), III (31)	15
	Fe(CO)$_5$ (0.57)	1. DME, reflux 2. 140°, 5 min	X = OCH$_3$ II (51), III (25)	14
C$_9$ p-CH$_3$C$_6$H$_4$COCH$_2$Br	Ni(CO)$_4$ (1.2)	THF, 50°, 10 hr	(p-CH$_3$C$_6$H$_4$COCH$_2$)$_2$ (15)	68
"		DMF, 10 hr	(50)	68
C$_6$H$_5$COCHBrCH$_3$	Fe(CO)$_5$ (0.57)	1. DME, reflux 2. 140°, 5 min	(C$_6$H$_5$COCHCH$_3$)$_2$ (46), C$_6$H$_5$COC$_2$H$_5$ (31)	14
C$_{10}$ C$_6$H$_5$COCBr(CH$_3$)$_2$	"		[C$_6$H$_5$COC(CH$_3$)$_2$]$_2$ (47), C$_6$H$_5$COC$_3$H$_7$-i (14)	14
	Co$_2$(CO)$_8$ (0.05), C$_6$H$_5$CH$_2$N(C$_2$H$_5$)$_3$Cl, NaOD/D$_2$O	—	(—)	51
C$_{12}$	Fe(CO)$_5$ (0.57)	1. DME, reflux 2. 140°, 5 min	(6), (24)	14

	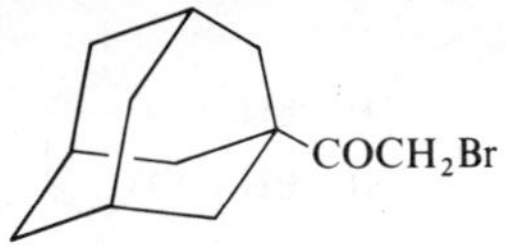	"	"	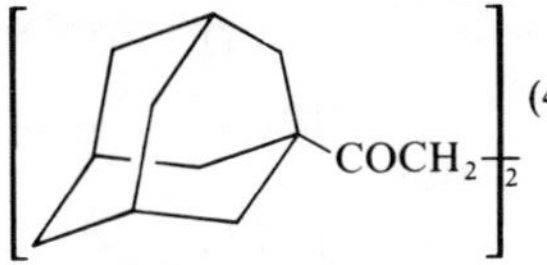 (4.6),	14
				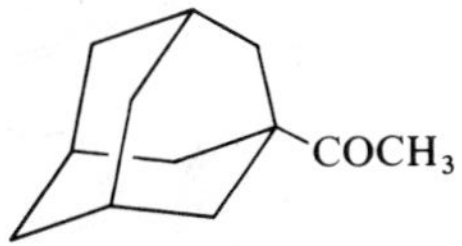 (13)	
C_{13}	$n\text{-}C_6H_{13}COCHBrC_5H_{11}\text{-}n$ (0.18*M*)	Zn	$DMSO/C_6H_6$ (1/10), 8 hr	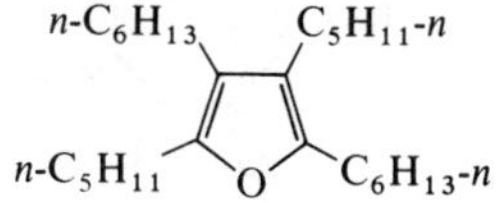 (30)	57
C_{14}	$[(CH_3)_2CCOCBr(CH_3)_2]_2$	Zn/Cu, $LiBF_4$	DME, 24 hr	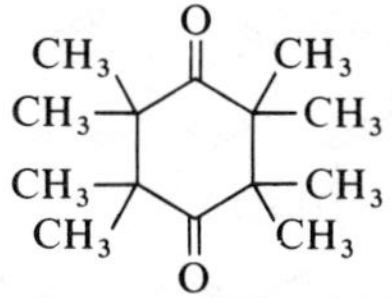 (3.6)	151
	$p\text{-}C_6H_5C_6H_4COCH_2Br$			$(p\text{-}C_6H_5C_6H_4COCH_2)_2$ (I), $p\text{-}C_6H_5C_6H_4COCH_3$ (II)	
		$Fe(CO)_5$ (0.57)	1. DME, reflux 2. 140°, 5 min	I (63), II (30)	14
		"	$CH_3CON(CH_3)_2$	I (—), II (—)	14
		"	$[(CH_3)_2N]_2CO$	I (—), II (—)	14
C_{16}	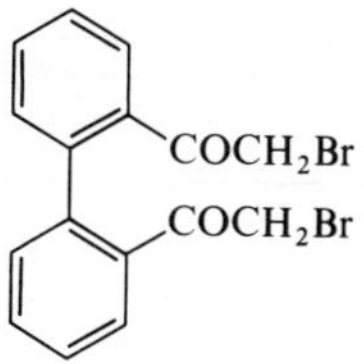	Zn/Cu, NaI, $NaHCO_3$ (2)	DMSO, 60°, 30 min	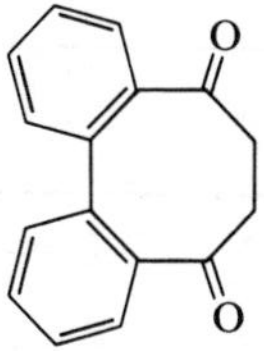 (48)	152

TABLE IV. REDUCTIVE ALKYLATION OF α-MONOHALO KETONES (*Continued*)

F. *With α-Bromo Ketones (Dimerization)* (*Continued*)

Halo Ketone	Reactants (equiv.)	Conditions	Products and Yields (%)	Refs.
C$_{16}$ (*Contd.*)	Fe(CO)$_5$ (0.57)	1. DME, reflux 2. 140°, 5 min	(15), (13)	14
	Zn/Cu, NaI, NaHCO$_3$ (2)	DMSO, 70°, 30 min	(56)	152
C$_{18}$	"	"	(64),	152

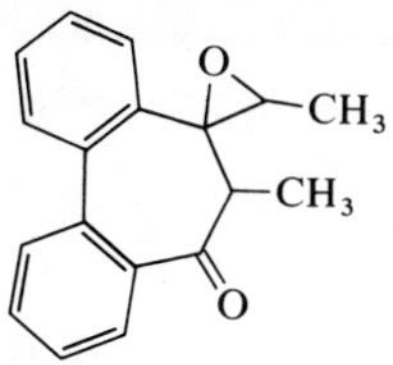

(17)

G. With α,β-Unsaturated Ketones

C$_5$ 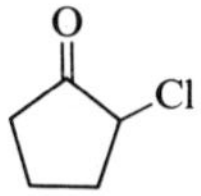		(CH$_3$)$_2$CuLi, MVK	Ether	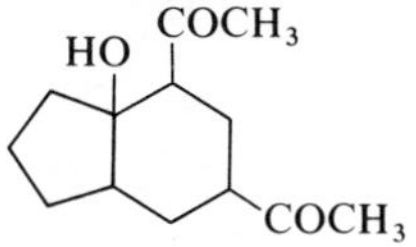 (—), 65
				(—)
C$_6$		"	"	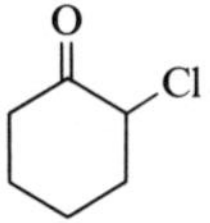 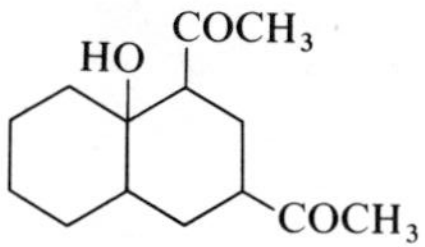 (50) 65
C$_7$		"	"	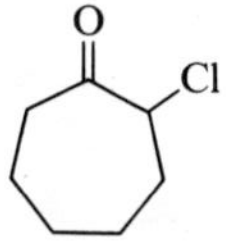 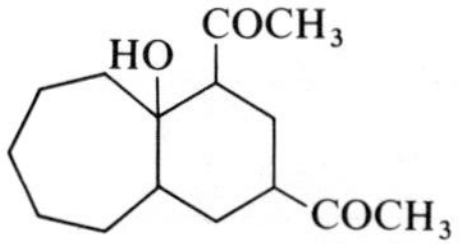 (—) 65
C$_9$	(CH$_3$)$_2$CHCH$_2$COCHBrC$_3$H$_7$-i	Mg, C$_6$H$_5$COCH=CHC$_6$H$_5$	"	(CH$_3$)$_2$CHCH$_2$COCHC$_3$H$_7$-i 69 C$_6$H$_5$COCH$_2$CHC$_6$H$_5$ (—)

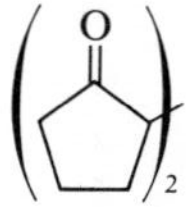

271

TABLE IV. REDUCTIVE ALKYLATION OF α-MONOHALO KETONES (*Continued*)

G. *With α,β-Unsaturated Ketones* (*Continued*)

Halo Ketone	Reactants (equiv.)	Conditions	Products and Yields (%)	Refs.
C₁₀	Zn, MVK	DME	(CH₂)₂COCH₃ (17), (35)	58
C₁₁	"	"	(CH₂)₂COCH₃ (25), (10)	58

C_{12}

(bromocyclododecanone structure: cyclododecane ring with O and Br)

" , " , "

H. With Arenesulfenyl Chlorides

cyclododecanone (45)

(cyclododecanone-derived structure with $(CH_2)_2COCH_3$) (20), 58

	Reactant	Conditions	Product	Ref.
C_5	$CH_3COCBr(CH_3)_2$	Zn, ArSCl (I), $HgCl_2$ (cat), Ether, reflux	$CH_3COC(SAr)(CH_3)_2$ (II)	
	I, Ar = C_6H_5	"	II, Ar = C_6H_5 (71)	70
	I, Ar = $p\text{-}ClC_6H_4$	"	II, Ar = $p\text{-}ClC_6H_4$ (68)	70
	I, Ar = $p\text{-}n\text{-}C_4H_9C_6H_4$	"	II, Ar = $p\text{-}n\text{-}C_4H_9C_6H_4$ (77)	70
	$C_2H_5COCHBrCH_3$	Zn, C_6H_5SCl, $HgCl_2$ (cat)	$C_2H_5COCH(SC_6H_5)CH_3$ (70)	70
C_9	$n\text{-}C_4H_9COCHBrC_3H_7\text{-}i$	"	$n\text{-}C_4H_9COCH(SC_6H_5)C_3H_7\text{-}i$ (69)	70
	$C_6H_5COCHBrCH_3$	Zn, ArSCl (I), $HgCl_2$ (cat)	$C_6H_5COCH(SAr)CH_3$ (II)	
	I, Ar = C_6H_5	"	II, Ar = C_6H_5 (73)	70
	I, Ar = $p\text{-}ClC_6H_4$	"	II, Ar = $p\text{-}ClC_6H_4$ (72)	70
C_{10}	$C_6H_5COCHBrC_2H_5$	Zn, C_6H_5SCl, $HgCl_2$ (cat)	$C_6H_5COCH(SC_6H_5)C_2H_5$ (75)	70
	$C_6H_5COCBr(CH_3)_2$	Zn, ArSCl (I), $HgCl_2$ (cat)	$C_6H_5COC(SAr)(CH_3)_2$ (II)	
	I, Ar = C_6H_5	"	II, Ar = C_6H_5 (71)	70
	I, Ar = $p\text{-}ClC_6H_4$	"	II, Ar = $p\text{-}ClC_6H_4$ (73)	70
C_{11}	$C_6H_5COCHBrC_3H_7\text{-}n$	Zn, ArSCl (I), $HgCl_2$ (cat)	$C_6H_5COCH(SAr)C_3H_7\text{-}n$ (II)	
	I, Ar = C_6H_5	"	II, Ar = C_6H_5 (68)	70
	I, Ar = $p\text{-}ClC_6H_4$	"	II, Ar = $p\text{-}ClC_6H_4$ (71)	70

TABLE IV. REDUCTIVE ALKYLATION OF α-MONOHALO KETONES (Continued)

1. With Miscellaneous Substrates

Halo Ketone	Reactants (equiv.)	Conditions	Products and Yields (%)	Refs.
C_{10}	Zn, $CH_2=CHCN$	DME	(4)	58
	Zn, CO_2	1. $DMSO/C_6H_6$ (1:10) 2. CH_3OH, H^+	(I) $R = CH_3$ (—)	58
	Zn, $(CH_3O)_2CO$	$DMSO/C_6H_6$ (1:10)	I, $R = CH_3$ (5)	58
	Zn, $(C_2H_5O)_2POCO_2C_2H_5$	1. DME 2. C_2H_5OH, H_2SO_4	I, $R = C_2H_5$ (50),	58
			(40)	

[a] Product was obtained by treatment of the crude aldol with p-toluenesulfonic acid in benzene.

[b] Ratio of *erythro* to *threo* isomer was not determined.

[c] A 1:1 mixture of *erythro* and *threo* isomers was obtained.

[d] A 4:1 mixture of *erythro* and *threo* isomers was obtained.

274

TABLE V. REACTIONS OF α,α-DIHALO KETONES

Halo Ketone	Reactants (equiv.)	Conditions	Products and Yields (%)	Refs.
C₇	(CH₃)₂CuLi	THF or ether, −78°	(—)	73
	Zn, NH₄Cl	CH₃OH	(poor)	153
	Zn	—	(I)	154
			$R^1 = H, R^2 = Cl$ (—)	
			I, $R^1 = R^2 = H$ (56)	
C₈	"	AcOH		154
	"	AcOH, reflux, 17 hr	(66)	155
	"	AcOH, 40°, 1 hr	(78)	156
C₉	"	AcOH, reflux	(I),	155

275

TABLE V. REACTIONS OF α,α-DIHALO KETONES (*Continued*)

Halo Ketone	Reactants (equiv.)	Conditions	Products and Yields (%)	Refs.
C_9 (*Contd.*) [structure: bicyclic Cl,Cl ketone]	Zn	AcOH, reflux	[structure] Cl (II); I/II (70:9) (90)	155
[structure: bicyclic O, Cl,Cl ketone]	"	"	[structure] (90)	73
[structure: H, Cl, Cl bicyclic ketone]	"	AcOH, 70°	[structure] O (I); R^1R^2, $R^1 = R^2 = H$ (90); I, $R^1 = H$, $R^2 = Cl$ (ca. 100)	73
	$(CH_3)_2CuLi$ (1–2)	1. THF, $-78°$; 2. AcOH or H_2O		73
	1. $(CH_3)_2CuLi$ (1–2); 2. LiI·2H$_2$O	THF or ether, $-78°$; HMPA, $-78°$ to room temp.	[structure] O (60)	73
	1. $(CH_3)_2CuLi$ (1–2); 2. CH$_3$I	THF, $-78°$; HMPA, -78 to $-40°$	[structure] O (I); R Cl; $R = CH_3$ (78)	73

276

Starting material	Reagents	Conditions	Product(s) (yield %)	Ref.
C_{10} (chlorocyclopentanone fused bicyclic structure)	1. $(CH_3)_2CuLi$ (1–2) 2. $CH_2=CHCH_2Br$	THF, -78 to $-40°$ HMPA, -78 to $-40°$	**I**, $R = CH_2CH=CH_2$ (71)	73
	1. $(CH_3)_2CuLi$ (1–2) 2. CH_3I	THF, $-78°$ HMPA, $-78°$ to room temp.	**(I)** $R = CH_3$ (78)	73
	1. $(CH_3)_2CuLi$ (1–2) 2. $CH_2=CHCH_2I$	THF, $-78°$ HMPA, $-78°$ to room temp.	**I**, $R = CH_2CH=CH_2$ (68)	73
(bicyclic chloro ketone)	Zn	Aq AcOH	R^1R^2 ... =O **(I)** $R^1 = H, R^2 = Cl$ (67)	71
(dibromo structure) **(I)**	1. $(CH_3)_2CuLi$ (1–2) 2. CH_3I	THF, $-78°$ HMPA, -78 to $-40°$	$R^1 = CH_3, R^2 = Cl$ (70)	73
	1. $(CH_3)_2CuLi$ (1–2) 2. CH_3I	THF, $-78°$ HMPA, $-78°$ to room temp.	CH_3-substituted azulenone (65); **(II)**, **(III)**	73

TABLE V. REACTIONS OF α,α-DIHALO KETONES (*Continued*)

Halo Ketone	Reactants (equiv.)	Conditions	Products and Yields (%)	Refs.
C_{10} (*Contd.*)	$(C_2H_5)_2Zn$	C_6H_6	II (100)	16
	"	C_6H_6, reflux, 22 hr	II (10), III (80)	16
	$(C_2H_5)_2Zn$ (0.5)	"	I (5), II (11), III (60)	16
	Zn	DMF	III (ca. 75)	16
	Zn/Cu	"	III (ca. 35)	16
C_{11}	Zn	Aq AcOH	(I)	71, 72
	"	AcOH, 70°, 15 hr	$R^1 = H, R^2 = Cl$ (>64)	71
	$(CH_3)_2CuLi$ (1–2)	THF or ether, −78°	I, $R^1 = R^2 = H$ (82)	73
	$(CH_3)_2CuLi$ (1–2)	"	I, $R^1 = H, R^2 = Cl$ (—)	73
	1. $(CH_3)_2CuLi$ (1–2) 2. CH_3I	HMPA, −78 to −40°	I, $R^1 = CH_3, R^2 = H$ (>49)	73
C_{12}	Zn	AcOH, 70°	(74)	73

278

TABLE VI. REDUCTIVE ALKYLATION OF α,α'-DIHALO KETONES

A. α-Monoalkylation

	Halo Ketone	Reactants (equiv.)	Conditions	Products and Yields (%)	Refs.
C_5	$(CH_3CHBr)_2CO$	$(CH_3)_2CuLi$	Pentane, $-50°$, 30 min	$C_2H_5COC_3H_7$-i (54)	60, 78
	$(CH_3)_2CBrCOCH_2Br$	$R(t\text{-}C_4H_9)CuLi$ (I)		i-$C_3H_7COCH_2R$ (II), $(CH_3)_2CRCOCH_3$ (III)	
		I, R = n-C_4H_9	THF, $-78°$, 30 min	R = n-C_4H_9 II (55), III (25)	60, 78
		I, R = $CH(CH_3)C_2H_5$	'', $-78°$,	R = $CH(CH_3)C_2H_5$ II (42), III (8)	60, 78
		I, R = t-C_4H_9	THF, $-78°$,	R = t-C_4H_9 II (56), III (10)	60, 78
	(I)	t-C_4H_9OK (2), $(C_2H_5)_3B$	THF, $20°$, 2 hr	(31), cyclohexanone (37)	76
C_6		R^1R^2CuLi (II)		(III)	
		II, $R^1 = n$-C_4H_9, $R^2 = n$-C_4H_9	Ether, $-78°$, 30 min	III, R = n-C_4H_9 (81)	78, 60
		II, $R^1 = n$-C_4H_9, $R^2 = t$-C_4H_9O	THF, $-78°$, 30 min	III, R = n-C_4H_9 (77), cyclohexanone (—)	78, 60
	cis- and trans-I	II, $R^1 = CH(CH_3)C_2H_5$, $R^2 = CH(CH_3)C_2H_5$	Ether, $-78°$, 30 min	III, R = $CH(CH_3)C_2H_5$ (38)	78, 60
		II, $R^1 = CH(CH_3)C_2H_5$, $R^2 = t$-C_4H_9O	THF, $-78°$, 30 min	III, R = $CH(CH_3)C_2H_5$ (75), cyclohexanone (—)	78, 60
		II, $R^1 = t$-C_4H_9, $R^2 = CH_3$	Ether, $-78°$, 30 min	III, R = t-C_4H_9 (20)	78, 60

279

TABLE VI. REDUCTIVE ALKYLATION OF α,α'-DIHALO KETONES (*Continued*)

A. α-Monoalkylation (*Continued*)

Halo Ketone	Reactants (equiv.)	Conditions	Products and Yields (%)	Refs.
C_6 (*Contd.*)	II, $R^1 = t\text{-}C_4H_9$, $R^2 = t\text{-}C_4H_9$	Ether, $-78°$, 30 min	III, $R = t\text{-}C_4H_9$ (18)	78, 60
	II, $R^1 = t\text{-}C_4H_9$, $R^2 = t\text{-}C_4H_9O$	THF, $-78°$, 30 min	III, $R = t\text{-}C_4H_9$ (65), cyclohexanone (—)	78, 60
	II, $R^1 = t\text{-}C_4H_9$, $R^2 = C_6H_5S$	"	III, $R = t\text{-}C_4H_9$ (63)	78, 60
cis-I	CH_3MgI (2)	Ether, 25°, 30 min	III, $R = CH_3$ (<10)	78
	II, $R^1 = R^2 = CH_3$	Ether, $-78°$, 30 min	III, $R = CH_3$ (70)	60
	"	1. Ether, $-78°$, 5 min 2. 25°, 1 hr	III, $R = CH_3$ (95)	78, 60
trans-I	$t\text{-}C_4H_9OK$ (2), $(C_2H_5)_3B$	THF, 20°, 2 hr	III, $R = C_2H_5$ (76), cyclohexanone (12)	76, 77
	CH_3MgI	Ether, $-78°$, 30 min	III, $R = CH_3$ (<1)	60
	$(CH_3)_2CuLi$	"	III, $R = CH_3$ (98)	78, 60
C_7 $n\text{-}C_4H_9CHBrCOCH_2Br$	$R(t\text{-}C_4H_9O)CuLi$ (I)		$n\text{-}C_5H_{11}COCH_2R$ (II), $n\text{-}C_4H_9CHRCOCH_3$ (III)	
	I, $R = n\text{-}C_4H_9$	THF, $-78°$, 30 min	$R = n\text{-}C_4H_9$, II (37), III (11)	78, 60
	I, $R = CH(CH_3)C_2H_5$	"	$R = CH(CH_3)C_2H_5$, II (49), III (8)	78, 60
	I, $R = t\text{-}C_4H_9$	THF, $-78°$	$R = t\text{-}C_4H_9$, II (53), III (6)	78, 60
$[(CH_3)_2CBr]_2CO$	CH_3MgI (2)	Ether, 25°, 30 min	$i\text{-}C_3H_7COC_4H_9\text{-}t$ (I) (70)	78
	$(CH_3)_2CuLi$ (2)	Pentane, $-50°$, 30 min	I (54)	78, 60
	1. $(CH_3)_2CuLi$ 2. D_2O	Ether, 0°	$(CH_3)_2CDCOC_4H_9\text{-}t$ (87)	80
	$R((t\text{-}C_4H_9O)CuLi$ (I)			

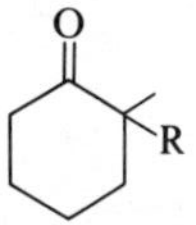

I, R = n-C₄H₉
I, R = CH(CH₃)C₂H₅
I, R = t-C₄H₉

THF, −78°, 30 min
"
"

(III)

R = n-C₄H₉ II (48), III (16)
R = CH(CH₃)C₂H₅ II (61), III (8)
R = t-C₄H₉ II (31), III (2)

78, 60
78, 60
78, 60

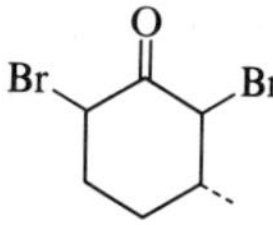

R₃COK (I) (2),
R₃B (II)

(III),

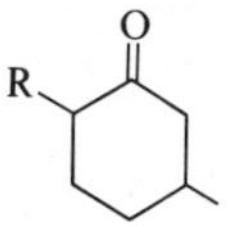

(IV)

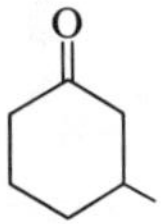

I, R = CH₃,
II, R = C₂H₅
I, R = C₂H₅,
II, R = C₂H₅
I, R = CH₃,
II, R = n-C₄H₉

THF, 20°
"
"

III, R = C₂H₅ (35), IV (52)
III, R = C₂H₅ (61), IV (12)
III, R = n-C₄H₉ (65), IV (10)

76
76
76, 77

R(t-C₄H₉O)CuLi (I)

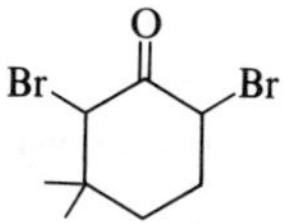

(II),

(III)

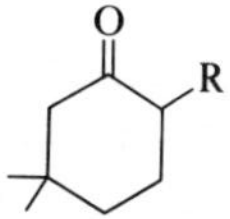

C₈

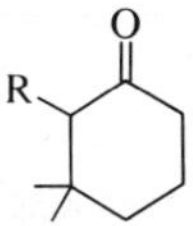

I, R = n-C₄H₉
I, R = t-C₄H₉

THF, −78°, 30 min
"

R = n-C₄H₉ II (31), III (5)
R = t-C₄H₉ II (33), III (10)

78, 60
78, 60

281

TABLE VI. REDUCTIVE ALKYLATION OF α,α'-DIHALO KETONES (Continued)

A. α-Monoalkylation (Continued)

Halo Ketone	Reactants (equiv.)	Conditions	Products and Yields (%)	Refs.
C₈ (Contd.)	CH₃MgI (2)	Ether, 25°, 30 min	(I) (65)	78
	(CH₃)₂CuLi (2)	1. Ether, −78°, 5 min 2. 25°, 1 hr	I (80)	78, 60
C₉ (n-C₃H₇CHBr)₂CO	CH₃MgI (2) R¹R²CuLi (I)	Ether, 25°, 30 min	n-C₄H₉COCH(CH₃)C₃H₇-n (17) n-C₄H₉COCHRC₃H₇-n (II)	78 78, 60
	I, R¹ = R² = CH₃		II, R = CH₃ (70)	78, 60
	I, R¹ = n-C₄H₉, R² = t-C₄H₉	Ether, −78°, 30 min THF, −78°, 30 min	II, R = n-C₄H₉ (75)	78, 60
	I, R¹ = CH(CH₃)C₂H₅, R² = t-C₄H₉	"	II, R = CH(CH₃)C₂H₅ (67)	78, 60
	I, R¹ = t-C₄H₉, R² = t-C₄H₉	"	II, R = t-C₄H₉ (60)	78, 60
	(t-C₄H₉)₂CuLi			
(i-C₃H₇CHBr)₂CO	Pentane, −78°	i-C₃H₇CH(C₄H₉-t)COCH₂CH(CH₃)₂ (60)	80
		—	(—)	157

282

C$_{10}$

(I)

RONa (II) (2),
R$_3$B (III)

(IV),

(V)

	Conditions	Products	Refs.
cis-I			
II, R = C(CH$_3$)$_2$C$_2$H$_5$ III, R = n-C$_4$H$_9$	THF, 20°	IV, R = n-C$_4$H$_9$ (75), V (4)	76
II, R = C(C$_2$H$_5$)$_3$ III, R = n-C$_4$H$_9$	"	"	76
t-C$_4$H$_9$OK (1.5), (n-C$_4$H$_9$)$_3$B	"	IV, R = n-C$_4$H$_9$ (67), V (8)	76
ROK (II) (2), R$_3$B (III)			
II, R = t-C$_4$H$_9$ III, R = C$_2$H$_5$	"	IV, R = C$_2$H$_5$ (85), V (10)	76, 77
II, R = t-C$_4$H$_9$ III, R = n-C$_3$H$_7$	"	IV, R = n-C$_3$H$_7$ (83), V (10)	76
II, R = t-C$_4$H$_9$ III, R = n-C$_4$H$_9$	THF, 0°, 0.2 hr	IV, R = n-C$_4$H$_9$ (4), V (—)	76
"	THF, 0°, 1 hr	IV, R = n-C$_4$H$_9$ *cis*-IV (15), V (—)	76
"	THF, 0°, 1.5 hr	IV, R = n-C$_4$H$_9$ *cis*-IV (25), V (—)	76
"	THF, 0°, 4 hr	IV, R = n-C$_4$H$_9$ (63)	76
"	THF, 20°, 0.1 hr	R = n-C$_4$H$_9$ *cis*-IV (25), *trans*-IV (6), V (6)	76
"	THF, 20°, 0.5 hr	R = n-C$_4$H$_9$ *cis*-IV (50), *trans*-IV (12), V (9)	76
"	THF, 20°, 1–24 hr	R = n-C$_4$H$_9$ *cis*-IV (70), *trans*-IV (15), V (10)	76
"	THF, 65°, 0.2–4 hr	IV, R = n-C$_4$H$_9$ (40), V (—)	76
"	THF, 70°	IV, R = n-C$_4$H$_9$ (51), V (—)	76

283

TABLE VI. REDUCTIVE ALKYLATION OF α,α'-DIHALO KETONES (*Continued*)

A. α-*Monoalkylation* (*Continued*)

Halo Ketone	Reactants (equiv.)	Conditions	Products and Yields (%)	Refs.
C_{10} (*Contd.*)	II, R = $C(CH_3)_2C_2H_5$ III, R = $n\text{-}C_4H_9$	THF, 20°	IV, R = $n\text{-}C_4H_9$ (95), V (5)	76
	II, R = $C(C_2H_5)_3$ III, R = $n\text{-}C_4H_9$	"	IV, R = $n\text{-}C_4H_9$ (100)	76, 77
	II, R = $t\text{-}C_4H_9$ III, R = $n\text{-}C_4H_9$ $t\text{-}C_4H_9OH$ (0.5)	"	IV, R = $n\text{-}C_4H_9$ (80), V (20)	76
	II, R = $t\text{-}C_4H_9$ III, R = $n\text{-}C_4H_9$ $t\text{-}C_4H_9OH$ (2)	"	IV, R = $n\text{-}C_4H_9$ (20), V (26)	76
	II, R = $t\text{-}C_4H_9$ III, R = $n\text{-}C_4H_9$	DME, 20°	IV, R = $n\text{-}C_4H_9$ (trace)	76
	"	C_6H_6, 20°	"	76
	II, R = $t\text{-}C_4H_9$ III, R = $CH_2CH(CH_3)_2$	THF, 20°	IV, R = $CH_2CH(CH_3)_2$ (trace)	76
	II, R = $C(C_2H_5)_3$ III, R = $CH_2CH(CH_3)_2$	"	IV, R = $CH_2CH(CH_3)_2$ (5)	76
	II, R = $C(C_2H_5)_3$ III, R = $CH(CH_3)C_2H_5$	"	IV, R = $CH(CH_3)C_2H_5$ (8), V (1)	76
	II, R = $C(C_2H_5)_3$, [$CH(CH_3)C_2H_5$]B-bbn	"	IV, R = $CH(CH_3)C_2H_5$ (6), V (27)	76
	II, R = $C(C_2H_5)_3$, ($c\text{-}C_6H_{11}$)B-bbn	"	IV, R = $c\text{-}C_6H_{11}$ (7)	76
trans-I	II, R = $t\text{-}C_4H_9$ III, R = C_2H_5	"	V (10)	76

284

Substrate	Reagent	Conditions	Product(s) and Yield(s) (%)	Refs.
C₁₁ (t-C₄H₉CHBr)₂CO	$(\triangleright)_2$CuLi	—	(—)	157
	1. R₂CuLi (I) 2. D₂O I, R = CH₃ I, R = t-C₄H₉	Ether, 0° Pentane, −78°	t-C₄H₉CHRCOCHDC₄H₉-t (II) II, R = CH₃ (85) II, R = t-C₄H₉ (50)	80 80
C₁₂	CH₃MgI (2)	Ether, 25°, 30 min	(I) (86) I (97)	78
	(CH₃)₂CuLi (2.5)	1. Ether, −78°, 30 min 2. H₂O		78, 60
	"	1. Ether, −78°, 30 min 2. D₂O	I-d^a (91)	78, 60
C₁₅ (C₆H₅CHBr)₂CO	CH₃MgI (CH₃)₂CuLi	Ether, 25°, 30 min Ether, −78°, 30 min	C₆H₅CH₂COCH(CH₃)C₆H₅ (I) (<20) I (72), (C₆H₅CH₂COCHC₆H₅)₂ (18)	78, 60 78, 60
C₂₇	"	"	(99)^b	78, 60

TABLE VI. REDUCTIVE ALKYLATION OF α,α'-DIHALO KETONES (*Continued*)

A. α-Monoalkylation (*Continued*)

Halo Ketone	Reactants (equiv.)	Conditions	Products and Yields (%)	Refs.
C_{27} (*Contd.*)	t-C_4H_9OK (2), (n-$C_4H_9)_3B$	THF, 20°	(15), (20)	76

B. α,α'-Dialkylation

C_5 $(CH_3CHBr)_2CO$	1. R_2CuLi (I) 2. RX (II)		$(CH_3CHR)_2CO$ (III), $C_2H_5COCHRCH_3$ (IV)	
	R = CH_3, I, II	Ether, 0°	R = CH_3, III (48), IV (22)	80, 55
	R = C_2H_5, I, II	"	R = C_2H_5, III (60), IV (25)	80
	R = i-C_3H_7, I, II	Pentane, −78°	R = i-C_3H_7, III (8), IV (86)	80, 55
C_6	(n-$C_4H_9)RCuLi$ (I)		(II),	

286

	Reagent (and conditions of step)	Conditions	Product(s) (%)	Yield (%)
	1. I, R = n-C$_4$H$_9$ 2. HMPA 3. CH$_3$I	THF, −78°, 30 min −78°, 30 min 25°, 30 min	R = CH$_3$ cis-II (40), trans-II (10)[c]	81
	1. I, R = t-C$_4$H$_9$O 2. CH$_3$I 3. NH$_4$Cl	THF, −78°, 30 min 0°, 1 hr H$_2$O/ether, 1 hr	R = CH$_3$ cis-II (18), trans-II (18), III (1)	81
	1. I, R = n-C$_4$H$_9$ 2. n-C$_4$H$_9$I, HMPA	THF, −78°, 30 min 25°, 2 hr	R = n-C$_4$H$_9$ II/III (total 31)	81
	1. I, R = t-C$_4$H$_9$O 2. n-C$_4$H$_9$I, HMPA 3. NH$_4$Cl	THF, −78°, 30 min 25°, 2 hr H$_2$O/ether, 1 hr	R = n-C$_4$H$_9$ II (28), III (17)[d]	81
	1. R = t-C$_4$H$_9$O 2. n-C$_3$H$_7$CHO 3. NH$_4$Cl	THF, −78°, 30 min THF, −78°, 5 min H$_2$O/ether, 1 hr	(43), (4)	81
C$_7$ (C$_2$H$_5$CHBr)$_2$CO	1. R$_2$CuLi (I) 2. CH$_3$I I, R = CH$_3$ I, R = C$_2$H$_5$	Ether, 0° "	C$_2$H$_5$CHR^{1}COCHR^{2}C$_2$H$_5$ (II) II, R^1 = R^2 = CH$_3$ (87) II, R^1 = CH$_3$, R^2 = C$_2$H$_5$ (74)	80 80
[(CH$_3$)$_2$CBr]$_2$CO	1. R$_2$CuLi (I) 2. RX (II) R = CH$_3$,I, II R = i-C$_3$H$_7$,I, II	Ether, 0° "	(CH$_3$)$_2$CRCOCR(CH$_3$)$_2$ (III) III, R = CH$_3$ (33) III, R = i-C$_3$H$_7$ (11), (CH$_3$)$_2$C(i-C$_3$H$_7$)COC$_3$H$_7$-i (29)	55 80

TABLE VI. REDUCTIVE ALKYLATION OF α,α'-DIHALO KETONES (*Continued*)

B. α,α'-*Dialkylation* (*Continued*)

Halo Ketone	Reactants (equiv.)	Conditions	Products and Yields (%)	Refs.
C$_9$ $(n\text{-}C_3H_7CHBr)_2CO$	1. $(CH_3)_2CuLi$ 2. CH_3I	Ether, $-78°$, 5 min $25°$, 12 hr	$n\text{-}C_3H_7\text{-}CHR^1COCHR^2C_3H_7\text{-}n$ (I) I, $R^1 = R^2 = CH_3$ (62)	78, 60
	1. $(CH_3)_2CuLi$ 2. $C_6H_5CH_2Br$	— —	I, $R^1 = CH_3$, $R^2 = CH_2C_6H_5$ (—)	78, 60
$(i\text{-}C_3H_7CHBr)_2CO$	1. R_2CuLi (I) 2. RX (II) $R = CH_3, I, II$ $R = i\text{-}C_3H_7, I, II$	Ether, $0°$ Pentane, $-78°$	$(i\text{-}C_3H_7CHR)_2CO$ (III), $i\text{-}C_3H_7CHRCOCH_2C_3H_7\text{-}i$ (IV) $R = CH_3$, III (16), IV (76) $R = i\text{-}C_3H_7$ III (12), IV (78)	80 80, 55
C$_{11}$ $(t\text{-}C_4H_9CHBr)_2CO$	1. R_2CuLi (I) 2. RX (II) $R = CH_3, I, II$ $R = C_2H_5, I, II$	Ether, $0°$ ”	$(t\text{-}C_4H_9CHR)_2CO$ (III), $t\text{-}C_4H_9CHRCOCH_2C_4H_9\text{-}t$ (IV) III, $R = CH_3$ (80) $R = C_2H_5$ III (18), IV (42)	80 80
C$_{12}$	1. $(CH_3)_2CuLi$ 2. CH_3I	THF, $-78°$, 30 min $25°$, 12 hr	 I, $R = CH_3$ (97)	78, 60
	1. $(CH_3)_2CuLi$ 2. $C_6H_5CH_2Br$	— —	I, $R = CH_2C_6H_5$ (—)	78, 60

288

C. Reductive Dimerization

		Reagent, conditions	Product(s) (yield %)	Refs.
C$_5$	(CH$_3$CHBr)$_2$CO	Zn/Cu, HCONHCH$_3$, 25°	(C$_2$H$_5$COCHCH$_3$)$_2$ (55)	82
	(CH$_3$)$_2$CBrCOCH$_2$Br	", HCONHCH$_3$, 25°, 24 hr	i-C$_3$H$_7$COCH$_2$C(CH$_3$)$_2$COCH$_3$ (I), (i-C$_3$H$_7$COCH$_2$)$_2$ (II), [CH$_3$COC(CH$_3$)$_2$]$_2$ (III) I/II/III (39:38:23) (36)	82
C$_6$	(CH$_3$)$_2$CBrCOCHBrCH$_3$	", HCONHCH$_3$, $-5°$, 2 hr	i-C$_3$H$_7$COCH(CH$_3$)C(CH$_3$)$_2$COC$_2$H$_5$ (I), [C$_2$H$_5$COC(CH$_3$)$_2$]$_2$ (II), [i-C$_3$H$_7$COCHCH$_3$]$_2$ (III) I/II/III (56:25:18) (65)	82
C$_7$	[(CH$_3$)$_2$CBr]$_2$CO	"	[i-C$_3$H$_7$COC(CH$_3$)$_2$]$_2$ (71), CH$_2$=C(CH$_3$)COC$_3$H$_7$-i (10)	82
		Zn/Cu, LiClO$_4$, DME, overnight	(2.5)	151
C$_{12}$		Zn/Cu, HCONHCH$_3$, 50°, 24 hr	(25)	82

TABLE VI. REDUCTIVE ALKYLATION OF α,α'-DIHALO KETONES (*Continued*)

C. *Reductive Dimerization* (*Continued*)

Halo Ketone	Reactants (equiv.)	Conditions	Products and Yields (%)	Refs.
C_{15} $(C_6H_5CHBr)_2CO$			$(C_6H_5CH_2COCHC_6H_5)_2$ (I),	
	Zn	CH_3CN, reflux	I (60)	84
	Zn	$CH_2{=}CHCN$, reflux	I (—), II (—)	84
	Zn/Cu	$HCONHCH_3$, 25°, 12 hr	I (43)	82
	"	$CH_2{=}CHCN$, reflux, 6 hr	I (22)	85

290

85

"   CH$_3$CN, reflux, 1 week   II (60)

(CH$_3$)$_2$CuLi,   Ether, −78°, 30 min   III (18), C$_6$H$_5$CH$_2$COCH(CH$_3$)C$_6$H$_5$ (72)   78, 60
NaI   Acetone, 1 week   III (41)   85
"   CS$_2$/acetone, reflux, 0.5 hr   III (60),   85

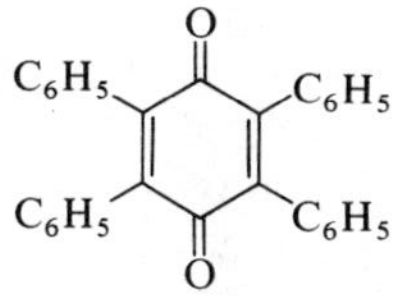

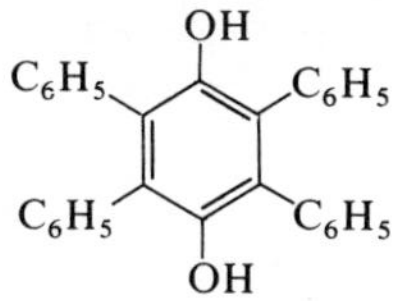

(structures: 2,3,5,6-tetraphenyl-1,4-benzoquinone (—), and 2,3,5,6-tetraphenyl-1,4-dihydroxybenzene (—))

---

[a] The product contained 95 % deuterium at C-12.

[b] The product was a mixture of 2α- and 4α-methyl-5α-cholestan-3-one.

[c] 2-Methylcyclohexanone (3 %) was also formed.

[d] 2,2,6-Tri-n-butylcyclohexanone was also formed in 13 % yield.

TABLE VII. [3 + 4] CYCLOCOUPLING OF α,α'-DIHALO AND POLYHALOGENATED KETONES

A. With Open-Chain 1,3-Dienes

Halo Ketone	Reactants (equiv.)	Conditions	Products and Yields (%)	Refs.
C$_5$ (CH$_3$CHBr)$_2$CO	Fe$_2$(CO)$_9$, CH$_2$=CR^{1}CR^2=CH$_2$ (I)		(II)	
	I, R^1 = R^2 = H	C$_6$H$_6$, 80°, 16 hr	II, R^1 = R^2 = H (30)	74, 86
	I, R^1 = H, R^2 = CH$_3$	C$_6$H$_6$, 25°, 36 hr, hv[a]	II, R^1 = H, R^2 = CH$_3$ (36)	74, 86
	I, R^1 = R^2 = CH$_3$	C$_6$H$_6$, 60°, 34 hr	II, R^1 = R^2 = CH$_3$ (46)	74, 86
	I, R^1—R^2 = (CH$_2$)$_4$ (1)	C$_6$H$_6$, 63°, 46 hr	II, R^1—R^2 = (CH$_2$)$_4$ (80)	74
	(I)[b]			
	I, R^1 = R^2 = H	C$_6$H$_6$, 80°, 9 hr	II, R^1 = R^2 = H (40)	74, 86
	I, R^1 = H, R^2 = CH$_3$	C$_6$H$_6$, 120°, 12 hr	II, R^1 = H, R^2 = CH$_3$ (51)	74
	I, R^1 = R^2 = CH$_3$	C$_6$H$_6$, 120°, 4 hr	II, R^1 = R^2 = CH$_3$ (33)	74
(CH$_3$)$_2$CBrCOCH$_2$X (I)	CH$_2$=C(CH$_3$)CH=CH$_2$ (II)		(III), (IV),	

292

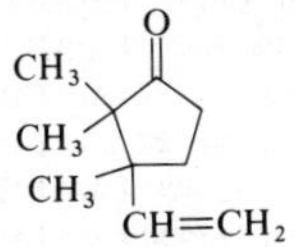

(V),

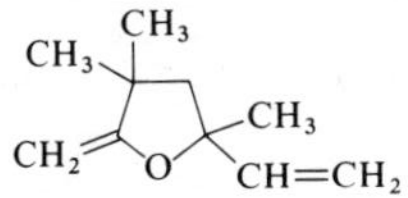

(VI),

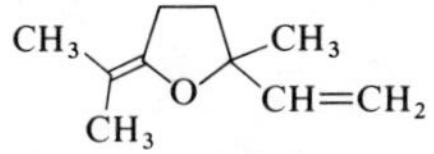

(VII)

I, X = Br	Cu, II, LiI	CH_3CN, $<30°$, 9 hr	III/IV/V (total 2), VI/VII (19 : 7) (—)[c]	158
	Cu, II, NaI	"	III/IV/V (total 2), VI/VII (35 : 8) (—)[c]	158
	"	DME, $<30°$, 9 hr	III/IV/V (total 2), VI/VII (25 : 8) (—)[c]	158
	Cu, II, $(n\text{-}C_4H_9)_4NI$	CH_3CN, $<30°$, 9 hr	III/IV/V (total 2), VI/VII (11 : 5) (—)[c]	158
	Zn/Cu, $RCH{=}CHCH{=}CH_2$ (II)			

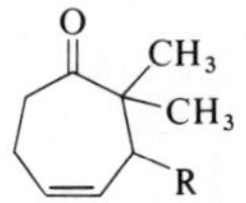

(III),

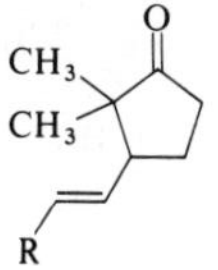

(IV),

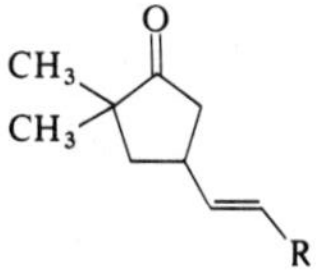

(V)

TABLE VII. [3 + 4] CYCLOCOUPLING OF α,α′-DIHALO AND POLYHALOGENATED KETONES (Continued)

A. With Open-Chain 1,3-Dienes (Continued)

Halo Ketone	Reactants (equiv.)	Conditions	Products and Yields (%)	Refs.
C₅ I, X = Br (Contd.)	II, R = H	CH₃CN/DME	R = H III/IV (93:7) (1)	87
	II, R = CH₃	1. CH₃CN/DME, 10°, 1.5 hr 2. 25°, 24 hr	R = CH₃ III/IV (62:38) (8)	87
I, X = I	"	Isopentane	R = CH₃ III/IV/V (26:65:9) (20)	87
	CH₂=C(CH₃)CH=CH₂ (II)		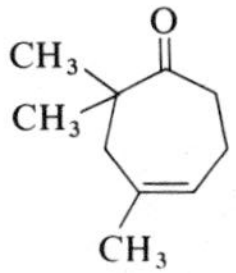 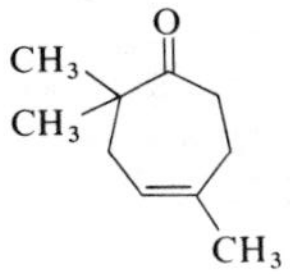 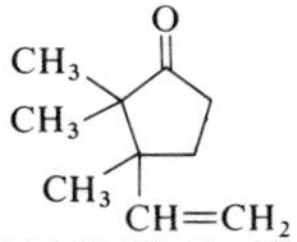	
	Zn/Cu, II	C₆H₆/isopentane or DME	III/IV/V (23:54:23) (12)	88, 87
	"	CH₃CN/DME, 10°, 24 hr	III/IV/V (48:43:9) (5)	88
	Hg, II, NaI	CH₃CN	III/IV/V (total 2)	86
	Fe₂(CO)₉, II	C₆H₆, 57°, 10 hr	III/IV/V (24:46:30) (—)	88
	"	C₆H₆, 57°, 20 hr	III/IV/V (22:43:30) (—)	88
	"	C₆H₆, 60 to 80°, 12 hr	III/IV/V (11:28:61) (—)	88
	"ᵈ	C₆H₆, reflux, 4.5 hr	III/IV/V (—:1:3.6) (—)	88

Structures (in Products column):

(III) — 7,7-dimethyl-4-methyl-cyclohept-4-en-1-one type: CH₃, CH₃ quaternary carbon adjacent to C=O; ring with CH₃ substituent.

(IV) — CH₃, CH₃ quaternary carbon adjacent to C=O; seven-membered ring with CH₃ substituent.

(V) — CH₃, CH₃ quaternary carbon with C(CH₃)(CH=CH₂) substituent on cyclopentanone ring.

294

C7 [(CH3)2CBr]2CO

Fe2(CO)9,
$CH_2=CR^1CR^2=CH_2$ (I)

I, $R^1 = R^2 = H$
I, $R^1 = H, R^2 = CH_3$
I, $R^1 = R^2 = CH_3$

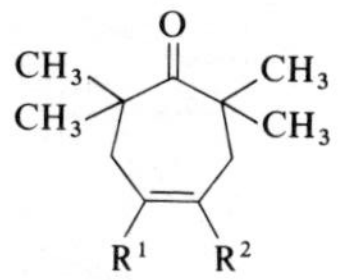

$(I)^b$

I, $R^1 = R^2 = H$
I, $R^1 = H, R^2 = CH_3$
I, $R^1 = R^2 = CH_3$

$C_6H_6, 60°, 38$ hr
$C_6H_6, 65°, 38$ hr
$C_6H_6, 60°, 38$ hr

$C_6H_6, 80°, 4$ hr
$C_6H_6, 87°, 12$ hr
$C_6H_6, 80°, 12$ hr

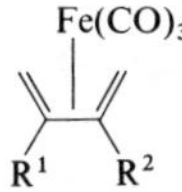

(II)

II, $R^1 = R^2 = H$ (33)
II, $R^1 = H, R^2 = CH_3$ (47)
II, $R^1 = R^2 = CH_3$ (71)

II, $R^1 = R^2 = H$ (90)
II, $R^1 = H, R^2 = CH_3$ (70)
II, $R^1 = R^2 = CH_3$ (100)

74, 86
74, 86
74, 86

74, 86
74, 86
74

C9 $(i\text{-}C_3H_7CHBr)_2CO$

Fe2(CO)9,
$CH_2=CR^1CR^2=CH_2$ (I)

I, $R^1 = R^2 = H$
I, $R^1 = H, R^2 = CH_3$
I, $R^1 = R^2 = CH_3$

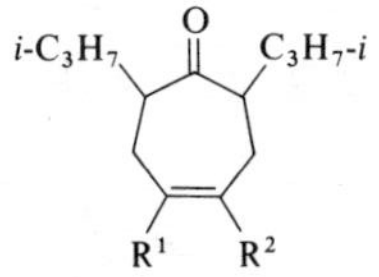

$(I)^b$

I, $R = H$
I, $R = CH_3$

$C_6H_6, 90°, 12$ hr
$C_6H_6, 25°, 36$ hr, hv^a
$C_6H_6, 60°, 38$ hr

$C_6H_6, 70°, 40$ hr
$C_6H_6, 120°, 20$ hr

(II)

II, $R^1 = R^2 = H$ (44)
II, $R^1 = H, R^2 = CH_3$ (31)
II, $R^1 = R^2 = CH_3$ (36)

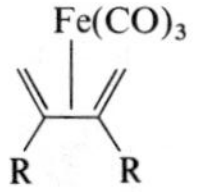

II, $R^1 = R^2 = H$ (77)
II, $R^1 = R^2 = CH_3$ (55)

74, 86
74, 86
74

74, 86
74, 86

295

TABLE VII. [3 + 4] CYCLOCOUPLING OF α,α′-DIHALO AND POLYHALOGENATED KETONES (*Continued*)

B. *With Carbocyclic Dienes*

Halo Ketone	Reactants (equiv.)	Conditions	Products and Yields (%)	Refs.
C₃ (CHBr₂)₂CO	Zn, cyclopentadiene, (C₂H₅O)₃B	THF, overnight	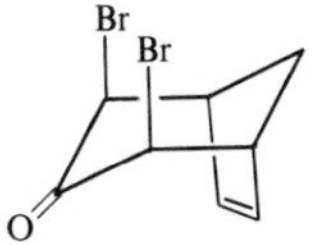 (12), (25), (1) (8)	47
	1. Fe(CO)₅, cyclopentadiene 2. Zn/Cu, NH₄Cl	THF/C₆H₆ (1/5), 80°, 30–40 min CH₃OH, 15 min	I (60, 47)	74, 96, 111
	1.  ⬠Fe(CO)₃ 2. Zn/Cu, NH₄Cl	Ether, 14 hr CH₃OH, 18 hr	I (56)	97
C₄ CH₃CHBrCOCH₂Br	Zn, cyclopentadiene, (C₂H₅O)₃B	THF, overnight	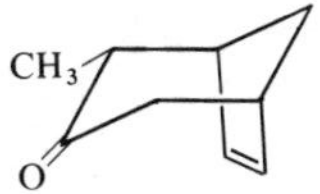 (8), 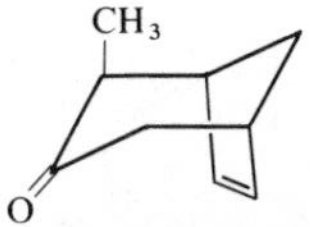 (6)	47

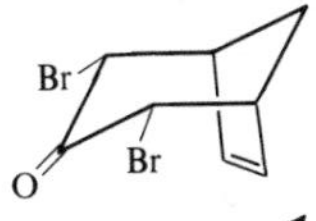

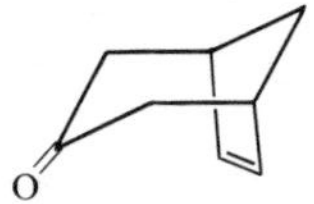

296

C$_5$ (CH$_3$CHBr)$_2$CO   Cyclopentadiene   (I)

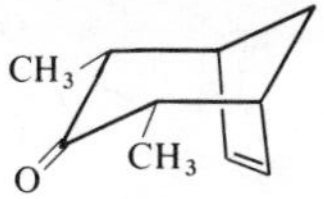

(II),

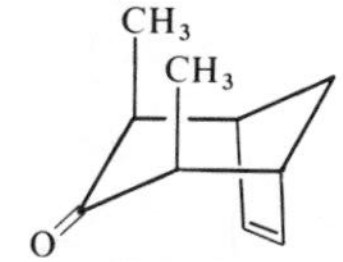

(III)

Cu, I, NaI   (2)   CH$_3$CN, 4.5 hr   II/III (6.4 : 1)   (91)   92

"   CH$_3$CN, 50°, 2 hr   II/III (6.4 : 1)   (82)   91

Zn/Cu, I   DME, −10°   II/III (1.67 : 1) (—)   93

Fe$_2$(CO)$_9$, I   C$_6$H$_6$, 60°, 25 hr II/III (47 : 53)   (86)   74, 86

Cu, NaI (2),   CH$_3$CN, 20°, 18 hr   92
C$_2$H$_5$O   OC$_2$H$_5$

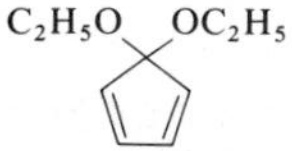

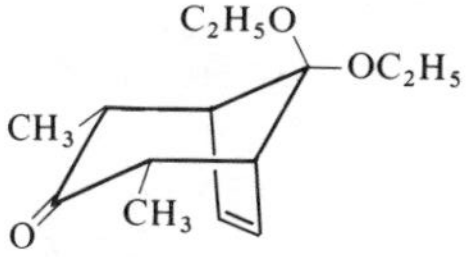

(I)

(11)

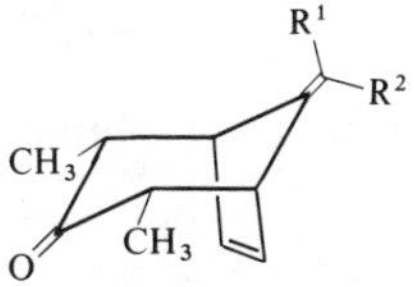

(II),

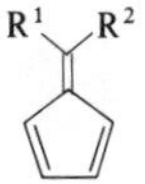

(III),

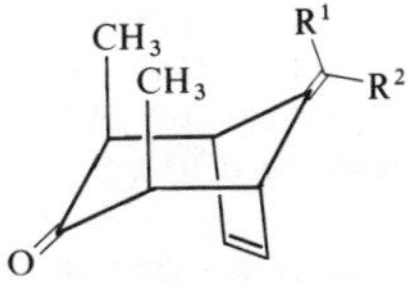

(IV)

297

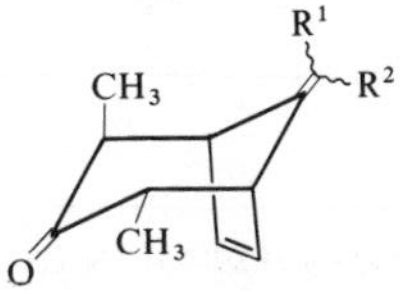

TABLE VII. [3 + 4] CYCLOCOUPLING OF α,α′-DIHALO AND POLYHALOGENATED KETONES (*Continued*)

B. With Carbocyclic Dienes (*Continued*)

Halo Ketone	Reactants (equiv.)	Conditions	Products and Yields (%)	Refs.
C₅ (CH₃CHBr)₂CO (*Contd.*)	Cu, NaI (2), I, R¹ = R² = CH₃	CH₃CN, 20°, 24.5 hr	R¹ = R² = CH₃, II/III (4.6:1) (80)	92
	Cu, NaI (2), I, R¹ = H, R² = OAc	CH₃CN, 20°, 18 hr	R¹ = H, R² = OAc II/IV (5.2:1) (70,ᵉ 33ᶠ)	92
	Zn/Cu, I, R¹ = R² = CH₃	DME, −15°, 2 hr	R¹ = R² = CH₃, II/III/IV (12.2:10:1) (−)	92
	"	—	R¹ = R² = CH₃ II/III (ca. 5:1) (60)	98
	Zn/Cu, I, R¹ = CH₃, R² = *t*-C₄H₉	—	R¹ = CH₃, R² = *t*-C₄H₉, II/III/IV (ca. 20:1:1) (60)	98
	Zn/Cu, I, R¹ = CH₃, R² = C₆H₅	—	R¹ = CH₃, R² = C₆H₅ II/III/IV (ca. 5:1:trace) (60)	98
	Fe₂(CO)₉, 1,3-cyclohexadiene	C₆H₆, 30°, 24 hr	(52), (34)	100
(CH₃)₂CBrCOCH₂X (I)	Cyclopentadiene (II)		(III)	
I, X = Br	Zn, II, (C₂H₅O)₃B	THF, overnight	III (55)	47
	Zn/Cu, II	—	III (30)	93
I, X = I	Zn, II, (C₂H₅O)₃B	THF, overnight	III (76)	47
(CH₃)₂CBrCOCHBr₂	1. Fe(CO)₅, cyclopentadiene	THF/C₆H₆ (1:10), 80°, 70 min	(83, 66)	48, 74
	2. Zn/Cu, NH₄Cl	CH₃OH, 20 min		

Substrate	Reagents and Conditions	Product(s) (Yield %)	Refs.
$\text{Br} \cdots \text{cyclopentanone} \cdots \text{Br}$ (O)	Fe₂(CO)₉, cyclopentadiene	(—), (—), (—)	95
C₆ (CH₃)₂CBrCOCHBrCH₃	Cyclopentadiene (I)	(II),	
	Zn, I, (C₂H₅O)₃B; THF, overnight	II/III (2.2 : 1) (78)	47
	Zn/Cu, I; DME, −5 to 0°	II/III (3.8 : 1) (60)	93
C₇ [(CH₃)₂CBr]₂CO	Zn/Cu, cyclopentadiene (I); DME, −5° to room temp.	(III)	93
	Fe₂(CO)₉ · I; C₆H₆, 60°, 38 hr	(II) (65); II (82, 71)	74, 86

299

TABLE VII. [3 + 4] CYCLOCOUPLING OF α,α'-DIHALO AND POLYHALOGENATED KETONES (*Continued*)

B. With Carbocyclic Dienes (Continued)

Halo Ketone	Reactants (equiv.)	Conditions	Products and Yields (%)	Refs.
C_7 $[(CH_3)_2CBr]_2CO$ (*Contd.*)	$Fe_2(CO)_9$, (I)		(II)	
	I, R = CH_3	C_6H_6, 50°, 35 hr	II, R = CH_3 (35)	99
	I, R = C_2H_5	C_6H_6, 60°, 16 hr	II, R = C_2H_5 (29)	99
	I, R = n-C_3H_7	C_6H_6, 50°, 34 hr	II, R = n-C_3H_7 (26)	99
	I, R = n-C_4H_9	C_6H_6, 60°, 32 hr	II, R = n-C_4H_9 (22)	99
	I, R = C_6H_5	—	II, R = C_6H_5 (0)	99
C_9 $(i$-$C_3H_7CHBr)_2CO$	$Fe_2(CO)_9$, cyclopentadiene	C_6H_6, 90°, 12 hr	(47), (43)	74, 86
C_{15} $(C_6H_5CHBr)_2CO$	NaI, cyclopentadiene (I)	CH_3CN, reflux, 15 min	(II) (40),	94

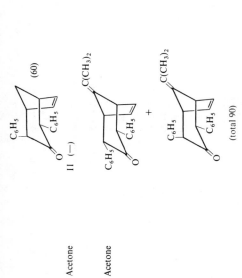

NaI, I

Acetone

(60)   84

Acetone

II (—)   98

(total 90)

C. With Pyrrole Derivatives

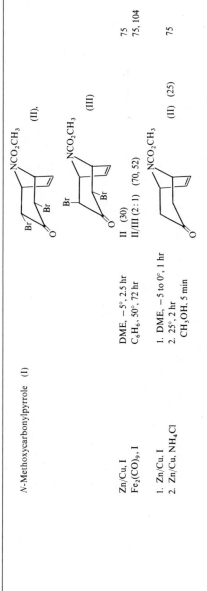

C$_3$ (CHBr$_2$)$_2$CO   N-Methoxycarbonylpyrrole   (I)

Zn/Cu, I   DME, −5°, 2.5 hr   II (30)   75
Fe$_2$(CO)$_9$, I   C$_6$H$_6$, 50°, 72 hr   II/III (2:1) (70, 52)   75, 104

1. Zn/Cu. I   1. DME, −5 to 0°, 1 hr   II (25)   75
2. Zn/Cu, NH$_4$Cl   2. 25°, 2 hr
                  CH$_3$OH. 5 min

TABLE VII. [3 + 4] CYCLOCOUPLING OF α,α′-DIHALO AND POLYHALOGENATED KETONES (*Continued*)

C. With Pyrrole Derivatives (*Continued*)

Halo Ketone	Reactants (equiv.)	Conditions	Products and Yields (%)	Refs.
C₅ (CH₃CHBr)₂CO	1. Fe₂(CO)₉, I 2. Zn/Cu, NH₄Cl R^3 R^3 R^2 N R^2 R^1 (I)	C₆H₆, 50°, 72 hr CH₃OH, 3 hr	II (57) (II), (III), (IV), (V) C₂H₅COCH(CH₃)	75
	Cu, NaI (2), I, R¹ = R² = R³ = H	CH₃CN, 3–4 hr	R¹ = R² = R³ = H II/III (6:1) (65)	101
	Cu, NaI (2), I, R¹ = CH₃, R² = R³ = H	CH₃CN, 3–4 hr	II, R¹ = CH₃, R² = R³ = H (89)	101

302

Substrate	Reagents	Conditions	Product (yield)	Refs.
C_6 $(CH_3)_2CBrCOCHBrCH_3$	Cu, NaI (2), I, $R^1 = CH_3$, $R^2 = R^3 = H$	CH_3CN, 8 hr	III, $R^1 = CH_3$, $R^2 = R^3 = H$ (52)	102
	Cu, NaI (2), I, $R^1 = C_6H_5$, $R^2 = R^3 = H$	CH_3CN, 20°, 4.25 hr	III, $R^1 = C_6H_5$, $R^2 = R^3 = H$ (—)	92
	Cu, NaI (2), I, $R^1 = R^3 = H$, $R^2 = CH_3$	CH_3CN, 3–4 hr	II, $R^1 = R^3 = H$, $R^2 = CH_3$ (66)	101
	Cu, NaI (2), I, $R^1 = H$, $R^2 = R^3 = CH_3$	"	II, $R^1 = H$, $R^2 = R^3 = CH_3$ (50)	101
	Cu, NaI (2), I, $R^1 = R^2 = CH_3$, $R^3 = H$	"	II, $R^1 = R^2 = CH_3$, $R^3 = H$ (74)	101
	Zn/Cu, I, $R^1 = CH_3$, $R^2 = R^3 = H$	—	α-V, $R^1 = CH_3$, $R^2 = R^3 = H$ (60)	102
	$Fe_2(CO)_9$, I, $R^1 = CH_3$, $R^2 = R^3 = H$	C_6H_6, 40°, 70 hr	$R^1 = CH_3$, $R^2 = R^3 = H$, α-V (48), β-V (33)	74, 103, 159
	$Fe_2(CO)_9$, I, $R^1 = CO_2CH_3$, $R^2 = R^3 = H$	C_6H_6, 10 hr, hv^g	$R^1 = CO_2CH_3$, $R^2 = R^3 = H$, II/III/IV (42:29:29) (60)^h	74
	Cu, NaI (2), N-methylpyrrole	CH_3CN, 3–4 hr	(structure, 70)	101
	"	—	(structure, 40)	102
C_7 $n\text{-}C_3H_7CHBrCOCHBrCH_3$	"	—	(structure, 57)	102
$[(CH_3)_2CBr]_2CO$	$Fe_2(CO)_9$, N-methylpyrrole	C_6H_6, 40°, 18 hr	(structures, 31), (51)	74, 159

303

TABLE VII. [3 + 4] CYCLOCOUPLING OF α,α'-DIHALO AND POLYHALOGENATED KETONES (*Continued*)

C. *With Pyrrole Derivatives* (*Continued*)

Halo Ketone	Reactants (equiv.)	Conditions	Products and Yields (%)	Refs.
C_7 [(CH₃)₂CBr]₂CO (*Contd.*)	Fe₂(CO)₉, *N*-acetylpyrrole	C_6H_6, 40 to 50°, 18.5 hr	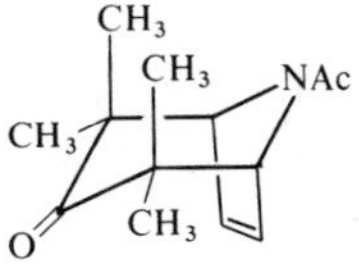 (68)	74, 103
C_9 (*i*-C₃H₇-CHBr)₂CO	Fe₂(CO)₉, *N*-methylpyrrole	C_6H_6, 40°, 60 hr	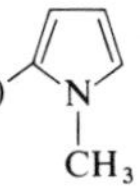 (CH₃)₂CHCH₂COCH(C₃H₇-*i*) (62), (CH₃)₂CHCH₂COCH(C₃H₇-*i*) (15)	74, 159
C_{12} (I)	Cu, NaI (2), II, R = CH₃ Cu, NaI (2), II, R = TMS	CH₃CN/C₆H₆, 50°, 48 hr CH₃CN/C₆H₆, 24 hr	α-IV, R = CH₃ (34) α-IV, R = H (−), β-IV, R = H (−) R = TMS (−)	83 83

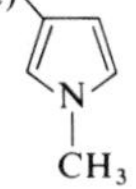

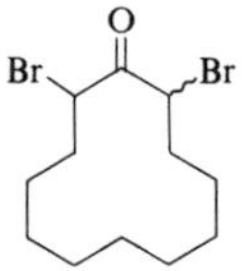

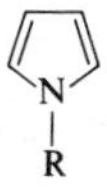

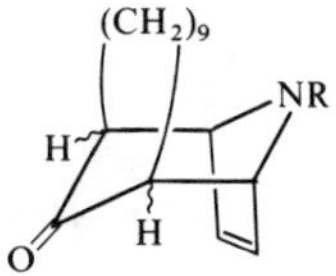

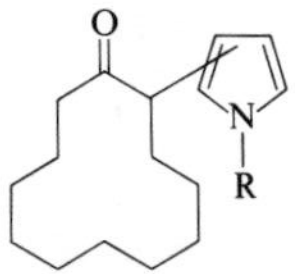

(III), (IV)

cis-I

	$Fe_2(CO)_9$, II, R = CO_2CH_3	C_6H_6, 50°, 48 hr	III, R = CO_2CH_3 (77)	83
	$Fe_2(CO)_9$, II, R = $CO_2CH_2CCl_3$	C_6H_6, 64°, 2 hr	R = $CO_2CH_2CCl_3$ III (58), α-IV (19)	83
	$Fe_2(CO)_9$, II, R = $CO_2C_4H_9$-t	C_6H_6, 50°, 21 hr	III, R = $CO_2C_4H_9$-t (5), α-IV, R = H (2), R = $CO_2C_4H_9$-t (25)	83
	$Fe_2(CO)_9$, II, R = $CO_2CH_2C_6H_5$	C_6H_6, 91°, 2 hr	R = $CO_2CH_2C_6H_5$ III (58), α-IV (14)	83
	Cu, NaI (2), II, R = CH_3	CH_3CN/C_6H_6, 50°, 45 hr	α-IV, R = CH_3 (5)	83

trans-I				
C_{15}	$(C_6H_5CHBr)_2CO$	Cu, NaI (2), N-methylpyrrole	80°	$C_6H_5CH_2COCH(C_6H_5)$ (50)	102

D. With Furan Derivatives

C_3	$(CHBr_2)_2CO$	Furan (I)		
		Zn, I, $(C_2H_5O)_2B$	THF, overnight	(II), II (30)^j	47, 102
		Zn/Ag, I	1. THF, −10°, 2 hr 2. 25°, 12 hr	II (65)	45
		$Fe_2(CO)_9$, I	Reflux, 48 hr	(III), II/III (9 : 1) (63)	74, 111
		Zn/Ag, CH_2OAc	1. THF, −10°, 10 min 2. 20°, 12 hr	CH_2OAc (52)	83

305

D. With Furan Derivatives (*Continued*)

Halo Ketone	Reactants (equiv.)	Conditions	Products and Yields (%)	Refs.
			II, $R^1 = R^2 = R^3 = H$ (55)	
C$_3$ (CHBr$_2$)$_2$CO (*Contd.*)	1. Zn/Ag, I, $R^1 = R^2 = R^3 = H$	1. THF, $-10°$, 2 hr 2. $25°$, 12 hr	II, $R^1 = R^2 = R^3 = H$ (55)	45
	2. Zn/Cu, NH$_4$Cl	CH$_3$OH, 15 min		
	1. Zn/Ag, I, $R^1 = R^3 = H$, $R^2 = OCH_2C_6H_5$	1. THF, $0°$, 1 hr 2. $20°$, 48 hr	II, $R^1 = R^3 = H$, $R^2 = OCH_2C_6H_5$ (14)	160
	2. Zn/Cu, NH$_4$Cl	CH$_3$OH, $25°$, 1 hr		
	1. Zn/Ag, I, $R^1 = R^3 = H$, $R^2 = OTHP$	1. THF, $0°$, 1 hr 2. $20°$, 60 hr	II, $R^1 = R^3 = H$, $R^2 = OTHP$ (31)	83, 114l
	2. Zn/Cu, NH$_4$Cl	CH$_3$OH, $20°$, 2 hr		
	1. Zn/Ag, I, $R^1 = R^2 = H$ $R^3 = OCH_2C_6H_5$	1. THF, $-10°$, 1 hr 2. $30°$, 49 hr	II, $R^1 = R^2 = H$, $R^3 = OCH_2C_6H_5$ (35)	83
	2. Zn/Cu, NH$_4$Cl	1. CH$_3$OH, $0°$, 2 hr 2. $25°$, 1 hr		
	1. Zn/Ag, I, $R^1 = R^2 = H$, $R^3 = OAc$	THF, $20°$, 12 hr	II, $R^1 = R^2 = H$, $R^3 = OAc$ (46)	114k
	2. Zn/Cu, NH$_4$Cl	CH$_3$OH, $20°$, 1 hr		
	1. Zn/Ag, I, $R^1 = R^3 = H$, $R^2 = CH_3$	THF, $20°$	II, $R^1 = R^3 = H$, $R^2 = CH_3$ (60)	161
	2. Zn/Cu, NH$_4$Cl	CH$_3$OH, $20°$		
	1. Zn/Ag, I, $R^1 = R^2 = H$, $R^3 = CH_3$	THF, $20°$, 14 hr	II, $R^1 = R^2 = H$, $R^3 = CH_3$ (86)	114i
	2. Zn/Cu, NH$_4$Cl	CH$_3$OH, $20°$, 1 hr		
	1. Zn/Ag, I, $R^1 = R^3 = CH_3$, $R^2 = H$	THF, $20°$, 12 hr	II, $R^1 = R^3 = CH_3$, $R^2 = H$ (66)	114j

306

Reactants	Conditions	Product (Yield %)	Refs.
2. Zn/Cu, NH₄Cl	1. CH₃OH, 0°, 2 hr; 2. 20°, 1 hr		
1. Zn/Ag, I, R¹ = R³ = H, R² = C₄H₉-t; 2. Zn/Cu, NH₄Cl	1. THF, 0°; 2. 20°, 60 hr; CH₃OH, 20°, 1 hr	II, R¹ = R³ = H, R² = C₄H₉-t (23)	83, 160
1. Zn/Ag, I, R¹ = R³ = H, R² = C₅H₁₁-n; 2. Zn/Cu, NH₄Cl	1. THF, 0°, 1 hr; 2. 30°, 48 hr; CH₃OH, 15°, 2 hr	II, R¹ = R³ = H, R² = C₅H₁₁-n (20)	83, 160
1. Zn/Ag, I, R¹ = R² = H, R³ = C₅H₁₁-n; 2. Zn/Cu, NH₄Cl	1. THF, −10°, 1 hr; 2. 25°, 12 hr; CH₃OH, 1 hr	II, R¹ = R² = H, R³ = C₅H₁₁-n (50)	83
1. Zn/Ag, I, R¹ = R² = H, R³ = C₆H₅; 2. Zn/Cu, NH₄Cl	1. THF, −10 to 25°, 1 hr; 2. 25°, 40 hr; CH₃OH, 2 hr	II, R¹ = R² = H, R³ = C₆H₅ (62)	83
1. Zn/Ag, I, R¹ = R³ = H, R² = C₆H₅; 2. Zn/Cu, NH₄Cl	1. THF, 0°; 2. 19°, 48 hr; CH₃OH, 2 hr	II, R¹ = R³ = H, R² = C₆H₅ (22)	83
1. Zn/Ag, I, R¹ = R³ = CH₃, R² = H; 2. Zn/Cu, NH₄Cl	1. THF, −10°, 30 min; 2. 20°, 12 hr; CH₃OH, 20°, 1 hr	II, R¹ = R³ = CH₃, R² = H (66)	83, 114j
1. Zn/Ag, I, R¹ = R³ = C₅H₁₁-n, R² = H; 2. Zn/Cu, NH₄Cl	1. THF, −10°, 1 hr; 2. 20°, 24 hr; CH₃OH, 0°, 4 hr	II, R¹ = R³ = C₅H₁₁-n, R² = H (41)	83
1. Zn/Ag, I, R¹ = R³ = C₆H₅, R² = H; 2. Zn/Cu, NH₄Cl	1. THF, −10°, 10 min; 2. 20°, 12 hr; CH₃OH, 20°, 1 hr	II, R¹ = R³ = C₆H₅, R² = H (53)	83
1. Fe₂(CO)₉, I, R¹ = R² = R³ = H; 2. Zn/Cu, NH₄Cl	Reflux, 48 hr; CH₃OH, 15 min	II, R¹ = R² = R³ = H (63)	74, 111
1. Fe₂(CO)₉, I, R¹ = R² = H, R³ = Br; 2. Zn/Cu, NH₄Cl	C₆H₆, 60°, 3 hr; CH₃OH, 15 min	II, R¹ = R² = H, R³ = Br (44)	83
1. Fe₂(CO)₉, I, R¹ = R² = H, R³ = CH₃; 2. Zn/Cu, NH₄Cl	C₆H₆, 60°, 5 hr; CH₃OH, 25°, 1 hr	II, R¹ = R² = H, R³ = CH₃ (70)	161
1. Fe₂(CO)₉, I, R¹ = R² = H, R³ = i-C₃H₇; 2. Zn/Cu, NH₄Cl	C₆H₆, 60°, 2 hr; CH₃OH, 30 hr	II, R¹ = R² = H, R³ = i-C₃H₇ (47)	89, 111

TABLE VII. [3 + 4] Cyclocoupling of α,α'-Dihalo and Polyhalogenated Ketones (*Continued*)

D. With Furan Derivatives (*Continued*)

Halo Ketone	Reactants (equiv.)	Conditions	Products and Yields (%)	Refs.
C_3 $(CHBr_2)_2CO$ (*Contd.*)	1. $Fe_2(CO)_9$, I, $R^1 = R^3 = H$, $R^2 = i\text{-}C_3H_7$ 2. Zn/Cu, NH_4Cl	C_6H_6, 60°, 2.5 hr	II, $R^1 = R^3 = H$, $R^2 = i\text{-}C_3H_7$ (71)	89, 110
	1. $Fe_2(CO)_9$, I, $R^1 = R^2 = H$, $R^3 = C_5H_{11}\text{-}n$ 2. Zn/Cu, NH_4Cl	CH_3OH, 20 min C_6H_6, 60°, 5 hr	II, $R^1 = R^2 = H$, $R^3 = C_5H_{11}\text{-}n$ (67)	83, 161
	1. $Fe_2(CO)_9$, I, $R^1 = R^2 = H$, $R^3 = C_6H_5$ 2. Zn/Cu, NH_4Cl	CH_3OH, 25°, 1 hr C_6H_6, 60°, 5 hr CH_3OH, 25°, 1 hr	II, $R^1 = R^2 = H$, $R^3 = C_6H_5$ (63)	114h
C_4 $CH_3CXBrCOCHYBr$ (I)	Furan (II)		(III), (IV)	
I, X = Y = H	Zn, II, $(C_2H_5O)_3B$	THF, overnight	X = Y = H III/IV (total 10)	47
I, X, Y = H, Br^j	"	"	X = Y = H III/IV (total 9), III, X = H, Y = Br (23), III and/or IV, X = Br, Y = H (14)	47
I, X = Y = Br	Zn/Ag, II	1. THF, −10°, 2.5 hr 2. 25°, 17 hr	III and/or IV, X = Y = Br (61)	45, 114a,b
	1. Zn/Ag, II	1. THF, −10°, 2.5 hr 2. 25°, 17 hr	III, X = Y = H (53)	45
	2. Zn/Cu, NH_4Cl	CH_3OH, 1 hr		
	1. $Fe_2(CO)_9$, II	24 hr		
	2. Zn/Cu, NH_4Cl	CH_3OH, 15 min	X = Y = H III/IV (55 : 45) (63)	74

C₅ C₂H₅CHBrCOCHBr₂

Zn/Cu, furan	—	(30)		102

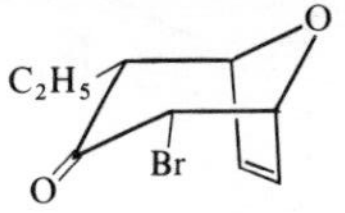

(CH₃CHBr)₂CO

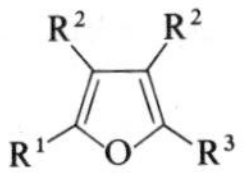

(I)

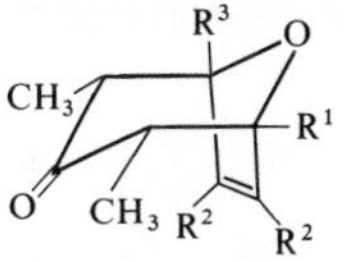

Reagent	Solvent/Conditions	Product	Yield (ratio)	Ref.
Cu, I, R¹ = R² = R³ = H, NaI (2)	CH₃CN, 45 to 60°, 4 hr	II, R¹ = R² = R³ = H (48)		107, 102
"	CH₃CN	R¹ = R² = H	II/III/IV (91:3:6) (—)	92
Zn/Cu, I, R¹ = R² = R³ = H	—	R¹ = R² = R³ = H	II/III/IV (total 70)	102
"	DME, −10°, 1.5 hr	R¹ = R² = R³ = H	II/III/IV (81:10:9) (85)	93, 105
"	DME	R¹ = R² = R³ = H	II/III/IV (74:9:7) (—)	92
"	CH₃CN	R¹ = R² = R³ = H	II/III/IV (75:10:15) (—)	92
Zn/Cu, I, R¹ = R² = R³ = H, (n-C₄H₉)₄NClO₄	DME	R¹ = R² = R³ = H	II/III/IV (79:8:13) (—)	92

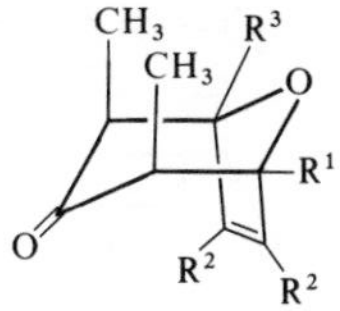

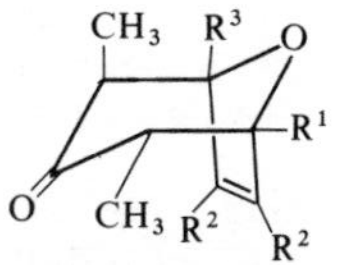

(II),

(III),

(IV)

TABLE VII. [3 + 4] CYCLOCOUPLING OF α,α'-DIHALO AND POLYHALOGENATED KETONES (Continued)

D. With Furan Derivatives (Continued)

Halo Ketone	Reactants (equiv.)	Conditions	Products and Yields (%)	Refs.
C$_5$ $(CH_3CHBr)_2CO$ (Contd.)	Zn/Cu, I, R^1 = R^2 = H, LiClO$_4$	"	R^1 = R^2 = R^3 = H II/III/IV (92 : 4 : 4) (—)	92
	Zn/Cu, I, R^1 = R^2 = H	DME, 0 to 25°, 16 hr	II, R^1 = R^2 = H, R^3 = C(CH$_3$)O(CH$_2$)$_2$O (53)	105
	R^3 = C(CH$_3$)O(CH$_2$)$_2$O			
	Zn/Cu, I, R^1 = R^3 = H, R^2 = CH$_3$	DME	R^1 = R^3 = H, R^2 = CH$_3$ II/III/IV (45 : 34 : 21) (—)	92
	Zn/Ag, I, R^1 = R^2 = R^3 = H	1. THF, −10°, 1 hr 2. 25°, 12 hr	II, R^1 = R^2 = R^3 = H (80)	45
	Fe$_2$(CO)$_9$, I, R^1 = R^2 = R^3 = H	C$_6$H$_6$, 40°, 53 hr	R^1 = R^2 = R^3 = H II/IV (44 : 56) (90)	74
	Fe$_2$(CO)$_9$, I, R^1 = R^3 = CH$_3$, R^2 = H	C$_6$H$_6$, 80°, 16 hr	II, R^1 = R^3 = CH$_3$, R^2 = H (78)	74

(III)

Halo Ketone	Reactants (equiv.)	Conditions	Products and Yields (%)	Refs.
$(CH_3)_2CBrCOCHXBr$ (I)	Furan (II)			
I, X = H	Zn/Cu, II	Acetone	III, X = H (20)	93
I, X = Br	Zn/Ag, II	1. THF, −10°, 1.5 hr 2. 25°, 16 hr	III, X = Br (55)	45, 114a,b
	1. Zn/Ag, II	1. THF, −10°, 1.5 hr 2. 25°, 19 hr	III, X = H (50)	45
	2. Zn/Cu, NH$_4$Cl	CH$_3$OH, 1 hr		
	1. Fe$_2$(CO)$_9$, II	Reflux, 40 hr	III, X = H (93)	74
	2. Zn/Cu, NH$_4$Cl	CH$_3$OH, 15 min		

Halo Ketone	Reactants (equiv.)	Conditions	Products and Yields (%)	Refs.
C$_6$ i-C$_3$H$_7$CHBrCOCHBr$_2$	1. Fe$_2$(CO)$_9$, furan	Reflux, 16 hr	(35)	89, 111
	2. Zn/Cu, NH$_4$Cl	CH$_3$OH, 1.5 hr		

310

(CH₃)₂CBrCOCHBrCH₃	Zn/Cu, furan	Acetone, −5° to room temp.	(I) (35), (II) (<5)	93
	Fe₂(CO)₉, furan	40°, 45 hr	(35)	74
	"	Reflux, 30 hr	I and/or II (84) (35)	74, 159
			(II)	
C₇ [(CH₃)₂CBr]₂CO	Furan (I)	Acetone, −5° to room temp.	II (43)	93
	Zn/Cu, I	40°, 38 hr	II (96)	74
	Fe₂(CO)₉, I	C₆H₆, 40°, 38 hr	II (89)	74
	"	CH₃CN, 14°	II (—),	20
	2e⁻, I, (n-C₄H₉)₄NBF₄		i-C₃H₇COC(CH₃)₂ (—)	
	Fe₂(CO)₉, furan	Reflux, 62 hr	(54),	74, 159

D. With Furan Derivatives   (*Continued*)

Halo Ketone	Reactants (equiv.)	Conditions	Products and Yields (%)	Refs.
C$_7$ (*Contd.*)	Fe$_2$(CO)$_9$, furan	Reflux, 62 hr	(21)	
C$_8$ n-C$_5$H$_{11}$CBr$_2$COCHBr$_2$	1. Zn/Ag, furan 2. Zn/Cu,NH$_4$Cl	1. THF, $-10°$, 2 hr 2. 25°, 39 hr CH$_3$OH	(47)	83, 114a,b
	Fe$_2$(CO)$_9$, furan	Reflux, 48 hr	(27), (10), (10)	74, 159
C$_9$ (i-C$_3$H$_7$CHBr)$_2$CO	Fe$_2$(CO)$_9$, furan (I)		(11),	

312

			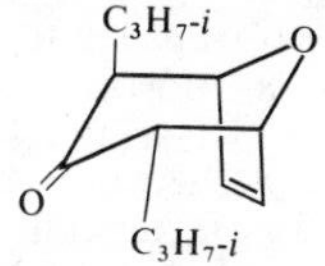 (III)	
	I	40°, 80 hr	II/III (77 : 23) (96)	74
	"	C_6H_6, 40°, 80 hr	II/III (75 : 25) (89)	74
	"	DMF, 22 hr	II (20), 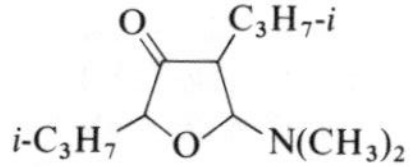 (49)	34
$C_6H_5CHBrCOCH_2Br$	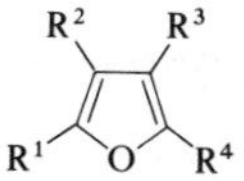 (I)		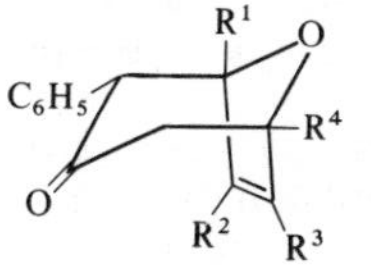 (II), 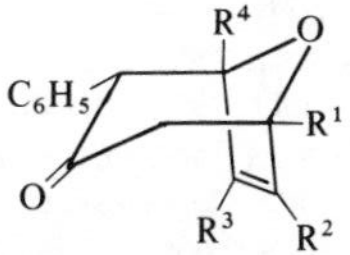 (III), 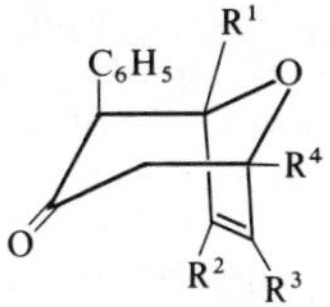 (IV), 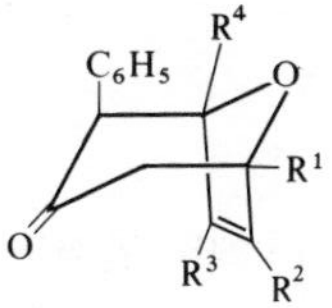 (V)	

TABLE VII. [3 + 4] CYCLOCOUPLING OF α,α'-DIHALO AND POLYHALOGENATED KETONES (Continued)

D. With Furan Derivatives (Continued)

Halo Ketone	Reactants (equiv.)	Conditions	Products and Yields (%)	Refs.
C₉ C₆H₅CHBrCOCH₂Br (Contd.)	Zn/Cu, I, R¹ = R³ = R⁴ = H, R² = CH₃	DME, −10°	R¹ = R³ = R⁴ = H, R² = CH₃ II/III (65:35) (36), IV/V (64:36) (18)	83
	Zn/Cu, I, R¹ = R³ = R⁴ = H, R² = CO₂C₂H₅	"	R¹ = R³ = R⁴ = H, R² = CO₂C₂H₅ II/III (67:33) (41)	83
	Zn/Cu, I, R¹ = CO₂C₂H₅, R² = R³ = R⁴ = H	"	R¹ = CO₂C₂H₅, R² = R³ = R⁴ = H II/III (88:12) (40)	83
	Zn/Cu, I, R¹ = CO₂CH₃, R² = CH₃, R³ = R⁴ = H	"	R¹ = CO₂CH₃, R² = CH₃, R³ = R⁴ = H II/III (95:5) (20)	83
	Zn/Cu, I, R¹ = CO₂CH₃, R² = R³ = H, R⁴ = CH₃	"	R¹ = CO₂CH₃, R² = R³ = H, R⁴ = CH₃ II/III (79:21) (23)	83
	Zn/Cu, I, R¹ = CO₂CH₃, R² = R⁴ = H, R³ = CH₃	"	R¹ = CO₂CH₃, R² = R⁴ = H, R³ = CH₃ II/III (80:20) (20)	83
	Zn/Ag, I, R¹ = R² = R³ = R⁴ = H	1. THF, −10°, 1.5 hr 2. 25°, 19 hr	II, R¹ = R² = R³ = R⁴ = H (77)	83
	Fe₂(CO)₉, I, R¹ = R³ = R⁴ = H, R² = Br	C₆H₆, 30°	R¹ = R³ = R⁴ = H, R² = Br II/III (65:35) (44), IV/V (66:34) (29)	36
	Fe₂(CO)₉, I, R¹ = R³ = R⁴ = H, R² = CH₃	C₆H₆, 30°	R¹ = R³ = R⁴ = H, R² = CH₃ II/III (56:44) (29), IV/V (56:44) (19)	36
	Fe₂(CO)₉, I, R¹ = R³ = R⁴ = H, R² = CO₂C₂H₅	"	R¹ = R³ = R⁴ = H, R² = CO₂C₂H₅ II/III (55:45) (52)	36
	Fe₂(CO)₉, I, R¹ = CO₂C₂H₅, R² = R³ = R⁴ = H	"	R¹ = CO₂C₂H₅, R² = R³ = R⁴ = H II/III (90:10) (56)	36
	Fe₂(CO)₉, I, R¹ = CO₂CH₃, R² = CH₃, R³ = R⁴ = H	"	R¹ = CO₂CH₃, R² = CH₃, R³ = R⁴ = H II/III (92:8) (63)	36
	Fe₂(CO)₉, I, R¹ = CO₂CH₃, R² = R³ = H, R⁴ = CH₃	"	R¹ = CO₂CH₃, R² = R³ = H, R⁴ = CH₃ II/III (65:35) (65)	36
	Fe₂(CO)₉, I, R¹ = CO₂CH₃, R² = R⁴ = H, R³ = CH₃	"	R¹ = CO₂CH₃, R² = R⁴ = H, R³ = CH₃ II/III (76:24) (68)	36

			83, 114a,b
			162
			117
			162

$C_6H_5CBr_2COCHBr_2$

1. Zn/Ag, furan
2. Zn/Cu, NH_4Cl

1. THF, $-10°$, 2 hr
2. 25°, 12 hr
CH_3OH, 1 hr

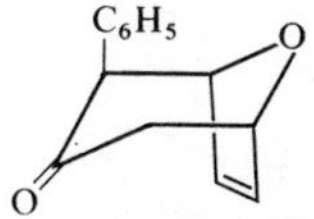

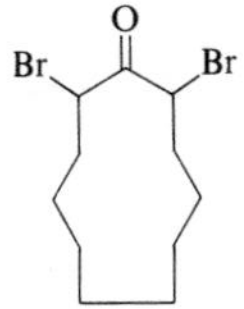

(48)

Zn/Cu, furan — (—)

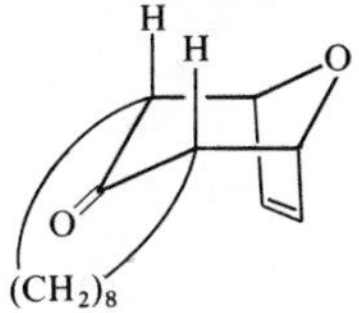

$Fe_2(CO)_9$ C_6H_6, 80°, 3 hr   (41)

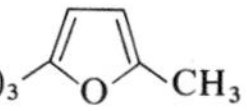

$BrCH_2COCHBr(CH_2)_3$ — CH_3

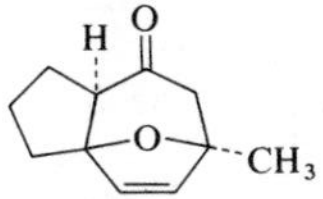

Zn/Cu, furan (I) — (II) (—), (—)

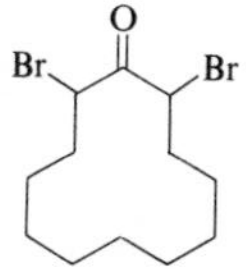

(III) (—)

C_{11}

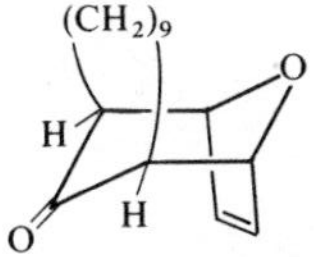

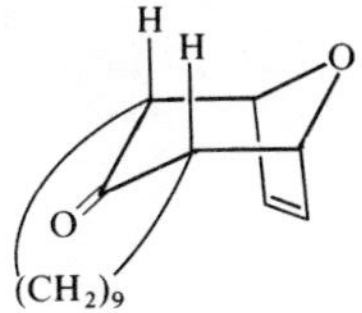

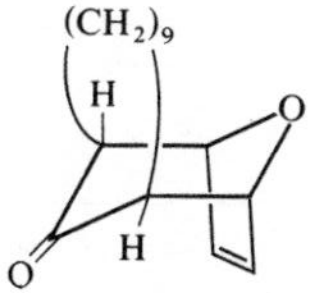

C_{12}

315

TABLE VII. [3 + 4] CYCLOCOUPLING OF α,α'-DIHALO AND POLYHALOGENATED KETONES (*Continued*)

D. With Furan Derivatives (*Continued*)

Halo Ketone	Reactants (equiv.)	Conditions	Products and Yields (%)	Refs.
C$_{12}$ (*Contd.*)	Fe$_2$(CO)$_9$, I	Reflux, 40 hr	II (7), III (45), (34)	74, 159
BrCH$_2$COCBr(CH$_3$)(CH$_2$)$_3$	Fe$_2$(CO)$_9$	C$_6$H$_6$, 80°, 3 hr	(38)	117
C$_{13}$	Zn/Cu, furan	—	(–), (–),	162

316

C$_{15}$ (C$_6$H$_5$CHBr)$_2$CO   Furan   (I)

Reagent	Conditions	Product(s) (%)	Refs.
Cu, I, NaI (2)	CH$_3$CN, 80°	II and/or III (80)	102
Zn/Cu, I	CH$_3$CN, reflux, 2 days	II (29), (C$_6$H$_5$CH$_2$COCHC$_6$H$_5$)$_2$ (49)	94
Hg, I	—	II (35)	94
Fe$_2$(CO)$_9$, I	25°, 6 hr	II (45), III (45)	74
"	C$_6$H$_6$, 25°, 12 hr	II/III (1:1) (90)	29, 28
(CH$_3$)$_2$CuLi, I	−78°	II and/or III (20)	78
NaI, I	—	II (65)	84
"	CH$_3$CN, reflux,	II (30), III (10),	94

C$_6$H$_5$CH$_2$COCH(C$_6$H$_5$)$_2$ (45)

E. With Thiophene

C$_5$ (CH$_3$CHBr)$_2$CO	Fe$_2$(CO)$_9$, thiophene	C$_6$H$_6$, 35 hr, hv^g	C$_2$H$_5$COCH(CH$_3$)- (37) 74

TABLE VII. [3 + 4] CYCLOCOUPLING OF α,α'-DIHALO AND POLYHALOGENATED KETONES (*Continued*)

F. *With Anthracene*

Halo Ketone	Reactants (equiv.)	Conditions	Products and Yields (%)	Refs.
C₅ (CH₃CHBr)₂CO	Anthracene (I)		(II),	
			(III)	
	Zn, I, CuCl (0.4)	Dioxane, 80 to 85°, 5 hr	II/III (ca. 90:10) (22)	116
	Zn, I, CuCl (0.4), TMSCl (2.1)	C₆H₆	II/III (ca. 1:1) (93)	116
(CH₃)₂CBrCOCH₂Br	Zn, anthracene (I), CuCl (0.4)	Dioxane, 80 to 85°, 5 hr	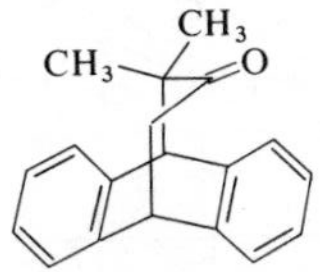 (II) (12)	116

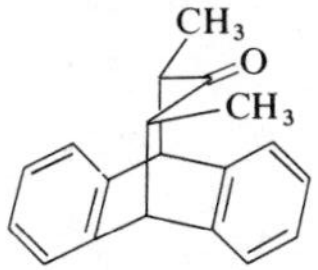

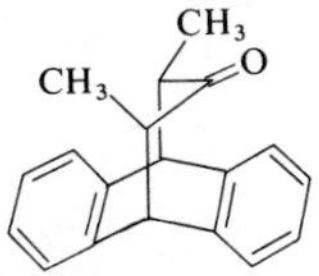

318

	Reactant	Conditions	Solvent	Product(s) (%)	Ref.
		Zn, I, CuCl (0.4), TMSCl (2.1)	C₆H₆	II (76) (II) (25)	116
		Zn, anthracene (I), CuCl (0.4)	Dioxane, 80 to 85°, 5 hr		116
C₆	(CH₃)₂CBrCOCHBrCH₃	Zn, I, CuCl (0.4), TMSCl (2.1)	C₆H₆	II (97) (II) (3)	116
		Zn, anthracene (I), CuCl (0.4)	Dioxane, 80 to 85°, 5 hr		116
C₇	[(CH₃)₂CBr]₂CO	Zn, I, CuCl (0.4), TMSCl (2.1)	C₆H₆	II (71)	116

[a] The mixture of dibromide, diene, and Fe₂(CO)₉ was irradiated with visible light.

[b] The complex was prepared *in situ* by irradiation of a mixture of the diene and Fe(CO)₅.

[c] A major product was 3-bromo-3-methylbutan-2-one.

[d] The iron carbonyl was added batchwise to the refluxing reaction mixture.

[e] The yield represents crude product.

[f] The products were hydrolyzed under workup conditions to give small amounts of *syn*- (6.3%) and *anti*-8-aldehydes (3.2%).

[g] The reaction mixture was irradiated with visible light.

[h] The yield was determined by ¹H NMR.

[i] 8-Oxabicyclo[3.2.1]oct-6-en-3-one (2%) and the 2,2,4-tribromo derivative (2%) were also formed.

[j] A mixture of the two tribromides was used.

TABLE VIII. [3 + 2] CYCLOCOUPLING OF α,α'-DIBROMO KETONES

A. With Simple Olefins

Halo Ketone	Reactants (equiv.)	Conditions	Products and Yields (%)	Refs
C$_9$ C$_6$H$_5$CHBrCOCH$_2$Br	Fe$_2$(CO)$_9$, (CH$_3$)$_2$C=CH$_2$	C$_6$H$_6$, 55°	(I), [cyclopentanone structure with C$_6$H$_5$, CH$_3$, CH$_3$]; C$_6$H$_5$CH$_2$CO(CH$_2$)$_2$C(CH$_3$)=CH$_2$ (II), CH$_3$COCH(C$_6$H$_5$)CH$_2$C(CH$_3$)=CH$_2$ (III), (IV) [lactone structure]; I/II/III/IV (14:52:5:29) (35)	33
C$_{10}$ [structure: BrCH$_2$CO–C(CH$_3$)(Br)CH$_2$CH$_2$CH=C(CH$_3$)$_2$]	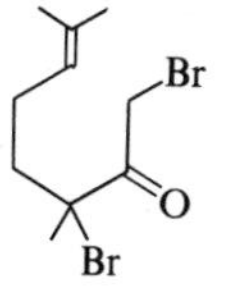 Fe$_2$(CO)$_9$	C$_6$H$_6$, 100 to 115°, 1.5 hr	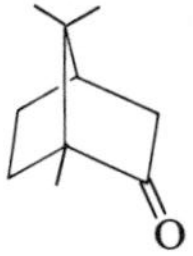 (I), 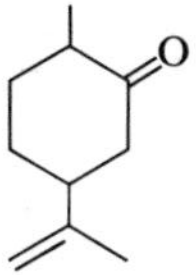 (II), 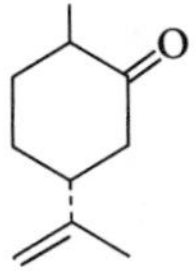 (III), 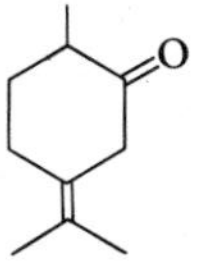 (IV), 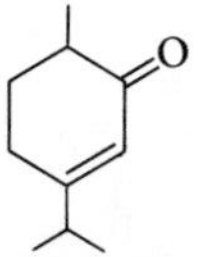 (V) [cyclohexanone/bicyclic structures]; I/II/III/IV/V (54:6:14:4:10) (70)	117

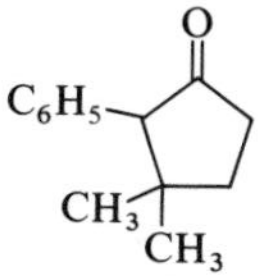

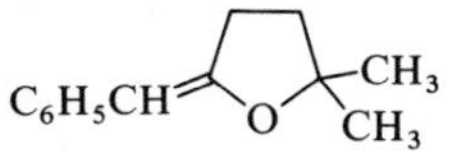

320

C$_{11}$ C$_6$H$_5$CHBrCOCBr(CH$_3$)$_2$ Fe$_2$(CO)$_9$, (CH$_3$)$_2$C=CH$_2$ C$_6$H$_6$, 55°

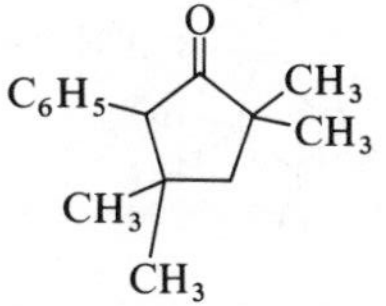

(I),

C$_6$H$_5$CH$_2$COC(CH$_3$)$_2$CH$_2$C(CH$_3$)=CH$_2$ (II)

I/II (91:9)   (5.5)

C$_{15}$

(1)

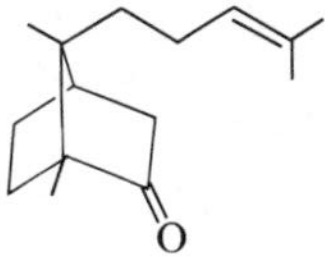

(II),

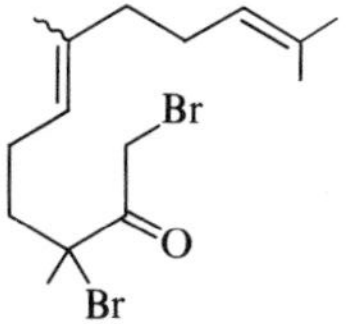

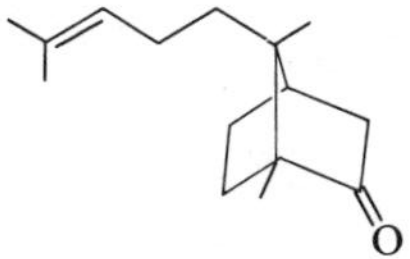

(III)

(E)-I   Fe$_2$(CO)$_9$ C$_6$H$_6$, 100°   II/III (2:1) (58)   117

(Z)-I   "   "   II/III (1:2) (—)   117

B. With Aryl-Substituted Olefins

C$_5$ (CH$_3$CHBr)$_2$CO   Fe$_2$(CO)$_9$,   C$_6$H$_5$CR^1=CHR^2 (I)

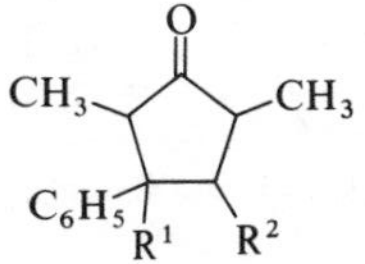

(II)

I, R^1 = R^2 = H   C$_6$H$_6$, 50°, 14 hr   II, R^1 = R^2 = H   (65)[a]   118, 119

I, R^1 = H, R^2 = CH$_3$ C$_6$H$_6$, 55°, 14 hr   II, R^1 = H, R^2 = CH$_3$ (20)[b]   118, 119

I, R^1 = CH$_3$, R^2 = H   C$_6$H$_6$, 60°, 12 hr   II, R^1 = CH$_3$, R^2 = H   (70)[a]   118, 119

I, R^1 = cyclopropyl, R^2 = H   C$_6$H$_6$, 65°, 24 hr   II, R^1 = cyclopropyl, R^2 = H   (95)[b]   118, 119

TABLE VIII. [3 + 2] Cyclocoupling of α,α'-Dibromo Ketones (Continued)

B. With Aryl-Substituted Olefins (Continued)

Halo Ketone	Reactants (equiv.)	Conditions	Products and Yields (%)	Refs.
C$_5$ (CH$_3$CHBr)$_2$CO (Contd.)	I, R^1 = H, R^2 = C$_6$H$_5$	C$_6$H$_6$, 50°, 14 hr	II, R^1 = H, R^2 = C$_6$H$_5$ (30)^a	118, 119
	I, R^1 = C$_6$H$_5$, R^2 = H	C$_6$H$_6$, 60°, 14 hr	II, R^1 = C$_6$H$_5$, R^2 = H (70),^{a} C$_2$H$_5$COCH(CH$_3$)CH=C(C$_6$H$_5$)$_2$ (7)	118, 119
	C$_6$H$_5$CH=CH$_2$^{c} Fe(CO)$_4$	C$_6$H$_6$, 50°, 19 hr	(16)^a	118, 119
	Fe$_2$(CO)$_9$, indene	C$_6$H$_6$, 50°, 14 hr	(45)^b	118, 119
	Fe$_2$(CO)$_9$,	C$_6$H$_6$, 65°, 14 hr	(30)^b	118, 119
(CH$_3$)$_2$CBrCOCH$_2$Br	Fe$_2$(CO)$_9$, p-CH$_3$C$_6$H$_4$C(CH$_3$)=CH$_2$	C$_6$H$_6$, 55°, 17 hr	(18),	121

322

Substrate	Conditions	Product	Reference
$(CH_3)_2CBrCOCHBr_2$	1. $Fe_2(CO)_9$, $p\text{-}CH_3C_6H_4C(CH_3)=CH_2$ 2. Zn/Cu, NH_4Cl	C_6H_6; CH_3OH	(1.2) CH_3, CH_3, $C_6H_4CH_3\text{-}p$, CH_3	121
C_6			
$(CH_3)_2CBrCOCHBrCH_3$	$Fe_2(CO)_9$, $C_6H_5CH=CH_2$	C_6H_6, 55°	(35) CH_3, CH_3, $C_6H_4CH_3\text{-}p$, CH_3 (I), CH_3, CH_3, C_6H_5 (II) CH_3, CH_3, C_6H_5 (II) I/II (94:6) (31)	36
C_7			
$[(CH_3)_2CBr]_2CO$	$Fe_2(CO)_9$, $ArCR^1=CHR^2$ (I)	C_6H_6, 55°, 12 hr	CH_3, CH_3, CH_3 (II), Ar, R^1, R^2 $(E)\text{-}i\text{-}C_3H_7COC(CH_3)_2CR^2=CR^1Ar$ (III) $Ar = C_6H_5$, $R^1 = R^2 = H$ II (23), III (23)	118, 119
	$I, Ar = C_6H_5$, $R^1 = R^2 = H$		
	$I, Ar = C_6H_5$, $R^1 = CH_3$, $R^2 = H$	C_6H_6	$Ar = C_6H_5$, $R^1 = CH_3$, $R^2 = H$ II (5), III (37), $i\text{-}C_3H_7COC(CH_3)_2CH_2C(C_6H_5)=CH_2$ (11),	118, 119

323

TABLE VIII. [3 + 2] CYCLOCOUPLING OF α,α′-DIBROMO KETONES (*Continued*)

B. With Aryl-Substituted Olefins (*Continued*)

Halo Ketone	Reactants (equiv.)	Conditions	Products and Yields (%)	Refs.
C$_7$ [(CH$_3$)$_2$CBr]$_2$CO (*Contd.*)	I, Ar = C$_6$H$_5$, R^1 = CH$_3$, R^2 = H	C$_6$H$_6$	CH$_3$ CH$_3$ CH$_3$ C$_6$H$_5$ (CH$_3$)$_2$ lactone (16)	
	(*E*)-I, Ar = *p*-CH$_3$OC$_6$H$_4$, R^1 = H, R^2 = CH$_3$	C$_6$H$_6$, 55°, 12 hr	II, Ar = *p*-CH$_3$OC$_6$H$_4$, R^1 = H, R^2 = CH$_3$ (55), III (3), CH$_3$ CH$_3$ CH$_3$ *i*-C$_3$H$_7$ HO C$_6$H$_4$OCH$_3$-*p* (30)	118, 119
	1. Fe$_2$(CO)$_9$ 2. (*Z*)-C$_6$H$_5$CH=CHD	C$_6$H$_6$, 50°, 15 min 50°, 1 hr	O CH$_3$ CH$_3$ D C$_6$H$_5$ H H (6), (*E*)-(CH$_3$)$_2$CD$_{0.35}$CO C$_6$H$_5$CH=CD$_{0.65}$C(CH$_3$)$_2$ (3)	118, 119 120
C$_9$ (*i*-C$_3$H$_7$CHBr)$_2$CO	Fe$_2$(CO)$_9$, (C$_6$H$_5$)$_2$C=CH$_2$	C$_6$H$_6$, 50°, 18 hr	*i*-C$_3$H$_7$ C$_3$H$_7$-*i* *i*-C$_3$H$_7$ O C$_6$H$_5$ C$_6$H$_5$ (44),[a] *i*-C$_3$H$_7$CH$_2$CO (C$_6$H$_5$)$_2$C=CHCHC$_3$H$_7$-*i* (11)	118, 119

324

C₆H₅CHBrCOCH₂Br	$Fe_2(CO)_9$, $C_6H_5CH{=}CH_2$	C_6H_6, 55°	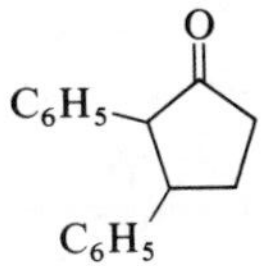 (44)	36
C₁₁ C₆H₅CHBrCOCBr(CH₃)₂	"	"	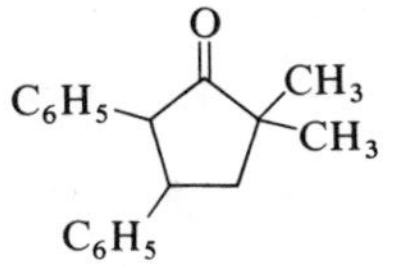 (49)	36

C. With Enamines[a]

$$\text{NCR}^1{=}\text{CR}^2\text{R}^3 \quad (I)$$

Zn/Cu, I, $R^1{-}R^2 = (CH_2)_4$, $R^3 = H$	1. DME, $-5°$, 30 min 2. 25°, 16 hr	II, $R^1{-}R^2 = (CH_2)_4$, $R^3 = H$ (39)	122
$Fe_2(CO)_9$, I, $R^1 = H$, $R^2 = R^3 = CH_3$	1. C_6H_6, 60°, 20 min 2. 25°, 12 hr	II, $R^1 = H$, $R^2 = R^3 = CH_3$ (79)	122, 123
$Fe_2(CO)_9$, I, $R^1 = C_2H_5$, $R^2 = H$, $R^3 = CH_3$	C_6H_6, 45°, 18 hr	II, $R^1 = C_2H_5$, $R^2 = H$, $R^3 = CH_3$ (73)	122
$Fe_2(CO)_9$, I, $R^1 = C_6H_5$, $R^2 = R^3 = H$	C_6H_6, 32°, 20 hr	II, $R^1 = C_6H_5$, $R^2 = R^3 = H$ (84,[e] 91[f])	124, 122
$Fe_2(CO)_9$, I, $R^1{-}R^2 = (CH_2)_3$, $R^3 = H$	C_6H_6, 25°, 12 hr	II, $R^1{-}R^2 = (CH_2)_3$, $R^3 = H$ (74)	122, 123
$Fe_2(CO)_9$, I, $R^1{-}R^2 = (CH_2)_4$, $R^3 = H$	C_6H_6, 25°, 24 hr	II, $R^1{-}R^2 = (CH_2)_4$, $R^3 = H$ (100)	122, 123
$Fe_2(CO)_9$, I, $R^1{-}R^2 = (CH_2)_4$, $R^3 = H$, furan	25°, 24 hr	II, $R^1{-}R^2 = (CH_2)_4$, $R^3 = H$ (4),	122

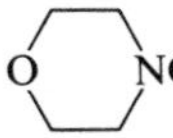

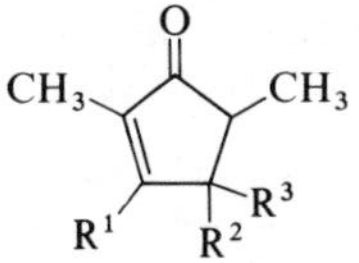

C₅ (CH₃CHBr)₂CO

TABLE VIII. [3 + 2] CYCLOCOUPLING OF α,α'-DIBROMO KETONES (*Continued*)

C. With Enamines[d] (*Continued*)

Halo Ketone	Reactants (equiv.)	Conditions	Products and Yields (%)	Refs.
C_5 $(CH_3CHBr)_2CO$ (*Contd.*)	$Fe_2(CO)_9$, I, $R^1-R^2 = (CH_2)_4$, $R^3 = H$, furan	$40°$, 24 hr	(III), (IV) III/IV (2:1) (35)	122
	$Fe_2(CO)_9$, I, $R^1 = H$, $R^2-R^3 = (CH_2)_5$	C_6H_6, $42°$, 18 hr	II, $R^1-R^2 = (CH_2)_4$, $R^3 = H$ (10), III/IV (Total 90)	122, 123
	$Fe_2(CO)_9$, I, $R^1-R^2 = (CH_2)_5$, $R^3 = H$	C_6H_6, $25°$, 14 hr	II, $R^1 = H$, $R^2-R^3 = (CH_2)_5$ (71)	122, 123
	$Fe_2(CO)_9$, $R^1-R^2 = (CH_2)_{10}$, $R^3 = H$	C_6H_6, $45°$, 19 hr	II, $R^1-R^2 = (CH_2)_5$, $R^3 = H$ (100)	122, 123
	$Fe_2(CO)_9$, I, $R^1 = H$, $R^2-R^3 = (CH_2)_{11}$	C_6H_6, $25°$, 12 hr	II, $R^1-R^2 = (CH_2)_{10}$, $R^3 = H$ (90)	122, 123
			II, $R^1 = H$, $R^2-R^3 = (CH_2)_{11}$ (65)	122 123
C_6 $(CH_3)_2CBrCOCHBrCH_3$	$Fe_2(CO)_9$, 1-morpholinocyclohexene	C_6H_6, $30°$	(I),[a]	36

326

	Carbonyl Compound	Reagent and Conditions	Product	Refs.
C_7	$(C_2H_5CHBr)_2CO$	$Fe_2(CO)_9$,	(II) structure (CH_3); I/II (59:41) (86)	122, 123
		morpholine–$NCR^1=CR^2R^3$ (I)	(II)[a]	
		I, $R^1 = C_2H_5$, $R^2 = H$, $R^3 = CH_3$; C_6H_6, 46°, 16 hr	II, $R^1 = C_2H_5$, $R^2 = H$, $R^3 = CH_3$ (70)	122, 123
		I, $R^1 = C_6H_5$, $R^2 = R^3 = H$; C_6H_6, 25°, 14 hr	II, $R^1 = C_6H_5$, $R^2 = R^3 = H$ (64)	122
		I, $R^1-R^2 = (CH_2)_4$, $R^3 = H$; C_6H_6, 25°, 12 hr, $h\nu^g$	II, $R^1-R^2 = (CH_2)_4$, $R^3 = H$ (89)	122
		$Fe_2(CO)_9$, α-morpholinostyrene; C_6H_6, 40°, 22 hr	morpholino–C_6H_5 lactone structure (>86)	118
	$[(CH_3)_2CBr]_2CO$	$Fe_2(CO)_9$, 1-morpholinocyclohexene; 1. C_6H_6, 60°, 10 hr, $h\nu^g$; 2. 25°, 12 hr, $h\nu^g$	bicyclic morpholino–CH_3 structure (87)	122, 123
C_9	$(i\text{-}C_3H_7CHBr)_2CO$	$Fe_2(CO)_9$, morpholine–$NCR^1=CR^2R^3$ (I)	(II) structure ($i\text{-}C_3H_7$, $C_3H_7\text{-}i$)	

TABLE VIII. [3 + 2] CYCLOCOUPLING OF α,α'-DIBROMO KETONES (Continued)

C. With Enamines[d] (Continued)

Halo Ketone	Reactants (equiv.)	Conditions	Products and Yields (%)	Refs.
C₉ (i-C₃H₇CHBr)₂CO (Contd.)	I, R¹ = C₆H₅, R² = R³ = H I, R¹—R² = (CH₂)₄, R₃ = H	C₆H₆, 30°, 15 hr C₆H₆, 32°, 12 hr	II, R¹ = C₆H₅, R² = R³ = H (72) I, R¹—R² = (CH₂)₄, R³ = H (73)[h] 	122, 123 122, 123
C₆H₅CHBrCOCH₂Br	Fe₂(CO)₉, 1-morpholinocyclohexene	C₆H₆, 30°	 I/II (76:24) (22)	36
C₁₁ C₆H₅CHBrCOCHBr(CH₃)₂	"	"	 I/II (75:25) (61)	36

328

D. With Carboxamides and Lactams

	Reactant	Reagents / Conditions	Product (yield)	Refs.
C_{15}	$(C_6H_5CHBr)_2CO$	$Fe_2(CO)_9$, 1-morpholinocyclopentene C_6H_6, 30°, 22 hr	(structure, C_6H_5, C_6H_5) (66)[h]	122
C_5	$(CH_3CHBr)_2CO$	Zn/Cu, DMF $-35°$, 10 min	(I), (II) with $N(CH_3)_2$; I/II (5:3) (15)	128
		$Fe_2(CO)_9$, $RCON(CH_3)_2$ (I)	(II) R	
		I, R = H; 12 hr	II, R = H (53)	34, 35
		I, R = CH_3; 14 hr	II, R = CH_3 (21)	34, 35
		I, R = C_6H_5; C_6H_6, 19 hr	II, R = C_6H_5 (25)	34, 35
	$(CH_3)_2CBrCOCH_2X$ (I)	Zn/Cu, $RCONC(CH_3)_2$ (II)	(III) R, $N(CH_3)_2$	
	I, X = Br	II, R = H; $-35°$, 10 min	III, R = H (5.2)	128
		II, R = CH_3; $-15°$, 55 min	III, R = CH_3 (—)	128
	I, X = I	II, R = H; $-35°$, 10 min	III, R = H (11)	128

329

TABLE VIII. [3 + 2] CYCLOCOUPLING OF α,α′-DIBROMO KETONES (Continued)

D. With Carboxamides and Lactams (Continued)

Halo Ketone	Reactants (equiv.)	Conditions	Products and Yields (%)	Refs.
C₆ $(CH_3)_2CBrCOCHBrCH_3$	Zn/Cu, DMF	−35°, 10 min	(structure) $N(CH_3)_2$, CH_3, CH_3, CH_3CH (50)	128
C₇ $(C_2H_5CHBr)_2CO$	$Fe_2(CO)_9$, $RCON(CH_3)_2$ (I)		(structure II) C_2H_5, R, C_2H_5	
	I, R = H	12 hr	II, R = H (92, 64)	34, 35
	I, R = CH₃	19 hr	II, R = CH₃ (51)	34, 35
	I, R = CH₃	24 hr, hv^i	II, R = CH₃ (41)	34, 35
	I, R = C₆H₅	C_6H_6, 14 hr	II, R = C₆H₅ (42)	34, 35
$[(CH_3)_2CBr]_2CO$	Zn/Cu, $RCON(CH_3)_2$ (I)		(structure II) CH_3, R, $N(CH_3)_2$, $(CH_3)_2C$, CH_3	
	I, R = H	−35°, 10 min	II, R = H (−)	128, 164
	I, R = CH₃	−15°, 55 min	II, R = CH₃ (30), $[i-C_3H_7COC(CH_3)_2]_2$ (3)	128
	$Fe_2(CO)_9$, DMF	18 hr	(structure) CH_3, CH_3, $N(CH_3)_2$, CH_3, CH_3 (3), $CH_2=C(CH_3)COC_3H_7-i$ (80)	34, 28, 35
C₉ $(i-C_3H_7CHBr)_2CO$	$Fe_2(CO)_9$, $RCON(CH_3)_2$ (I)		(structure II) $i-C_3H_7$, R, C_3H_7-i	

Reactant	Reagent	Conditions	Product(s)	Refs.
I, R = H, furan		2. 140°, 15 min 1. 25°, 12 hr 2. 140°	II, R = H (49), (20)[j]	34, 35
I, R = CH₃		12 hr	II, R = CH₃ (49), (39)	34, 35
I, R = CH₃		1. 25°, 12 hr 2. 110°, 15 min	II, R = CH₃ (87)	34, 35
Fe₂(CO)₉, N-methylpyrrolidone		48 hr	(26)	34, 35
C₁₀ CH₃CHBrCOCHBrC₆H₅	Zn/Cu, DMF	1. −35°, 2 hr 2. 25°, overnight	(—)	127
C₁₁ (t-C₄H₉CHBr)₂CO	Fe₂(CO)₉, DMF	17 hr	(98)	34
(CH₃)₂CBrCOCHBrC₆H₅	Zn/Cu, DMF	1. −35°, 2 hr 2. 25°, overnight	(—), (—)	127

TABLE VIII. [3 + 2] Cyclocoupling of α,α′-Dibromo Ketones (Continued)

E. With Other C₂ Substrates

Halo Ketone	Reactants (equiv.)	Conditions	Products and Yields (%)	Refs.
C₅ (CH₃CHBr)₂CO	Cu, NaI (4), CH₂=C(OCH₃)₂ (1)	CH₃CN, 8–10 hr	CH₃ OCH₃ OCH₃ CH₃CH₃ (90)	79
	Fe₂(CO)₉, 1-methoxycyclohexene	C₆H₆, 15.5 hr	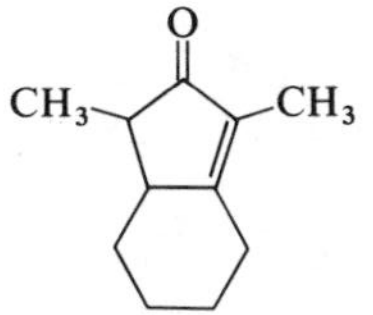 CH₃ O CH₃ (I) (12)	83
	Fe₂(CO)₉, 1-ethoxycyclohexene	C₆H₆, 7.5 hr	I (10)	83
	1. Fe₂(CO)₉ 2. CH₂=C(OC₆H₅)₂	C₆H₆, 1.5 hr 21 hr	CH₃ O OC₆H₅ CH₃ (8)	83
(CH₃)₂CBrCOCH₂Br	Zn/Cu, CH₃CN	DME	CH₃ N CH₃ CH₃ O CH₂ (—)	87
C₆ [structure: 2,6-dibromocyclohexanone]	Cu, CH₂=C(OCH₃)₂ (1), NaI (4)	CH₃CN, 8–10 hr	OCH₃ OCH₃ O (—)	79

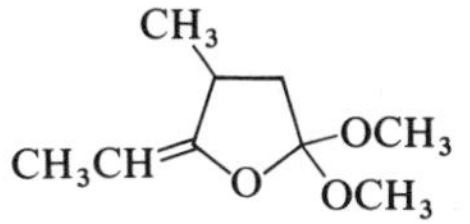

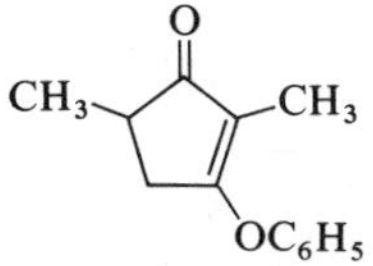

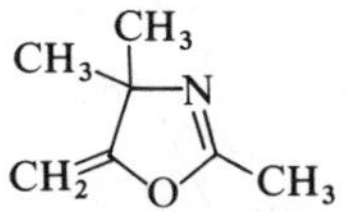

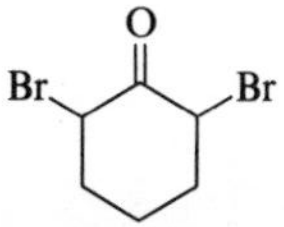

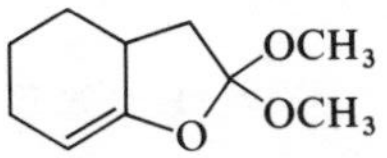

332

				Ref.
C_7 $(C_2H_5CBr)_2CO$	$Fe_2(CO)_9$, CH_3CN	40°, 20 hr	(structure) (I) (27)	83
	$Fe_2(CO)_9$, CH_3CN (100)	C_6H_6, 40°, 13 hr	I (44)	83
	$Fe_2(CO)_9$, CH_3CN (10)	C_6H_6, 40°, 20 hr	I (31)	83
	$Fe_2(CO)_9$, CH_3CN (1.2)	C_6H_6, 40°, 20 hr	I (0)	83
$CH_3CHBrCOCHBrC_3H_7\text{-}n$	Cu, $CH_2{=}C(OCH_3)_2$ (1), NaI (4)	CH_3CN, 8–10 hr	(structure) (—), $n\text{-}C_4H_9COCH(CH_3)CH_2CO_2CH_3$ (—)	79
$[(CH_3)_2CBr]_2CO$	$Hg,^k$ R^1COR^2 (I)		(structure) (II)	
	I, $R^1 = R^2 = CH_3$	25°, 1–2 days	II, $R^1 = R^2 = CH_3$ (49)	130
	I, $R^1 = CH_3$, $R^2 = C_2H_5$	"	II, $R^1 = CH_3$, $R^2 = C_2H_5$ (59)	130
	I, $R^1 = R^2 = C_2H_5$	"	II, $R^1 = R^2 = C_2H_5$ (25)	130
	I, $R^1 = CH_3$, $R^2 = C_3H_7\text{-}n$	"	II, $R^1 = CH_3$, $R^2 = C_3H_7\text{-}n$ (47)	130
	$Fe_2(CO)_9$, CH_3CN	14 hr	(structure) (50)	83
C_9 $(i\text{-}C_3H_7CHBr)_2CO$	"	30°, 40 hr	(structure) (64)	83
	"	CH_3CONH_2, 22 hr	(structure) (90, 76)	83

333

TABLE VIII. [3 + 2] Cyclocoupling of α,α'-Dibromo Ketones (*Continued*)

E. With Other C$_2$ Substrates (Continued)

Halo Ketone	Reactants (equiv.)	Conditions	Products and Yields (%)	Refs.
C$_{15}$ (C$_6$H$_5$CHBr)$_2$CO	NaI, (NC)$_2$C=C(CN)$_2$ (I)	Acetone, reflux, 1 hr	(II) (36)	84, 85
	NaI, I	CH$_3$CN, reflux, 1 hr	II (trace)	85
	NaI, C$_2$H$_5$O$_2$CN=NCO$_2$C$_2$H$_5$	1. CH$_3$CN, 1 day 2. Reflux, 1 hr	(15)	85

[a] The product was a mixture of diastereoisomers.

[b] The product was a mixture of diastereoisomers, from which a single epimer was obtained by thin-layer chromatography on silica gel or by distillation.

[c] Prepared according to reference 163.

[d] Generally, treatment of the initial products with silica gel or 3% ethanolic sodium hydroxide was required to obtain the cyclopentenones.

[e] This yield was obtained on a 100-mmol scale.

[f] This yield was obtained on a 10-mmol scale.

[g] The reaction mixture was irradiated with visible light through 10% aqueous copper sulfate.

[h] A single *trans* isomer was obtained.

[i] The reaction mixture was irradiated with visible light (>350 nm).

[j] A 1:1 mixture of *cis* and *trans* isomers.

[k] Ultrasonically dispersed mercury was used.

TABLE IX. MISCELLANEOUS CYCLOCOUPLINGS OF α,α'-DIBROMO KETONES

Halo Ketone	Reactants (equiv.)	Conditions	Products and Yields (%)	Refs.
C_5 $(CH_3CHBr)_2CO$	Cu, C_6H_5 (I)		(II), (III) structures with $CH_3CHCOC_2H_5$	
	I, R = H	CH_3CN	R = H II (—), III (—)	132
	I, R = CH_3	CH_3CN	R = CH_3 II (—), III (—)	132
C_7 $[(CH_3)_2CBr]_2CO$	Cu, RNC (I)		(II) structure	
	I, R = $C_4H_9\text{-}t$	C_6H_6, 12 hr	II, R = $C_4H_9\text{-}t$ (65)	108
	I, R = cyclohexyl	C_6H_6, 12 hr	II, R = cyclohexyl (99, 86)	108
	I, R = $C(CH_3)_2C_5H_{11}\text{-}n$	—	II, R = $C(CH_3)_2C_5H_{11}\text{-}n$ (60)	108
	I, R = $C_4H_9\text{-}t$, 2-methylfuran	C_6H_6, 10 hr	II, R = $C_4H_9\text{-}t$ (—), (40)	108
	Zn. I, R = $C_4H_9\text{-}t$	Py/C_6H_6, 3 hr	II, R = $C_4H_9\text{-}t$ (43)	108

335

TABLE IX. MISCELLANEOUS CYCLOCOUPLINGS OF α,α'-DIBROMO KETONES (*Continued*)

Halo Ketone	Reactants (equiv.)	Conditions	Products and Yields (%)	Refs.
C$_7$ [(CH$_3$)$_2$CBr]$_2$CO (*Contd.*)	Fe$_2$(CO)$_9$, 6-dimethylaminofulvene	C$_6$H$_6$, 45°, 6 hr	(17)	99
	Fe$_2$(CO)$_9$, NC$_6$H$_4$Cl-p	C$_6$H$_6$, 50°, 8 hr	(37)	131a
	Fe$_2$(CO)$_9$, NC$_6$H$_4$Cl-p ... Fe(CO)$_3$	C$_6$H$_6$, 45°, 8 hr	(22)	131a
	Fe$_2$(CO)$_9$, (I)		(II)	

I, R = H   C$_6$H$_6$, 50°, 30 hr   II, R = H (50)   99

I, R = Cl   C$_6$H$_6$, 40°, 70 hr   II, R = Cl (28)   99

I, R = C$_6$H$_5$ C$_6$H$_6$, 45°, 30 hr   II, R = C$_6$H$_5$ (65)   99

Fe$_2$(CO)$_9$, [Fe(CO)$_3$ cycloheptadiene complex]   C$_6$H$_6$, 50°, 30 hr   Fe(CO)$_3$ (20) [CH$_3$ CH$_3$ / CH$_3$ CH$_3$ / O]   131a

Fe$_2$(CO)$_9$, [N—CO$_2$C$_2$H$_5$ azepine Fe(CO)$_3$ complex]   C$_6$H$_6$, 50°, 8 hr   CO$_2$C$_2$H$_5$ Fe(CO)$_3$ (—), [CH$_3$ CH$_3$ / CH$_3$ CH$_3$ / O]   131a

   C$_6$H$_6$, 60°, 8 hr   CO$_2$C$_2$H$_5$ N—Fe(CO)$_3$ (—), i-C$_3$H$_7$COC(CH$_3$)$_2$

Cu, t-C$_4$H$_9$NC   C$_6$H$_6$, 12 hr   NC$_4$H$_9$-t (70) [O, bicyclic structure]   108

C$_{10}$ [Br, O, Br cyclohexanone structure]

337

TABLE IX. MISCELLANEOUS CYCLOCOUPLINGS OF α,α'-DIBROMO KETONES (*Continued*)

Halo Ketone	Reactants (equiv.)	Conditions	Products and Yields (%)	Refs.
C_{15} $(C_6H_5CHBr)_2CO$	(I)		(II), (III)	
	Cu, I, $R^1 = R^2 = R^3 = H$, NaI	—	$R^1 = R^2 = R^3 = H$ II $(-)$, III $(-)$	132
	Cu, I, $R^1 = R^2 = H$, $R^3 = CH_3$, NaI	—	II, $R^1 = R^2 = H$, $R^3 = CH_3$ $(-)$	132
	Cu, I, $R^1 = R^3 = H$, $R^2 = CH_3$, NaI	—	II, $R^1 = R^3 = H$, $R^2 = CH_3$ $(-)$	132
	Cu, I, $R^1 = CH_3$, $R^2 = R^3 = H$, NaI	—	$R^1 = CH_3$, $R^2 = R^3 = H$ II $(-)$, III $(-)$	132
	$Fe_2(CO)_9$, I, $R^1 = R^2 = R^3 = H$	C_6H_6, $60°$	$R^1 = R^2 = R^3 = H$ II (35), III (10)	132

TABLE X. MISCELLANEOUS REACTIONS OF α,α'-DIBROMO KETONES

	Halo Ketone	Reactants (equiv.)	Conditions	Products and Yields (%)	Refs.
C_5	$(CH_3CHBr)_2CO$	$HC{\equiv}CR$ (I) Cu, I, R = CH_2OH, NaI	CH_3CN, 8–10 hr	$RCH{=}C{=}C(CH_3)COC_2H_5$ (II) II, R = CH_2OH (25)	79
		Cu, I, R = $CH(OC_2H_5)_2$, NaI	"	II, R = $CH(OC_2H_5)_2$ (30)	79
		Cu, I, R = CH_2OCH_3, NaI	"	II, R = CH_2OCH_3 (35)	79
		$Fe_2(CO)_9$, 2,3-dihydropyrane	C_6H_6	$C_2H_5COCH(CH_3)$ (10)	83
C_6		1. $(CH_3)_2CuLi$ 2. TMSCl, $(C_2H_5)_3N$	Ether, −78°, 0.5 hr 1 hr	(46)	81
C_7	$[(CH_3)_2CBr]_2CO$	$Fe_2(CO)_9$, 1,3,5-cycloheptatriene	C_6H_6, 70°, 6 hr	$i\text{-}C_3H_7\text{-}COC(CH_3)_2$ (65)	131b
C_9	$(i\text{-}C_3H_7CHBr)_2CO$	$Fe_2(CO)_9$, 2,3-dihydropyrane	C_6H_6	$i\text{-}C_3H_7CH_2COCH(C_3H_7\text{-}i)$ (3)	83

REFERENCES

[1] (a) C. D. Gutsche and D. Redmore, *Carbocyclic Ring Expansion Reactions*, Academic Press, New York, 1968, Chap. 4; (b) C. D. Gutsche, *Org. React.*, **8**, 364 (1954).

[2] (a) A. T. Nielsen and W. J. Houlihan, *Org. React.*, **16**, 1 (1968); (b) R. A. Raphael, *Chemistry of Carbon Compounds*, E. H. Rodd, Ed., Vol. IIA, Elsevier, Amsterdam, 1953, Chap. 4; (c) M. Green, G. R. Knox, and P. L. Pauson, *Rodd's Chemistry of Carbon Compounds*, S. Coffey, Ed., Vol. IIA, Elsevier, Amsterdam, 1967, Chap. 4; (d) R. A. Ellison, *Synthesis*, **1973**, 397; (e) B. M. Trost and M. J. Bogdanowicz, *J. Am. Chem. Soc.*, **95**, 5311 (1973).

[3] (a) B. M. Trost, *Acc. Chem. Res.*, **7**, 85 (1974); (b) P. T. Lansbury, *ibid.*, **5**, 311 (1972).

[4] B. G. McFarland, *Steroid Reactions*, C. Djerassi, Ed., Holden-Day, San Francisco, 1963, p. 427.

[5] (a) R. Jacquier, *Bull. Soc. Chim. Fr.*, **1950**, 35; (b) A. S. Kende, *Org. React.*, **11**, 261 (1969); (c) D. Redmore and C. D. Gutsche, *Adv. Alicycl. Chem.*, **3**, 1 (1971).

[6] D. Fårcasiu and P. v. R. Schleyer, *J. Org. Chem.*, **38**, 3455 (1973).

[7] R. Noyori, *Acc. Chem. Res.*, **12**, 61 (1979).

[8] H. M. R. Hoffmann, *Angew. Chem., Int. Ed. Engl.*, **12**, 819 (1973).

[9] J. K. Kochi, *Organometallic Mechanisms and Catalysis*, Academic Press, New York, 1978, Chap. 7.

[10] J. R. Hanson and E. Premuzic, *Angew. Chem., Int. Ed. Engl.*, **7**, 247 (1968).

[11] H. Inoue, T. Nagata, H. Hata, and E. Imoto, *Bull. Chem. Soc. Jpn.*, **52**, 469 (1979).

[12] H. Alper and L. Pattee, *J. Org. Chem.*, **44**, 2568 (1979).

[13] P. J. Elving and J. T. Leone, *J. Am. Chem. Soc.*, **79**, 1546 (1957).

[14] H. Alper and E. C. H. Keung, *J. Org. Chem.*, **37**, 2566 (1972).

[15] E. Ghera, D. H. Perry, and S. Shoua, *J. Chem. Soc., Chem. Commun.*, **1973**, 858.

[16] L. T. Scott and W. D. Cotton, *J. Am. Chem. Soc.*, **95**, 2708 (1973).

[17] R. G. Doerr and P. S. Skell, *J. Am. Chem. Soc.*, **89**, 4684 (1967).

[18] R. Hoffmann, *J. Am. Chem. Soc.*, **90**, 1475 (1968).

[19] R. C. Bingham, M. J. S. Dewar, and D. H. Lo, *J. Am. Chem. Soc.*, **97**, 1302 (1975).

[20] J. P. Dirlam, L. Eberson, and J. Casanova, *J. Am. Chem. Soc.*, **94**, 240 (1972).

[21] H. M. R. Hoffmann, T. A. Nour, and R. H. Smithers, *J. Chem. Soc., Chem. Commun.*, **1972**, 963.

[22] A. J. Fry and J. J. O'Dea, *J. Org. Chem.*, **40**, 3625 (1975).

[23] A. J. Fry and D. Herr, *Tetrahedron Lett.*, **1978**, 1721.

[24] A. J. Fry and J. P. Bujanauskas, *J. Org. Chem.*, **43**, 3157 (1978).

[25] A. J. Fry and G. S. Ginsburg, *J. Am. Chem. Soc.*, **101**, 3927 (1979).

[26] A. J. Fry, W. A. Donaldson, and G. S. Ginsburg, *J. Org. Chem.*, **44**, 349 (1979).

[27] A. J. Fry and A. T. Lefor, *J. Org. Chem.*, **44**, 1270 (1979).

[28] R. Noyori, Y. Hayakawa, H. Takaya, S. Murai, R. Kobayashi, and N. Sonoda, *J. Am. Chem. Soc.*, **100**, 1759 (1978).

[29] R. Noyori, Y. Hayakawa, M. Funakura, H. Takaya, S. Murai, R. Kobayashi, and S. Tsutsumi, *J. Am. Chem. Soc.*, **94**, 7202 (1972).

[30] (a) H. E. Zimmerman and R. J. Pasteris, *J. Org. Chem.*, **45**, 4864 (1980); (b) *ibid.*, **45**, 4876 (1980).

[31] H. E. Zimmerman, D. S. Crumrine, D. Döpp, and P. S. Huyffer, *J. Am. Chem. Soc.*, **91**, 434 (1969).

[32] M. Nishizawa and R. Noyori, *Bull. Chem. Soc. Jpn.*, **54**, 2233 (1981).

[33] R. Noyori, F. Shimizu, and Y. Hayakawa, *Tetrahedron Lett.*, **1978**, 2091.

[34] Y. Hayakawa, H. Takaya, S. Makino, N. Hayakawa, and R. Noyori, *Bull. Chem. Soc. Jpn.*, **50**, 1990 (1977).

[35] R. Noyori, Y. Hayakawa, S. Makino, N. Hayakawa, and H. Takaya, *J. Am. Chem. Soc.*, **95**, 4103 (1973).

[36] (a) R. Noyori, F. Shimizu, K. Fukuta, H. Takaya, and Y. Hayakawa, *J. Am. Chem. Soc.*, **99**, 5196 (1977); (b) R. Noyori, *Ann. N.Y. Acad. Sci.*, **295**, 225 (1977).

[37] A. J. Fry and R. Scoggins, *Tetrahedron Lett.*, **1972**, 4079.

[38] W. J. M. van Tilborg, R. Plomp, R. de Ruiter, and C. J. Smit, *Recl. Trav. Chim. Pays-Bas*, **99**, 206 (1980).

[39] H. M. R. Hoffmann and T. A. Nour, *J. Chem. Soc., Chem. Commun.*, **1975**, 37.

[40] J.-E. Dubois, P. Fournier, and C. Lion, *C. R. Acad. Sci., Ser. C*, **279**, 965 (1974).

[41] J. J. Beereboom, C. Djerassi, D. Ginsburg, and L. F. Fieser, *J. Am. Chem. Soc.*, **75**, 3500 (1953).

[42] H. Inoue, H. Hata, and E. Imoto, *Chem. Lett.*, **1975**, 1241.

[43] H. Alper and D. D. Roches, *J. Org. Chem.*, **41**, 806 (1976).

[44] H. H. Inhoffen and G. Zühlsdorff, *Ber.*, **76**, 233 (1943).

[45] T. Sato and R. Noyori, *Bull. Chem. Soc. Jpn.*, **51**, 2745 (1978).

[46] H. E. Zimmerman and A. Mais, *J. Am. Chem. Soc.*, **81**, 3644 (1959).

[47] H. M. R. Hoffmann and M. N. Iqbal, *Tetrahedron Lett.*, **1975**, 4487.

[48] R. Noyori, T. Souchi, and Y. Hayakawa, *J. Org. Chem.*, **40**, 2681 (1975).

[49] E. J. Corey, S. W. Chow, and R. A. Scherrer, *J. Am. Chem. Soc.*, **79**, 5773 (1957).

[50] T.-Y. Luh, C. H. Lai, K. L. Lei, and S. W. Tam, *J. Org. Chem.*, **44**, 641 (1979).

[51] H. Alper, K. D. Logbo, and H. des Abbayes, *Tetrahedron Lett.*, **1977**, 2861.

[52] J. R. Bull and A. Tuinman, *Tetrahedron Lett.*, **1973**, 4349.

[53] J.-E. Dubois, C. Lion, and C. Moulineau, *Tetrahedron Lett.*, **1971**, 177.

[54] T.-L. Ho and C. M. Wong, *Synth. Commun.*, **3**, 237 (1973).

[55] T.-L. Ho and G. A. Olah, *Synthesis*, **1976**, 807.

[56] J. M. Townsend and T. A. Spencer, *Tetrahedron Lett.*, **1971**, 137.

[57] T. A. Spencer, R. W. Britton, and D. S. Watt, *J. Am. Chem. Soc.*, **89**, 5727 (1967).

[58] J. M. Townsend, Ph.D. Dissertation, Dartmouth College, Hanover, N.H., 1971, [*Diss. Abstr. Int. B*, **32**, 5118 (1972)].

[59] F. G. Saitkulova, G. G. Abashev, and I. I. Lapkin, *Zh. Org. Khim.*, **9**, 1405 (1973) [*C.A.*, **79**, 91529z (1973)].

[60] G. H. Posner and J. J. Sterling, *J. Am. Chem. Soc.*, **95**, 3076 (1973).

[61] E. Ghera and S. Shoua, *Tetrahedron Lett.*, **1974**, 3843.

[62] F. G. Saitkulova, G. G. Abashev, I. I. Lapkin, *Izv. Vyssh., Ucheb. Zaved., Khim. Khim. Tekhnol.*, **16**, 1458 (1973) [*C. A.*, **80**, 36667p (1973)].

[63] K. Maruoka, S. Hashimoto, Y. Kitagawa, H. Yamamoto, and H. Nozaki, *J. Am. Chem. Soc.*, **99**, 7705 (1977).

[64] I. I. Lapkin and F. G. Saitkulova, *Zh. Org. Khim.*, **7**, 2488 (1971) [*C.A.*, **76**, 71955v (1971)].

[65] C. Wakselman and M. Mondon, *Tetrahedron Lett.*, **1973**, 4285.

[66] T. Mitsudo, Y. Watanabe, T. Sasaki, H. Nakanishi, M. Yamashita, and Y. Takegami, *Tetrahedron Lett.*, **1975**, 3163.

[67] E. Yoshisato and S. Tsutsumi, *J. Am. Chem. Soc.*, **90**, 4488 (1968).

[68] E. Yoshisato and S. Tsutsumi, *J. Chem. Soc., Chem. Commun.*, **1968**, 33.

[69] L. Gorrichon-Guigon, Y. Maroni-Barnaud, and P. Maroni, *Bull. Soc. Chim. Fr.*, **1975**, 291.

[70] I. I. Lapkin, G. G. Abashev, and F. G. Saitkulova, *Zh. Org. Khim.*, **12**, 967 (1976) [*C.A.*, **85**, 46149p (1976)].

[71] A. E. Greene and J.-P. Deprés, *J. Am. Chem. Soc.*, **101**, 4003 (1979).

[72] C. Djerassi, *Steroid Reactions*, Holden-Day, San Francisco, 1963, Chap. 4, and references cited therein.

[73] J.-P. Deprés and A. E. Greene, *J. Org. Chem.*, **45**, 2036 (1980).

[74] H. Takaya, S. Makino, Y. Hayakawa, and R. Noyori, *J. Am. Chem. Soc.*, **100**, 1765 (1978).

[75] Y. Hayakawa, Y. Baba, S. Makino, and R. Noyori, *J. Am. Chem. Soc.*, **100**, 1786 (1978).

[76] (a) R. H. Prager and J. M. Tippett, *Aust. J. Chem.*, **27**, 1457 (1974); (b) *ibid.*, **27**, 1467 (1974).

[77] R. H. Prager and J. M. Tippett, *Tetrahedron Lett.*, **1973**, 5199.

[78] G. H. Posner, C. E. Whitten, and J. J. Sterling, *J. Am. Chem. Soc.*, **95**, 7788 (1973).

[79] A. P. Cowling and J. Mann, *J. Chem. Soc., Chem. Commun.*, **1978**, 1006.

[80] C. Lion and J.-E. Dubois, *Tetrahedron*, **31**, 1223 (1975).

[81] G. H. Posner, J. J. Sterling, C. E. Whitten, C. M. Lentz, and D. J. Brunelle, *J. Am. Chem. Soc.*, **97**, 107 (1975).

[82] C. Chassin, E. A. Schmidt, and H. M. R. Hoffmann, *J. Am. Chem. Soc.*, **96**, 606 (1974).

[83] Y. Hayakawa and R. Noyori, Nagoya University, Nagoya, Japan, unpublished results.

[84] R. C. Cookson and M. J. Nye, *Proc. Chem. Soc., London*, **1963**, 129.

[85] R. C. Cookson and M. J. Nye, *J. Chem. Soc.*, **1965**, 2009.

[86] R. Noyori, S. Makino, and H. Takaya, *J. Am. Chem. Soc.*, **93**, 1272 (1971).

[87] H. M. R. Hoffmann and R. Chidgey, *Tetrahedron Lett.*, **1978**, 85.

[88] R. Chidgey and H. M. R. Hoffmann, *Tetrahedron Lett.*, **1977**, 2633.

[89] H. Takaya, Y. Hayakawa, S. Makino, and R. Noyori, *J. Am. Chem. Soc.*, **100**, 1778 (1978).

[90] R. Noyori, Y. Hayakawa, S. Makino, and H. Takaya, *Chem. Lett.*, **1973**, 3.

[91] A. Busch and H. M. R. Hoffmann, *Tetrahedron Lett.*, **1976**, 2379.

[92] D. I. Rawson, B. K. Carpenter, and H. M. R. Hoffmann, *J. Am. Chem. Soc.*, **101**, 1786 (1979).

[93] H. M. R. Hoffmann, K. E. Clemens, and R. H. Smithers, *J. Am. Chem. Soc.*, **94**, 3940 (1972).

[94] R. C. Cookson, M. J. Nye, and G. Subrahmanyam, *J. Chem. Soc. C*, **1967**, 473.

[95] R. K. Siemionko and J. A. Berson, *J. Am. Chem. Soc.*, **102**, 3870 (1980).

[96] W. Wierenga, B. R. Evans, and J. A. Woltersom, *J. Am. Chem. Soc.*, **101**, 1334 (1979).

[97] S. A. Monti and J. M. Harless, *J. Am. Chem. Soc.*, **99**, 2690 (1977).

[98] K. Kashman and A. Rudi, *Tetrahedron*, **30**, 109 (1974).

[99] T. Ishizu, M. Mori, and K. Kanematsu, *J. Org. Chem.*, **46**, 526 (1981).

[100] A. S. Narula, *Tetrahedron Lett.*, **1979**, 1921.

[101] G. Fierz, R. Chidgey, and H. M. R. Hoffmann, *Angew. Chem., Int. Ed. Engl.*, **13**, 410 (1974).

[102] (a) A. P. Cowling and J. Mann, *J. Chem. Soc., Perkin Trans. 1*, **1978**, 1564; (b) A. P. Cowling, J. Mann, and A. A. Usmani, *ibid.*, **1981**, 2116.

[103] R. Noyori, S. Makino, Y. Baba, and Y. Hayakawa, *Tetrahedron Lett.*, **1974**, 1049.

[104] R. Noyori, Y. Baba, and Y. Hayakawa, *J. Am. Chem. Soc.*, **96**, 3336 (1974).

[105] J. D. White and Y. Fukuyama, *J. Am. Chem. Soc.*, **101**, 226 (1979).

[106] M. J. Arco, M. H. Trammell, and J. D. White, *J. Org. Chem.*, **41**, 2075 (1976).

[107] M. R. Ashcroft and H. M. R. Hoffmann, *Org. Synth.*, **58**, 17 (1978).

[108] Y. Ito, M. Asada, K. Yonezawa, and T. Saegusa, *Synth. Commun.*, **4**, 87 (1974).

[109] R. Noyori, S. Makino, and H. Takaya, *Tetrahedron Lett.*, **1973**, 1745.

[110] Y. Hayakawa, M. Sakai, and R. Noyori, *Chem. Lett.*, **1975**, 509.

[111] R. Noyori, S. Makino, T. Okita, and Y. Hayakawa, *J. Org. Chem.*, **40**, 806 (1975).

[112] (a) S. Masamune, C. U. Kim, K. E. Wilson, G. O. Spessard, P. E. Georghiou, and G. S. Bates, *J. Am. Chem. Soc.*, **97**, 3512 (1975); (b) S. Masamune, H. Yamamoto, S. Kamata, and A. Fukuzawa, *ibid.*, **97**, 3513 (1975).

[113] (a) R. Noyori, T. Sato, and Y. Hayakawa, *J. Am. Chem. Soc.*, **100**, 2561 (1978); (b) T. Sato, R. Ito, Y. Hayakawa, and R. Noyori, *Tetrahedron Lett.*, **1978**, 1829.

[114] (a) T. Sato, M. Watanabe, and R. Noyori, *Tetrahedron Lett.*, **1978**, 4403; (b) T. Sato, M. Watanabe, and R. Noyori, *Chem. Lett.*, **1978**, 1297; (c) T. Sato, M. Watanabe, and R. Noyori, *Tetrahedron Lett.*, **1979**, 2897; (d) T. Sato, K. Marunouchi, and R. Noyori, *ibid.*, **1979**, 3669; (e) T. Sato and R. Noyori, *Nucleic Acids Research*, Symposium Series No. 6, 19 (1979); (f) T. Sato and R. Noyori, *Heterocycles*, **13**, 141 (1979); (g) T. Sato and R. Noyori, *Bull. Chem. Soc. Jpn.*, **53**, 1195 (1980); (h) T. Sato, M. Watanabe, and R. Noyori, *Heterocycles*, **14**, 761 (1980); (i) T. Sato, H. Kobayashi, and R. Noyori, *Tetrahedron Lett.*, **21**, 1971 (1980); (j) T. Sato, M. Watanabe, and R. Noyori, *Chem. Lett.*, **1980**, 679; (k) T. Sato and R. Noyori, *Tetrahedron Lett.*, **21**, 2535 (1980); (l) T. Sato, H. Kobayashi, and R. Noyori, *Heterocycles*, **15**, 321 (1981).

[115] (a) R. S. Glass, D. R. Deardorff, and L. H. Gains, *Tetrahedron Lett.*, **1978**, 2965; (b) S. R. Wilson and R. A. Sawicki, *ibid.*, **1978**, 2969.

[116] (a) R. J. Giguere, D. I. Rawson, and H. M. R. Hoffmann, *Synthesis*, **1978**, 902; (b) R. J. Giguere, H. M. R. Hoffmann, M. B. Hursthouse, and J. Trotter, *J. Org. Chem.*, **46**, 2868 (1981).

[117] R. Noyori, M. Nishizawa, F. Shimizu, Y. Hayakawa, K. Maruoka, S. Hashimoto, H. Yamamoto, and H. Nozaki, *J. Am. Chem. Soc.*, **101**, 220 (1979).

[118] Y. Hayakawa, K. Yokoyama, and R. Noyori, *J. Am. Chem. Soc.*, **100**, 1791 (1978).

[119] R. Noyori, K. Yokoyama, and Y. Hayakawa, *J. Am. Chem. Soc.*, **95**, 2722 (1973).

[120] Y. Hayakawa, K. Yokoyama, and R. Noyori, *Tetrahedron Lett.*, **1976**, 4347.

[121] Y. Hayakawa, F. Shimizu, and R. Noyori, *Tetrahedron Lett.*, **1978**, 993.

[122] Y. Hayakawa, K. Yokoyama, and R. Noyori, *J. Am. Chem. Soc.*, **100**, 1799 (1978).

[123] R. Noyori, K. Yokoyama, S. Makino, and Y. Hayakawa, *J. Am. Chem. Soc.*, **94**, 1772 (1972).

[124] R. Noyori, K. Yokoyama, and Y. Hayakawa, *Org. Synth.*, **58**, 56 (1978).

[125] (a) R. M. Coates and R. L. Sowerby, *J. Am. Chem. Soc.*, **93**, 1027 (1971); (b) G. Stork and J. Benaim, *ibid.*, **93**, 5938 (1971).

[126] (a) R. A. Raphael, *Chemistry of Carbon Compounds*, E. H. Rodd, Ed., Vol. IIA, Elsevier, Amsterdam, 1953, pp. 298–307; (b) N. A. J. Rogers, *Rodd's Chemistry of Carbon Compounds*, S. Coffey, Ed., Vol. IIC, 2nd ed., Elsevier, Amsterdam, 1969, pp. 20–31; (c) A. P. Krapcho, *Synthesis*, **1974**, 383.

[127] M.-A. Barrow, A. C. Richards, R. H. Smithers, and H. M. R. Hoffmann, *Tetrahedron Lett.*, **1972**, 3101.

[128] H. M. R. Hoffmann, K. E. Clemens, and E. A. Schmidt, *J. Am. Chem. Soc.*, **94**, 3201 (1972).

[129] (a) C. H. Eugster, *Adv. Org. Chem.*, **2**, 427 (1960); (b) S. Wilkinson, *Quart. Rev., Chem. Soc.*, **15**, 153 (1961).

[130] A. J. Fry, G. S. Ginsburg, and R. A. Parente, *J. Chem. Soc., Chem. Commun.*, **1978**, 1040.

[131] (a) T. Ishizu, K. Harano, M. Yasuda, and K. Kanematsu, *J. Org. Chem.*, **46**, 3630 (1981); (b) T. Ishizu, K. Harano, M. Yasuda, and K. Kanematsu, *Tetrahedron Lett.*, **22**, 1601 (1981).

[132] T. Nakao, K. Kurita, H. Awaya, Y. Tominaga, Y. Matsuda, and G. Kobayashi, "Abstracts of Papers," 11th Symposium on Nonbenzenoid Aromatic Compounds, Osaka, Oct. 1978, Chemical Society of Japan, Tokyo, 1978.

[133] C. Rappe, *Acta Chem. Scand.*, **16**, 2467 (1962).

[134] E. LeGoff, *J. Org. Chem.*, **29**, 2048 (1964).

[135] (a) R. D. Clark and C. H. Heathcock, *J. Org. Chem.*, **41**, 636 (1976); (b) J. M. Denis, C. Girard, and J. M. Conia, *Synthesis*, **1972**, 549.

[136] R. B. King, *Organometal. Synth.*, **1**, 128 (1965).

[137] R. M. Evans, J. C. Hamlet, J. S. Hunt, P. G. Jones, A. G. Long, J. F. Oughton, L. Stephenson, T. Walker, and B. M. Wilson, *J. Chem. Soc.*, **1956**, 4356.

[138] A. W. Fort, *J. Am. Chem. Soc.*, **84**, 2620 (1962).

[139] G. Rosenkranz, J. Pataki, St. Kaufmann, J. Berlin, and C. Djerassi, *J. Am. Chem. Soc.*, **72**, 4081 (1950).

[140] K. L. Williamson and W. S. Johnson, *J. Org. Chem.*, **26**, 4563 (1961).

[141] K. E. Harding, J. W. Trotter, and L. May, *Synth. Commun.*, **2**, 231 (1972).

[142] J.-E. Dubois, P. Fournier, and C. Lion, *Tetrahedron Lett.*, **1975**, 4263.

[143] K. Gump, S. W. Moje, and C. E. Castro, *J. Am. Chem. Soc.*, **89**, 6770 (1967).

[144] R. R. Sauers and C. K. Hu, *J. Org. Chem.*, **36**, 1153 (1971).

[145] P. A. Grieco, T. Oguri, and S. Gilman, *J. Am. Chem. Soc.*, **102**, 5886 (1980).

[146] G. Rosenkranz, O. Mancera, J. Gatica, and C. Djerassi, *J. Am. Chem. Soc.*, **72**, 4077 (1950).

[147] F. Sondheimer, G. Rosenkranz, O. Mancera, and C. Djerassi, *J. Am. Chem. Soc.*, **75**, 2601 (1953).

[148] S. Binns, J. S. G. Cox, E. R. H. Jones, and B. G. Ketcheson, *J. Chem. Soc.*, **1964**, 1161.

[149] R. E. Beyler, F. Hoffman, L. H. Sarett, and M. Tishler, *J. Org. Chem.*, **26**, 2426 (1961).

[150] J.-E. Dubois and J. Itzkowitch, *Tetrahedron Lett.*, **1965**, 2839.

[151] B. K. Carpenter, D. I. Rawson, and H. M. R. Hoffmann, *Chem. Ind. (London)*, **1975**, 886.

[152] E. Ghera, Y. Gaoni, and S. Shoua, *J. Am. Chem. Soc.*, **98**, 3627 (1976).

[153] P. R. Brook and J. G. Griffiths, *J. Chem. Soc. D, Chem. Commun.*, **1970**, 1344.

[154] M. Rey, U. A. Huber, and A. S. Dreiding, *Tetrahedron Lett.*, **1968**, 3583.

[155] D. A. Bak and W. T. Brady, *J. Org. Chem.*, **44**, 107 (1979).

[156] E. J. Corey, Z. Arnold, and J. Hutton, *Tetrahedron Lett.*, **1970**, 307.

[157] R. G. Carlson and W. S. Mardis, *J. Org. Chem.*, **40**, 817 (1975).

[158] R. Chidgey and H. M. R. Hoffmann, *Tetrahedron Lett.*, **1978**, 1001.

[159] R. Noyori, Y. Baba, S. Makino, and H. Takaya, *Tetrahedron Lett.*, **1973**, 1741.

[160] R. Noyori, H. Kobayashi, and T. Sato, *Tetrahedron Lett.*, **21**, 2573 (1980).

[161] R. Noyori, T. Sato, and H. Kobayashi, *Tetrahedron Lett.*, **21**, 2569 (1980).

[162] J. G. Vinter and H. M. R. Hoffmann, *J. Am. Chem. Soc.*, **95**, 3051 (1973).

[163] H. E. Hennis and W. B. Trapp, *J. Org. Chem.*, **26**, 4678 (1961).

[164] H. M. R. Hoffmann and R. H. Smithers, *Angew. Chem.*, **82**, 43 (1970).

CHAPTER 3

BASE-PROMOTED ISOMERIZATIONS OF EPOXIDES

JACK K. CRANDALL

Indiana University, Bloomington, Indiana

AND

MARCEL APPARU

Université Scientifique et Médicale de Grenoble, Grenoble, France

CONTENTS

ACKNOWLEDGMENT

We gratefully thank Mr. T. C. Johns for help with the literature search.

INTRODUCTION

Epoxides have long enjoyed popularity as synthetic intermediates because of their facile preparation, often with substantial stereochemical control, and their high chemical reactivity, a feature attributable to the ring strain of these small-ring heterocycles.[1,2] The preponderance of synthetic applications involves nucleophilic opening of the epoxide ring, and an enormous range of nucleophilic species has been utilized for this purpose.[1,3] The conversion of epoxides to isomeric compounds under the influence of acidic reagents has also been used preparatively for some time.[1] However, it is only relatively recently that such isomerizations have been effected by strong, non-nucleophilic bases such as lithium dialkylamides (LiNR$_2$). This chapter is concerned with the synthetic potential of these base-promoted reactions of epoxides.[4,5]

Although several types of isomeric products are observed on treatment of epoxides with strong bases, the most interesting conversion from a synthetic point of view is the formation of allylic alcohols as indicated by the generalized transformation of Eq. 1. A specific example illustrating the remarkable selectivity often found for these reactions is the isomerization of α-pinene oxide to *trans*-pinocarveol (Eq. 2).[6] This process, coupled with the stereoselective generation of the epoxide by peracid oxidation of α-pinene, converts a simple double bond into an allylic alcohol with both positional and stereochemical control. Isolated early reports[7,8] on the transformation of Eq. 1 have been followed by extensive studies leading to the definition of the structural features and reaction variables that determine the course of the epoxide to allylic alcohol rearrangement. This subject constitutes the major emphasis of the discussion that follows.

(Eq. 1)

(90–95%)   (Eq. 2)

A second, much less common transformation is typified by the conversion of *cis*-cyclooctene oxide to *endo*-2-bicyclo[3.3.0]octanol (**1**), along with some of the expected allylic alcohol (Eq. 3).[9] This early observation attracted considerable attention because of the unanticipated generation of a new carbon skeleton in the isomerization process and is largely responsible for the subsequent flurry of activity in the area of base isomerizations of epoxides.

(16%)   **1** (70%)   (Eq. 3)

A third type of conversion also occasionally observed leads to the production of an isomeric carbonyl compound.[10]

(46%)

The final isomerization pathway included in this survey involves the transformation of epoxides possessing allylic, propargylic, or benzylic methylene groups with a 1,3 relationship to an epoxide carbon into cyclopropylcarbinols by lithium amides in the presence of hexamethylphosphoramide (HMPA).[11]

(80%)

This review attempts to provide the necessary information for predicting the type of isomerization to be expected for a given epoxide under a specified set of conditions and, where possible, for manipulating these conditions to achieve the desired conversions. The survey is limited to the isomerizations of simple epoxides by powerful, relatively non-nucleophilic bases. Thus the significant literature concerning epoxides with proximate carbonyl functions and other anion-stabilizing substituents is not considered since the chemistry of such molecules is typical of the stabilized anions (intramolecular alkylations, Favorskii rearrangements, *etc.*) with the epoxide unit serving the trivial function of a good leaving group.[1,4,5] In addition to lithium amides and organolithium reagents, the recently developed dialkylaluminum amides[12] are systematically covered. Studies utilizing non-nucleophilic alkoxides (*e.g.*, potassium *tert*-butoxide), particularly in polar aprotic solvents, are also included. However, the vast literature concerning the interaction of epoxides with metal alkoxides in alcohol solvents is not reviewed since these conditions are generally conducive to nucleophilic addition and are only rarely effective in promoting the epoxide isomerizations of concern in this chapter.

MECHANISM

Most of the mechanistic information available regarding the title reactions has been deduced from product studies and deuterium-labeling experiments designed to show the site of proton removal by the basic reagents. Little useful kinetic data are available because of the complicated nature of the commonly utilized lithium reagents. These species undoubtedly exist as aggregates, the nature of which is likely to depend strongly on the solvent systems and other reaction conditions.[13,14] Moreover, the character of these reagents surely changes as the reaction proceeds to generate lithium alkoxides, which can themselves be involved in the aggregation process.

Allylic Alcohol Formation

The formation of allylic alcohols from the reaction of epoxides with lithium amide bases in relatively nonpolar solvents appears to proceed predominately, if not exclusively, by a β-elimination pathway. This is shown most clearly in a study of the isomerization of *cis*- and *trans*-4,5-epoxyoctanes labeled with deuterium at the two ring carbons (Eq. 4).[15] The allylic alcohol product retains almost all of the deuterium label, thereby demonstrating that the major pathway to product does not involve attack of base at a ring proton. The only reasonable mechanistic alternative consistent with this observation invokes proton removal from an adjacent carbon center (β elimination). Similar results are obtained with an appropriately deuterated *cis*-epoxycyclodecane (Eq. 5).[16] It has been noted that the loss of a small amount of the label in these experiments is consistent with a competing route to the allylic alcohol that proceeds by ring metalation, followed by α elimination and rearrangement of the resulting carbenoid species. This mechanism is illustrated in Eq. 6 for *cis*-epoxycyclodecane. However, the observation that unlabeled *cis*-epoxycyclodecane incorporates deuterium in the presence of lithium diethylamide and diethylamine-N-d_1 strongly suggests that the loss of label arises from reversible metalation of the epoxide ring in a process that is unrelated to its rearrangement to allylic alcohol.[16] Other instances of the reversible metalation at an epoxide ring under the conditions of the rearrangement are known.[17-20] Thus, whereas an α-elimination route to allylic alcohol cannot be excluded as a minor mechanistic contribution, this pathway has yet to be conclusively demonstrated. Consequently, a β-elimination mechanism is assumed for much of the discussion concerning allylic alcohol formation.

(Eq. 4)

(Eq. 5)

(Eq. 6)

A second, crucial mechanistic point concerns the stereochemistry of the β-elimination process. This has been shown to proceed by *syn* elimination in the isomerizations of *cis*- and *trans*-4-*tert*-butylepoxycyclohexanes by lithium diethylamide.[21] The *trans* isomer **2** reacts more readily and yields only allylic alcohol products with a high degree of regioselectivity for **3**. A rather complicated deuterium-labeling experiment demonstrates that epoxide **2d** is transformed to alcohol **3** with loss of deuterium.[17] Thus the proton *syn* to the epoxide function is preferentially abstracted by the base. The more slowly reacting *cis* epoxide **4** yields considerable amounts of two cyclohexanone products and, furthermore, shows little regioselectivity in its transformation to allylic alcohols **5** and **6**. Nonetheless, labeled epoxide **4d** gives alcohol **6d** with retention of the deuterium label, indicating that *syn* elimination is also operative in this transformation.

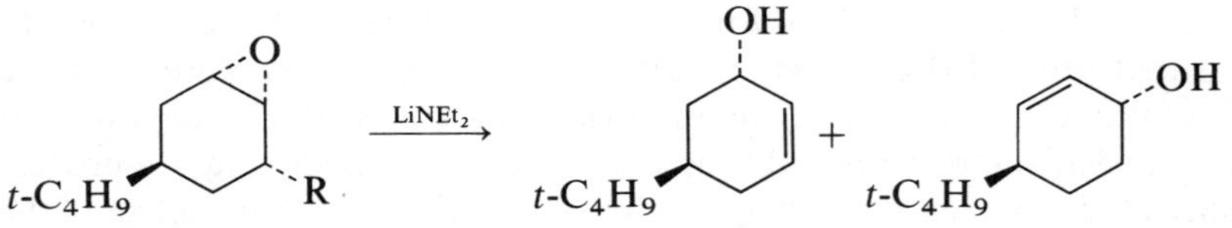

d:R = D

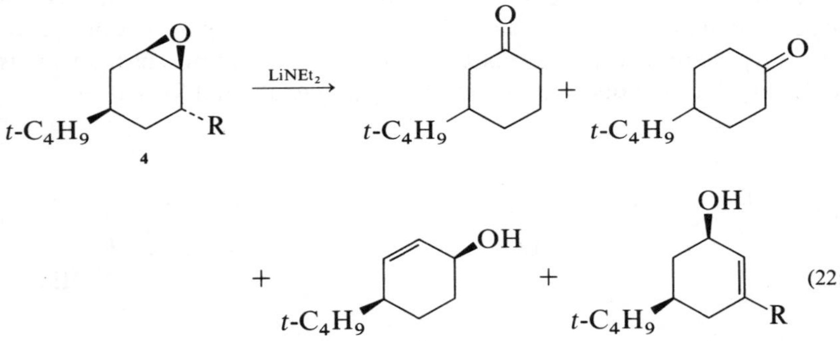

The hypothesis that reaction occurs from a complex of the epoxide with the lithium amide reagent is extremely useful in rationalization of these results and many others discussed in this chapter. Coordination of an oxygen lone pair with an electron-deficient lithium center of the reagent gives a species whose essential features are represented by structure **7**. (The lithium amide reagent is depicted here as monomeric for simplicity, although more than one base unit may actually be involved.) Decomposition of complex **7** in a cyclic, concerted manner provides a reasonable route to allylic alcohol products. Such a process requires the removal of a *syn* hydrogen in cyclic systems such as the rigid epoxycyclohexanes **2** and **4** discussed above. Furthermore, stereoelectronic considerations suggest that β elimination should be facilitated by the availability of a conformation that permits bond reorganization to occur with a maximum of orbital overlap and a minimum of molecular deformation in the transition from reactant to product. This is achieved best from a conformation in which the more acute dihedral angle of the epoxide ring subtends the dihedral angle of the adjacent proton-bearing carbon in the fashion indicated by the Newman projection structure **8**. This situation obtains for removal of a *syn* quasi-axial proton in a half-chair conformation of epoxycyclohexane, but not for abstraction of a *syn* quasi-equatorial hydrogen. This line of reasoning explains the regioselectivity found for the *trans* epoxide **2**, since base removal of the *syn* quasi-axial proton at C_6 in the favored half-chair conformation (equatorial *tert*-butyl group) leads to the predominant allylic alcohol **3** as illustrated by structure **9**.

According to these arguments, the *cis* epoxide **4** should react by removal of the quasi-axial hydrogen at C_3 as shown in **10** and, indeed, the appropriate alcohol **5** is the major product from this epoxide. However, the sluggishness of this reaction suggests that the nearby *tert*-butyl group provides some steric hindrance to this process, allowing the formation of other products in significant quantities. The *syn* elimination leading to the isomeric allylic alcohol **6** most probably proceeds by way of the accessible half-boat conformation **11**, thus satisfying the stereoelectronic requirements. This conformation can also give alcohol **5** since both adjacent *syn* hydrogens have the appropriate relationship to the epoxide ring. Alternatively, product **6** could arise by *syn* elimination of a quasi-equatorial hydrogen atom from the energetically more favorable half-chair conformation; steric interference by the *tert*-butyl group hinders the otherwise favorable abstraction to yield **5**.

The proposed involvement of epoxide–base complexes in epoxide isomerizations is also useful in understanding some of the other propensities of these reactions. In general, a given epoxide can give rise to two isomeric complexes, depending on which lone pair of the oxygen is used as the donor site. This is depicted by structures **12** and **13**. The relative stabilities of these complexes are largely determined by steric interactions between the base moiety and the *syn* substituents. The steric demands of the base are also important in this regard. Furthermore, these features are probably highly influential in determining the activation parameters for subsequent transformations of the complexes. Thus the number and steric bulk of the substituents exert significant control over the competition among potential isomerization pathways.

Trisubstituted epoxides $(R_1, R_2, R_3 = R; R_4 = H)$, for example, will normally show a preference for complex **13** in which the base unit is *syn* to the single hydrogen substituent on the epoxide ring. Consequently, in the absence of other overriding effects, β elimination into the alkyl group (R_3) *cis* to the hydrogen is expected to predominate.

A second situation of some importance concerns *cis*-disubstituted epoxides $(R_1, R_2 = R; R_3, R_4 = H)$. In this instance complex **13** with the base situated *anti* to the alkyl substituents is clearly preferred over complex **12**. However, allylic alcohol formation must proceed by way of the less stable complex if the mechanism previously elaborated is operative. Although β elimination is usually observed with this type of epoxide, it is frequently accompanied by formation of other products. This is especially true with epoxycycloalkanes whose carbocyclic rings provide additional steric destabilization of complex **12**. The other products are derived from metalated epoxide intermediates. Metalation presumably occurs from the more stable complex **13** in which the base is held in close proximity to the epoxide hydrogen atoms. Products derived from epoxide metalation are normally observed only with *cis*-disubstituted epoxides, apparently a result of slow β elimination by means of the destabilized complex **12**. Experimental illustrations of these ideas are deferred to the section entitled "Scope and Limitations."

The concept of a *syn* β elimination from an epoxide–base complex is a useful generalization for describing most epoxide isomerizations. This is particularly important when lithium amides are used in relatively nonpolar solvents such as hexane or ether. Alkyllithium reagents are used less frequently as bases but appear to behave similarly. The more recently introduced dialkylaluminum amides undoubtedly form complexes with even more avidity and seem to display analogous behavior.[12] However, metal alkoxides in polar, aprotic solvents probably utilize other mechanistic pathways that do not involve complexation to the degree suggested for the other bases. Kinetic studies[22,23] of several activated epoxides that isomerize to allylic alcohols under the influence of alkoxide bases in hydroxylic solvents seem to indicate typical E-2 elimination reactions.[24,25] Likewise, the use of more polar solvents or additives such as hexamethyl-phosphoramide profoundly influence certain reactions with lithium amides in a manner suggestive of mechanistic change.[26,27] The ability of hexamethyl-phosphoramide to break up the base aggregates and coordinate with the electro-philic lithium species surely precludes significant association with weakly basic epoxides under these conditions.[28]

An important feature of the β-elimination reactions is the very high degree of stereoselectivity for the *trans* double-bond isomer of the allylic alcohol product. This is again illustrated most conclusively with the *cis*- and *trans*-4,5-epoxy-octanes, both of which are transformed by lithium diethylamide into *trans*-oct-5-en-4-ol (Eq. 4, p. 349).[15] The isomeric *cis* alcohol is not formed in detectable quantities in these reactions. Likewise, this reagent converts the spiroepoxide **14** exclusively to the *trans* allylic alcohol **15**.[29] Similarly, both *n*-butyllithium[30] and diethylaluminum 2,2,6,6-tetramethylpiperidide (DATMP)[31,32] promote the conversion of either *cis*- or *trans*-epoxycyclododecane to the corresponding *trans* allylic alcohol. Of course, the *cis* isomer is formed exclusively when the double bond is endocyclic to an eight-membered or smaller ring. However, *trans*-cyclodec-2-en-1-ol is apparently the major allylic alcohol derived from *cis*- and *trans*-epoxycyclodecanes (Eq. 5).[33]

The distinct propensity for formation of a *trans* double bond in conforma-tionally unrestricted systems appears to be a result of the *syn*-elimination mechanism. Thus extensive studies on a wide variety of elimination reactions show that the production of *trans* double bonds is generally associated with *syn* elimination.[34]

This is understandable in terms of the transition states that lead to the *cis* and *trans* olefins. Clearly, the transition state that yields the *cis* double bond experiences substantial nonbonded interaction involving the side chain R. This is not the case with the transition state that gives the *trans* double bond.

Cyclic Alcohol Formation

In a number of instances where the β-elimination process is retarded or somehow excluded, a competing cyclization such as that of *cis*-epoxycyclooctane to bicyclic alcohol **1** (Eq. 3) is observed.[9] The available evidence supports a carbenoid mechanism for these reactions. The initial step in these transformations almost certainly involves metalation of the epoxide ring. The corresponding metalated epoxide is thought to undergo α elimination and insertion of the carbenoid center into a neighboring C–H bond.

The reversible metalation of epoxides under typical reaction conditions is supported by several deuterium-labeling studies.[16,17,19,20] Thus the treatment of *exo*-norbornene oxide (**16**) with lithium cyclohexylamide in the presence of an excess of cyclohexylamine-d_2 gives recovered starting material with a very substantial level (83 %) of deuterium incorporation.[18] In addition, the concomitantly formed rearrangement product nortricyclanol (**17**) also shows significant label incorporation (69 %), specifically at the carbinol center. The lack of appreciable amounts of deuterium elsewhere in alcohol **17** is not consistent with a mechanism triggered by base attack at another site of epoxide **16**, for example, a 1,3 elimination[35] resulting from proton removal from the unfunctionalized two-carbon bridge. The metalation of the epoxide ring most probably takes place within an epoxide–base complex similar to that proposed as an intermediate in β elimination.

Ample precedent for the postulated carbene insertion process in the norbornane skeleton and, indeed, in many of the other ring systems where this type of cyclization occurs, is available from studies of the analogous unfunctionalized carbenes generated by alternative methods.[36] However, a complication arises with the stepwise mechanism depicted for norbornene oxide when it is applied to a detailed description of the results with the stereoisomeric *cis*- and *trans*-epoxycyclodecanes.[33] In addition to some β-elimination product, the *cis* epoxide gives *endo*-2-*cis*-bicyclo[4.4.0]decanol (**18**) along with a little *endo*-2-*cis*-bicyclo[5.3.0]decanol (**19**), whereas the *trans* epoxide produces *exo*-2-*cis*-bicyclo[4.4.0]decanol (**20**), the epimer of alcohol **18**. Thus, whereas a transannular carbene insertion mechanism satisfactorily accounts for the products from each of the epoxides when taken individually, the stereospecificity of the conversion of the isomeric epoxides to epimeric *cis*-2-bicyclo[4.4.0]decanols clearly precludes common intermediates, most notably the free carbene species **21**. An analogous situation exists for *cis*- and *trans*-epoxycyclooctanes that stereospecifically produce the epimeric *endo*- and *exo*-2-*cis*-bicyclo[3.3.0] octanols, respectively.[9] Nonetheless, rearrangement of *cis*-epoxycyclodecane labeled with deuterium at both epoxide carbons proceeds with loss of approximately half of the label to give bicyclic alcohols **18** and **19** with the remaining deuterium at the carbinol carbons;[16] these observations complement those on the norbornene oxide rearrangement.

The results can be accommodated within the mechanistic framework shown for the latter process, provided α elimination and C–H insertion are compressed into a single, concerted reaction step. Structure **22** illustrates a hypothetical

transition state for the rather complicated bonding changes implicit in this one-step process. Such "carbenoid" mechanisms (carbenelike reactions that do not proceed by way of a free divalent carbon species) have been invoked to rationalize aspects of "carbene" chemistry where the data are inconsistent with a simple carbene intermediate.[37,38] In the present reaction a concerted α elimination–insertion mechanism allows for the absence of crossover products from epimeric epoxycycloalkanes. It is noteworthy that cyclizations by C–H insertion occur only with epoxides in which the appropriate C–H bond is held in very close proximity to the reactive center by structural features of the molecule. Finally, the stereospecificity of the rearrangements discussed above also demonstrates that the intermediate metalated epoxides are configurationally stable to the usual reaction conditions.

The involvement of carbenes or carbenoid intermediates in the isomerization of epoxides under strongly basic conditions is supported by consideration of a side reaction that occurs on treatment of certain monosubstituted and disubstituted epoxides with alkyllithium reagents.[19,39,40] The overall conversion results in transformation of an epoxide into an olefin with concomitant substitution of one of the original ring hydrogen atoms with an alkyl group from the reagent (Eq. 7). A particularly efficient example of this process is the reaction of tert-butyllithium with tert-butylethylene oxide to yield trans-di-tert-butylethylene.[40] This unusual reaction has been rationalized by a mechanism proceeding by metalation of the epoxide followed by α elimination and insertion of the carbenoid center into a C–Li bond of the alkyllithium reagent. This generates intermediate 23, which gives the observed olefin by the loss of lithium oxide. With the monoepoxide of 1,5-hexadiene, this reaction is accompanied by intramolecular trapping of the carbenoid center by the side-chain double bond to give endo-2-bicyclo[3.1.0]hexanol (24).[39] These observations supplement the evidence for the intervention of carbenoid reactions of epoxides under strongly basic conditions.

$$(Eq. 7)$$

The stereochemical features of the transannular carbenoid reactions of medium-ring epoxycycloalkanes follow from the conformational properties of the metalated epoxide intermediates and the known propensity for retention of configuration in C–H insertions.[36] Thus cis-fused bicyclic systems are formed exclusively in these transformations as, indeed, they are in other transannular carbene insertions.[36] This is a result of the relative stabilities of the ring conformations that allow a transannular C–H bond to interact with the carbenoid center. In principle, the C–H group can approach this site in two different fashions termed "backside" and "frontside." In the former, the C–H approaches the metalated epoxide center from a direction away from the departing oxygen; in the latter, attack is from the oxygen side of the reactive site. If the metalated epoxide moiety is so situated as to minimize steric interaction with the ring residue, the backside mode is clearly favored for these cyclic compounds as illustrated for cis-cyclooctene oxide in structure 25. This defines the stereochemistry of the hydroxyl group, namely, endo alcohols from cis epoxides and

25

exo alcohols from trans epoxides. However, backside attack is not inherently favored, but rather is a result of specific structural features of the medium-ring compounds. Thus frontside and backside modes compete in the conversion of the acyclic epoxide trans-di-tert-butylethylene oxide into the diastereomeric cyclopropylcarbinols 26 and 27, along with some tert-butyl neopentyl ketone.[39] In fact, frontside approach is observed exclusively with 2,3-epoxybicyclo[2.2.2] octane, a structurally unbiased substrate that yields tricyclic alcohol 28 as the only transannular insertion product. The major product from this epoxide is bicyclic ketone 29.[18]

Ketone Formation

In a number of base isomerizations of epoxides, ketones are generated as primary products. This type of transformation is generally found under circumstances where neither β elimination nor transannular insertion processes are effective. It should be noted that ketones can also arise from further isomerization of allylic alcohols under the reaction conditions. Consequently, it is not always clear whether the observed ketone products are formed directly from the epoxides or are secondary products. However, several unambiguous examples of direct ketone production from epoxides have been cited above. This isomerization is also initiated by metalation of the epoxide ring. Two pathways are considered as reasonable routes to the enolate anions, which are the actual products prior to hydrolysis. The first of these proceeds by α elimination and hydrogen migration from the adjacent ring carbon as illustrated by path a. The alternative is an electrocyclic ring opening, sometimes referred to as β *elimination*, which is depicted in path b.

The two mechanisms predict different products when the substituents on the metalated epoxide are not the same. Unfortunately, in most instances of ketone formation there is competitive metalation at the two ring sites, obscuring the relationship between the products and the mechanism. The clean conversion of epoxide **30** to ketone **31** indicates that the α-elimination mechanism occurs in this reaction, if the reasonable assumption of metalation at the benzylic epoxide position is made.[41] In several other instances, more involved arguments can be generated to support this mechanistic possibility.[42–44] At the present time, however, adequate information is not available to enable confident prediction of the mechanism that will be followed for rearrangement of a given epoxide to an isomeric ketone.

30

31

Formation of Cyclopropyl Carbinols

Relatively few examples of this type of conversion are known and little rigorous mechanistic information is available because of the rather specific structural requirements and reaction conditions necessary for such reactions to occur. However, it is most likely that this kind of transformation is not truly an epoxide isomerization since the reactive site appears to be a carbanion generated by proton abstraction from an activated (allylic, benzylic, or propargylic) center. This accounts for the required structural features in the reactant and also the need for hexamethylphosphoramide, which facilitates the metalation process. An appropriately situated epoxide function then alkylates the nucleophilic carbanion in the usual fashion. This mechanism is illustrated for the conversion of the phenyl-activated epoxide **32** to the indicated cyclopropyl carbinols.[11] The high

32

(86%)

(cis:trans = 6:94)

yield in this reaction is noteworthy, in view of the fact that many monosubstituted epoxides react with lithium amides to give nucleophilic addition to the terminal carbon. The high degree of selectivity for the *trans* product is rationalized on steric grounds. Analogous intramolecular alkylations of enolates and other stabilized carbanions by epoxides are well known as routes to cyclopropanes and larger carbocyclic rings.[4,5] A relevant example is the reaction of 4,5-epoxycyclooctanone (**33**).[45]

33

(90%)

The S_N2-type mechanism with its inherent inversion of configuration at the site of attack on the epoxide ring requires that the reactant be able to achieve an appropriate geometry in cyclic systems. Furthermore, this process defines the stereochemistry of the products. Thus the *trans* isomer of 3-allylepoxycyclohexane (**34**) is smoothly converted to a mixture of alcohols **35** (epimeric at the vinyl group) with a *trans* relationship between the cyclopropyl and hydroxy functions.[11] The corresponding *cis* epoxide gives no cyclopropane product because of the unfavorable geometric relationship between the allyl and epoxide functions that renders an intramolecular alkylation impossible.

SCOPE AND LIMITATIONS

The main focus of this section is on the isomerization of epoxides to allylic alcohols (Eq. 1) as commonly effected by lithium amide bases in relatively nonpolar solvent systems. Other reaction conditions that significantly modify the chemistry taking place are specifically indicated. The discussion is subdivided according to increasing complexity of the epoxide reactant. Thus saturated epoxides are considered first, followed by epoxides possessing unsaturation (double bonds, triple bonds, or aryl groups) and finally, epoxides with other functional groups in the molecule. Departure from this classification occurs in the selection of examples only where the additional functionality has no significant influence on the transformations observed.

Saturated Epoxides

These substrates are considered according to their substitution patterns.

Terminal Epoxides. Relatively few examples of the formation of allylic alcohols from epoxides with an unsubstituted ring carbon have been recorded. Nucleophilic additions of the base to these terminal epoxides is expected to be an important competing reaction, and such behavior has been noted.[11,39,40,46,47] However, epoxides with structural features that sterically retard this side reaction appear to be suitable precursors of allylic alcohols, as illustrated with β-pinene oxide (**36**).[48] The use of more sterically encumbered bases, such as lithium diisopropylamide, shows promise in expanding the range of substrates for epoxide isomerizations to include terminal epoxides. A relevant example is given by the conversion of epoxide **37** to the corresponding primary allylic alcohol.[49] Hindered aluminum amide **38** (diethylaluminum 2,2,6,6-tetramethylpiperidide) is reported to successfully isomerize terminal epoxides such as **39** in several instances.[31,32] However, organolithium reagents seldom are useful bases for isomerizations of such epoxides because of competing processes.[39,40,50] Further work designed to elaborate on the reaction conditions necessary for the synthesis of allylic alcohols from terminal epoxides is clearly warranted.

$$\mathbf{36} \xrightarrow{\text{LiNEt}_2} (81\%)$$

$$\mathbf{37} \xrightarrow{\text{LDPA*}} (64\%)$$

$$(\text{CH}_2)_{11}\ \overset{O}{\underset{}{C}}-\text{CH}_2 + (C_2H_5)_2\text{AlN} \longrightarrow (\text{CH}_2)_{10}\ \overset{\text{CH}}{\underset{}{C}}-\text{CH}_2\text{OH}$$

39   **38** (DATMP) $(85\%; E:Z = 69:31)$

2,3-Disubstituted Oxiranes. A large number of substrates in this category have been examined, including both acyclic and alicyclic epoxides. The concept of an epoxide–base complex is basic to understanding the significant selectivities observed with these epoxides. The steric requirements of the base moiety exert control over both the competition among the different types of isomerization and the regiochemistry of the β-elimination reaction.

The simple aliphatic epoxides normally undergo clean β elimination with an extraordinary degree of discrimination for proton removal from the least substituted site.[21] Thus both *cis* and *trans* isomers of epoxide **40** suffer base attack at the CH$_3$ group much more readily than at the CH$_2$ unit. Likewise, proton abstraction from a CH$_2$ group is preferred over that from a CH center, as illustrated with epoxide **41**. In each of these cases the indicated conversion is favored by a factor of at least 100 over the alternative mode of β elimination. In fact, removal of a methine proton is ordinarily quite difficult as indicated by the lack of reactivity of epoxide **42**.[29] As discussed previously, removal of a proton from a CH$_2$ group proceeds with high stereoselectivity for the *trans* allylic alcohol as observed in the isomerization of **41**.

* Throughout the text, LDPA is used to denote lithium diisopropylamide.

$$cis,trans\text{-}CH_3HC\overset{O}{\overbrace{}}CHCH_2CH_3 \quad \xrightarrow{\text{LiNEt}_2} \quad CH_2{=}CHCHOHCH_2CH_3$$

40

$$cis,trans\text{-}(CH_3)_2CHHC\overset{O}{\overbrace{}}CHCH_2CH_2CH_3$$

41

$$\xrightarrow{\text{LiNEt}_2} \quad trans\text{-}(CH_3)_2CHCHOHCH{=}CHCH_2CH_3$$

$$i\text{-}C_3H_7HC\overset{O}{\overbrace{}}CHC_3H_7\text{-}i \quad \xrightarrow{\text{LiNEt}_2} \quad \text{No reaction}$$

42

Epoxycycloalkanes show a more complicated pattern of reactivity because of the importance of transannular C–H insertion and ketone formation with this type of substrate. Because of the propensity for such epoxides to form complexes unsuitable for β elimination, plus the unfavorable stereoelectronic situation for β elimination in some medium-ring compounds, epoxide metalation and the ensuing transformations are significant processes in many instances. Both the epoxide structure and the reaction conditions can be important in determining the course of isomerization with these substrates.

As anticipated, large-ring epoxides behave much like their acyclic analogs. This is illustrated by the conversion of a mixture of *cis*- and *trans*-epoxycyclohexadecanes to the corresponding *trans* allylic alcohol.[51]

Cyclohexene oxide is smoothly converted to 2-cyclohexenol under a variety of conditions,[26,52,53] as are a number of more complicated analogs.[54–56] The high regioselectivity shown by the rigid *trans*-decalin system **43** is rationalized by preferential *syn* elimination of the quasi-axial β hydrogen (p. 350).[21] A mixture of *cis*- and *trans*-3-methylcyclohexene oxide (**44**) leads predominately to the epimeric 6-methyl-2-cyclohexenols, consistent with both proton abstraction from a CH_2 over a CH group and *syn* elimination.[21] However, ketone formation is important with certain cyclohexene oxides as shown for apopinene oxide (**45**).[52] In fact, the use of sterically bulky amides results in the formation of significant amounts of cyclohexanone from cyclohexene oxide itself.[53] Thus a 62 % conversion to ketone is observed with lithium 2,2,6,6-tetramethylpiperidide. This reagent favors epoxide metalation over β elimination because of steric destabilization of the requisite *syn* complex for the latter process.

An observation of obvious relevance for synthetic applications concerns the dramatic effect of hexamethylphosphoramide on epoxide rearrangements.[26] Whereas the bulky base lithium diisopropylamide converts cyclohexene oxide to a product mixture consisting of roughly equal amounts of 2-cyclohexenol, cyclohexanone, and the nucleophilic adduct **46** under the usual reaction conditions, the addition of two equivalents of hexamethylphosphoramide to this reaction results in the exclusive formation of 2-cyclohexenol. A further example of this dramatic effect is found in the rearrangement of bicyclic epoxide **47**. This substrate yields only the allylic alcohol **48** in hexamethylphosphoramide solvent as compared to a mixture of **48** and isomeric ketones in the usual ether–hexane system.[18,26]

Another report on cyclohexene oxide concerns the intriguing possibility of asymmetric induction during epoxide isomerizations.[57] The use of bases prepared from optically active amines results in enrichment of the 2-cyclohexenol in one of its enantiomers. Several amine enantiomers have been used, and, depending on their structure, either the R or the S alcohol is preferentially generated in optical yields in the range 3–31%. These results, particularly with such a simple epoxide, point to great potential for asymmetric induction in epoxide isomerizations. Further exploration of this subject can be anticipated in the near future.

A new competing reaction, namely, transannular C–H insertion, is a significant process with medium-ring epoxides of 7- to 10-membered carbocycles. Thus cycloheptene oxide yields bicyclic alcohol **49** in addition to cycloheptanone and 2-cycloheptenol on base isomerization.[52] The use of hexamethylphosphoramide as solvent does suppress the first two products, but significant amounts of 3-cycloheptenol are also formed.[26]

(23:36:41)   49

(82:18)

Some very useful information concerning epoxide rearrangements is provided by the transformations of *cis*-epoxycyclooctane, which have been studied extensively. Interestingly, the initial observation that bicyclic alcohol **1** is the major product of this epoxide, in fact, depends on the presence of lithium bromide in the lithium diethylamide reagent.[9,58] When *salt-free* lithium diethylamide is used, 2-cyclooctenol is the predominant product.[42,58] However, the bulkier base lithium diisopropylamide converts this epoxide almost exclusively to bicyclic alcohol **1**.[58] On the other hand, this same base in hexamethylphosphoramide results in the slow but complete isomerization to 2-cyclooctenol.[26] The allylic alcohol is also the major product when the reaction is conducted with potassium *tert*-butoxide in dimethyl sulfoxide.[59] Finally, *n*-butyllithium is an effective reagent for the generation of bicyclic alcohol **1**.[42]

LiNEt$_2$	20:80 (−)
LDPA	2:98 (80%)
LDPA-HMPA	100: 0 (70%)

The conformationally rigid *trans*-epoxycyclooctane yields 2-cyclooctenol, bicyclic alcohol **50**, and cycloheptanecarboxaldehyde on treatment with lithium diethylamide.[9] A favorable geometry for β elimination is difficult to achieve with this compound, allowing for competitive epoxide metalation and subsequent transformations. The high degree of stereospecificity in the rearrangements of the isomeric *cis*- and *trans*-epoxycyclooctanes to the epimeric bicyclic alcohols is discussed in connection with the analogous reactions of the epoxycyclo-decanes (p. 354).[33] Transannular C–H insertions are not reported for larger carbocycles.

$$\xrightarrow{\text{LiNEt}_2}$$

(10–15:55–60:32; 84%)

50

The second class of epoxides that give C–H insertions are bridged bicyclic derivatives that are precluded from β elimination by Bredt's rule restrictions. The structures of these compounds, like the medium-ring carbocycles, are often favorable for transannular carbenoid reactions. Norbornene oxide and its derivatives are particularly prone to such transformations.[18,60] Thus the *endo*-5-methyl compound **51** behaves analogously to the parent system in isomerizing to the tricyclic alcohol **52**.[18] However, bicyclic epoxide **53** yields ketone **54** as the only significant rearrangement product.[18] Ketone formation is also the principal process with epoxides in the bicyclo[2.2.2]octane and bicyclo[2.1.1]hexane systems.[10,18] The striking transformation of tetracyclic epoxide **55** into alcohol **56** is noteworthy in that insertion occurs into a proximate C–H unit that is several bonds removed from the carbenoid center.[61]

51 $\xrightarrow{\text{LiNEt}_2}$ **52**

53 $\xrightarrow{\text{LiNEt}_2}$ **54**   (91%)

55 $\xrightarrow{\text{LiNEt}_2}$ **56**   (82%)

These results emphasize the stringent geometric requirements for C–H insertion. This process is efficient only when the structure of the molecule holds a C–H bond in close proximity to the metalated epoxide carbon in a rigid or readily accessible conformation. With substrates that do not fulfill this prerequisite, metalation is usually followed by ketone formation. Isomerization of epoxides to ketones is observed with normal-ring and large-ring epoxycycloalkanes that are not susceptible to transannular carbenoid reactions. Significant ketone formation is normally confined to cyclic epoxides and is enhanced by conditions that favor epoxide metalation, namely, the use of bulky lithium amides or alkyllithiums as isomerizing agents. A notable exception to this generalization is found with the di-*tert*-butylethylene oxides. The *cis* isomer **57** is converted to *tert*-butyl neopentyl ketone under forcing conditions with *tert*-butyllithium as the base.[62] This ketone is also produced from the *trans* epoxide that gives carbenoid insertion into the C–H bonds of the adjacent *tert*-butyl group as the principal reaction.[39]

$$t\text{-}C_4H_9 \overset{O}{\underset{57}{\triangle}} C_4H_9\text{-}t \xrightarrow{t\text{-Bu Li}} t\text{-}C_4H_9COCH_2C_4H_9\text{-}t$$

Trisubstituted Epoxides. This class of substrate is also well studied, and a number of important applications in synthesis are recorded in the literature. Allylic alcohols are the only significant isomerization products normally observed from these epoxides, which apparently are not easily metalated. The major allylic alcohol is usually derived from the more stable epoxide–base complex.

For example, epoxide **58** gives the only possible β-elimination product in good yield.[52] The complications introduced by the availability of more than one β-elimination pathway are illustrated by the isomerization of trimethyloxirane with lithium diethylamide.[21] The low degree of regioselectivity in this reaction indicates that base attack at the *gem*-dimethyl and monomethyl centers is reasonably competitive. Interestingly, potassium *tert*-butoxide in dimethyl sulfoxide reacts slowly with this epoxide to give a product mixture that is

$$t\text{-}C_4H_9HC \overset{O}{\underset{58}{\diagup \diagdown}} C(CH_3)_2 \xrightarrow{\text{LiNEt}_2} t\text{-}C_4H_9CHOHC(CH_3)=CH_2$$

$$CH_3HC \overset{O}{\diagup \diagdown} C(CH_3)_2$$

$$\longrightarrow CH_3CHOHC(CH_3)=CH_2 + CH_2=CHC(CH_3)_2OH$$

LiNEt₂	(59:41)
t-BuOK–DMSO	(16:84)

enriched in the isomer derived from proton removal at the single methyl group.[46] Although this point has not received experimental attention, bulkier amide bases are expected to favor β elimination at the *gem*-dimethyl center since they should favor the less hindered complex.

The sterically hindered diethylaluminum 2,2,6,6-tetramethylpiperidide (**38**) offers some useful advantages as a base for the isomerization of trisubstituted epoxides. Thus, the Z epoxide **59** is converted cleanly to alcohol **60** by proton abstraction from the CH_3 substituent.[31,32] On the other hand, the E epoxide **61** gives only a minor amount of alcohol **60**, and the major product is the isomeric alcohol **62** formed by base attack at the CH_2 of the alkyl group *cis* to the hydrogen substituent. It is noteworthy that elimination toward the other CH_2 group is not observed for either of these epoxides. These results indicate a predisposition for reaction by means of the least crowded complex that overcomes the usual selectivity for elimination toward a CH_3 group.

59 → **60** (90%)

61

38 → **62** (78%) + **60** (12%)

An important aspect of the reaction of epoxide **61** is the high degree of stereoselectivity for the E form of the trisubstituted double bond in **62**. This suggests that isomerization proceeds by way of conformation **63** of the complex in which the side-chain residue is oriented away from the sterically demanding base moiety. Such high discrimination is unusual in the generation of trisubstituted olefins.

$(C_2H_5)_2AlN$

38 (DATMP)

O---$Al(C_2H_5)_2$

63

A number of terpenoid epoxides of partial structure **64** are converted to allylic alcohols of type **65** in good yields by the aluminum amide **38**.[32] A related transformation by use of lithium diisopropylamide proceeds equally selectively,[63] but an isomerization analogous with that for this base is reported to give some of the allylic alcohol derived from attack at the CH_2 substituent.[64]

$$(CH_3)_2C\overset{O}{\overbrace{\qquad}}CHCH_2R \quad \longrightarrow \quad CH_2{=}C(CH_3)CHOHCH_2R$$

64   **65**

A striking example illustrating the difference between lithium diethylamide and aluminum amide **38** is found in the reactions of the farnesol diepoxide **66**. Lithium diethylamide is reported to give exclusive proton abstraction from the CH_3 group at both sites to yield **67**.[65] However, the use of **38** results in the formation of the isomeric triol **68**.[32] Thus it appears that, whereas the aluminum amide reaction is controlled by the more stable complex, the less hindered lithium amide selects proton removal from a CH_3 group over that from a CH_2. This difference is of obvious interest in synthetic applications.

66

$$\xrightarrow{\text{LiNEt}_2}$$

67 ($>50\%$)

$$\textbf{66} + \textbf{38} \quad \longrightarrow$$

68 (41%)

The isomerizations of the spiroepoxides of structure **69** by lithium diethylamide represent a competition between two different CH_2 groups.[29] With one notable exception, it is the endocyclic CH_2 that participates preferentially in the elimination process. This is once again consistent with reaction by way of the sterically favored complex. The exception is the cyclohexyl derivative **69** ($n = 2$) that is slowly isomerized by elimination into the ethyl substituent to give **70**. In this reaction the conformational preference of the cyclohexane ring does not allow a readily accessible geometry that fulfills the stereoelectronic requirements for β elimination. Consequently, proton abstraction from the conformationally mobile ethyl group is the preferred pathway despite the necessity for proceeding by way of the more hindered epoxide–base complex. The other spiroepoxides have no serious problem in achieving an optimum geometry for elimination into the ring.

$$n = 0 \quad 83:17; 66\%$$
$$n = 1 \quad 100:0; 84\%$$
$$n = 2 \quad 5:95; 69\%$$
$$n = 3 \quad 98:2; 76\%$$
$$n = 4 \quad 100:0; 74\%$$
$$n = 8 \quad 100:0; 66\%$$

The lower homologs of **69**, with a methyl group in place of the ethyl side chain, show poor regioselectivity between elimination into the ring and into the methyl group.[29] The tendency for reaction from the less hindered complex is counterbalanced by that for proton removal from a CH_3 over a CH_2 group in this case. The cyclohexyl system is an exception again, yielding only elimination toward the CH_3 substituent.

The use of HMPA can influence the regioselectivity of β elimination with epoxides of this type. This is illustrated by the synthetically relevant example of steroid epoxides of partial structure **71**.[66] In this case a mixture of alcohols **72** and **73** is obtained on treatment with lithium diethylamide under the usual conditions, whereas only **72** is observed in the presence of two equivalents of hexamethylphosphoramide. The results with added hexamethylphosphoramide appear to implicate attack of the base at the most accessible site without prior complexation.

The 1-methylepoxycycloalkanes are an interesting class of cyclic epoxides whose base-promoted reactions are frequently utilized in synthesis. Proton removal from the CH_3 group by way of the more stable complex is normally anticipated for these substrates. Thus 1-methylepoxycycloheptane can be converted cleanly to 2-methylenecycloheptanol.[67] However, such conversions must be monitored closely to achieve good yields and purities of the methylenecycloalkanols since the intermediate alkoxides are especially susceptible to further isomerization. Further examples of relevant transformations are shown for epoxides **74** and **75**.[68,69]

Structural features in more complicated epoxides can, however, overcome the selectivity for elimination into the CH₃ group. This is demonstrated with epoxide **76** (a stereoisomer of **75**) that gives competitive base attack at the CH_3 and ring CH_2.[69] The use of the basic solvent tetrahydrofuran in this study may be a significant feature leading to these results. Other examples that show little selectivity are known.[70-73]

The nature of the base can exert a significant influence on the course of the isomerization of epoxides of this type. In the conversion of epoxide **77**, lithium diethylamide gives a mixture of alcohols **78** and **79** in a 17:83 ratio.[74] The use of lithium *tert*-butyltrimethylsilylamide, on the other hand, favors the desired compound **78** by an 80:20 ratio. Other amides of intermediate steric bulk result in roughly equal amounts of the two alcohols. These observations imply an important steric effect, although it is not clear in this case whether this operates on the regioselectivity of the β elimination or on the secondary isomerization of alcohol **78** to its endocyclic isomer **79**.

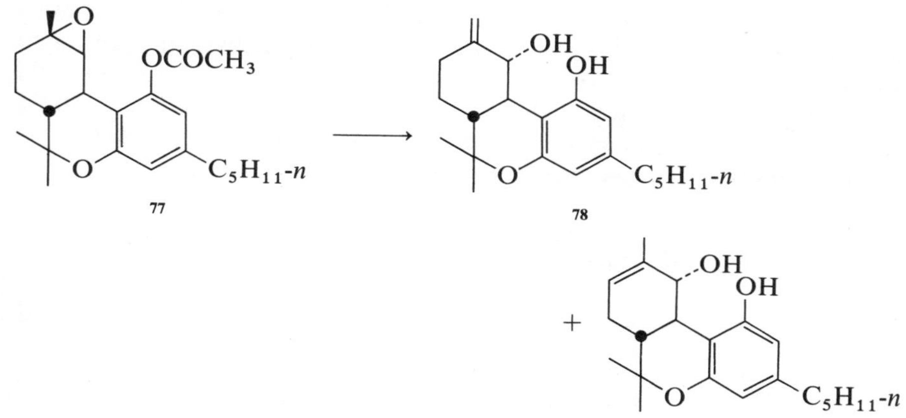

77 → **78**

+ **79**

Another informative study concerns the isomerization of **80**, an epoxide that gives complex product mixtures with lithium amides.[75,76] In this reaction the amide diethylaluminum 2,2,6,6-tetramethylpiperidide (**38**) yields the *exo*-methylene isomer **81** cleanly. This result suggests more widespread application of aluminum amides with trisubstituted epoxides that give complex product mixtures with lithium amides.

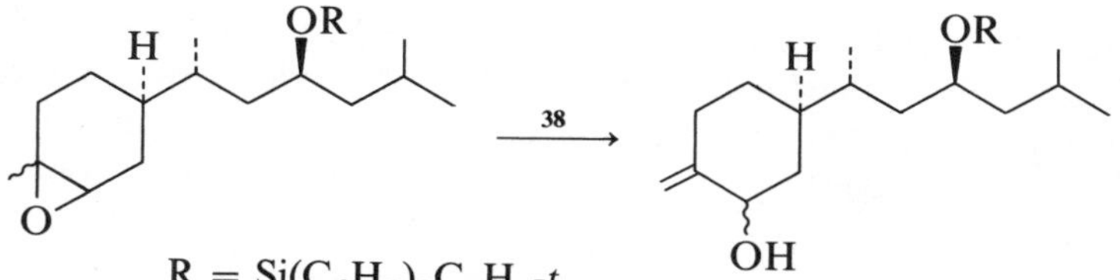

$$R = Si(C_6H_5)_2C_4H_9\text{-}t$$

80 → **38** → **81** (92%)

Little information is available concerning 1-alkylepoxycycloalkanes with alkyl groups other than methyl, despite the synthetic potential for isomerizations of such substrates. However, the reaction of 1-*tert*-butylepoxycyclooctane is interesting in that only products from β elimination are found.[67] This reversal of character of the eight-membered ring, which usually undergoes transannular insertion, is attributed to the difficulty of metalation as a result of steric retardation.

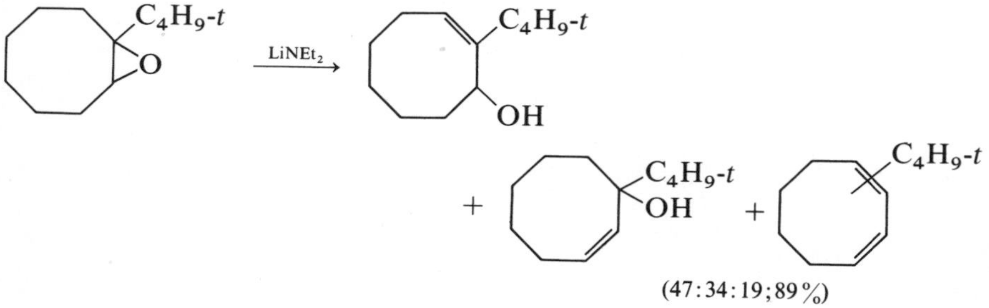

$$\xrightarrow{\text{LiNEt}_2}$$

(47:34:19; 89%)

Finally, bicyclic compound **82**, whose epoxide ring is fused to one cyclo-hexane unit and spiro to the second, is converted solely to the endocyclic olefin **83** rather than the other possible allylic alcohol.[77] Removal of a *syn* proton in the fused ring is preferred over β elimination into the spiro cyclohexane. This is reasonable on the basis of stereoelectronic considerations and parallels the behavior of the spiro epoxides discussed earlier.

Tetrasubstituted Epoxides.   Data are available for only a few substrates of this type, but the selectivities appear to follow the principles already elaborated for less highly substituted epoxides. Thus selective proton abstraction from a CH_3 group is observed with epoxide **84**.[78] Likewise, base attack is highly regioselective for a CH_3 over a CH_2 group with the interesting spiro epoxide **85**.[79] Excellent yields of the highly reactive cyclopropanol can be obtained if the product is converted directly into the trimethylsilyl derivative. The bis–spiro epoxides **86a** and **86c** undergo similar transformations to yield cyclopropanols **87a** and **87c**.[79] Elimination into the three-membered ring is not competitive because of the highly strained nature of the cyclopropene derivative that would result. Interestingly, the cyclohexane analog **86b** does yield some of this product, alcohol **88**, illustrating once again the resistance to elimination into the spiro cyclohexane moiety.

Side Reactions. Nucleophilic addition of the base to the epoxide ring probably occurs to some extent in most isomerization reactions. The addition products formed in this manner from amide bases are conveniently removed from the neutral products during processing of the reaction mixture as a result of their basic properties. As a consequence, these materials are rarely examined in synthetic conversions, where they pose no great inconvenience apart from decreasing the yields of the desired isomerization products. Base adducts have been characterized in a number of reactions using lithium amides.[11,15,44,47,52]

This side reaction is expected to be subject to the usual considerations that govern S_N2-type substitutions.[3] Thus an increase in the number and steric size of the substituent groups on the epoxide ring should retard nucleophilic attack. Likewise, an increase in the steric encumberment of the lithium amide should be detrimental to the nucleophilic addition process. It appears that these steric effects are usually more critical for nucleophilic processes than for the competing proton-abstraction events that lead to epoxide rearrangement.

In a number of instances the initially formed allylic alcoholates are subject to further isomerizations under the reaction conditions.[21,43,67,70] For reasons that remain obscure, this is especially true of the reactions of 1-methylepoxycyclo-alkanes. Thus lithium salts of the primary products, 2-methylenecycloalkanols, are transformed into the precursors of 2-methyl-2-cycloalkenols and 2-methyl-cycloalkanones.[67,70] This reaction pathway is indicated by studies of the evolution of product mixtures with time and by independent transformations of the allylic alcohols. The most obvious explanation of these secondary isomerizations invokes allylic metalation of the intermediate alkoxides followed by allylic rearrangement and reprotonation. This is illustrated for the 2-methylenecyclo-alkanols. A similar process accounts for the slow conversion of 1-penten-3-ol to 3-pentanone.[21] In a number of other reactions complex product mixtures probably result from secondary isomerizations.[67,71,74,76,80]

$$(n = 1, 2, 3, \text{ or } 4)$$

The formation of small amounts of 3-cyclohexenol during the rearrangement of cyclohexene oxide is another side reaction that must proceed by a related mechanism.[53] Thus metalation of the lithium salt of 2-cyclohexenol, followed by rearrangement and protonation, results in double-bond migration away from the alkoxide function. This reaction appears to be favored by hexamethyl-phosphoramide, a solvent that greatly enhances the prospects for metalation. This is illustrated with epoxycycloheptane, which gives 2-cycloheptenol con-taminated with significant amounts of 3-cycloheptenol when hexamethyl-phosphoramide is used.[26]

Analogous transformations of allylic alcohols can sometimes be performed efficiently on a synthetic scale by using the lithium salt of ethylenediamine as the base.[78,81] This is demonstrated in the reactions of alcohol **89**, the clean iso-merization product of the lithium diethylamide treatment of epoxide **90**. Reaction of alcohol **89** with the lithium salt of ethylenediamine for a brief period of time gives isomeric alcohol **91**, whereas longer reaction times yield alcohol **92**, which is presumably the thermodynamically favored product. These further isomerizations have considerable synthetic potential in their own right.

In a few instances conjugated dienes are found as minor products from epoxy-cycloalkanes, as in the formation of cycloheptadiene from cycloheptene oxide.[51,52] These dienes are secondary products derived from the allylic alcohols and are favored by high temperatures and prolonged reaction times. A reasonable mechanism invokes allylic metalation as the initial event, but decomposition of the key intermediate gives lithium oxide and the diene. The elimination of lithium oxide as a driving force for reactions is proposed in several other instances.[40]

Finally, it should be mentioned that certain terminal and cyclic epoxides react with organolithium reagents at higher temperatures to give substituted olefins in low to moderate yields according to the transformation indicated below.[20,39,40,50,82] Thus 1,2-epoxybutane gives olefins of the indicated structure on prolonged heating with various alkyllithium reagents. The yields appear to increase with the size of the alkyl group of the reagent; the *trans* isomers of the product are preponderant. In a similar fashion cyclopentene oxide is converted to the analogous trisubstituted olefins. This reaction is less prevalent for the medium-ring and bicyclic epoxides that undergo transannular insertions.[50] In certain situations this conversion constitutes a satisfactory synthesis of substituted olefins.

$$C_2H_5HC\underset{O}{\overset{O}{\diagup\!\!\diagdown}}CH_2 \xrightarrow{\text{RLi}} C_2H_5CH\!=\!CHR$$

R (yield): t-C$_4$H$_9$ (57%), n-C$_4$H$_9$ (35%), i-C$_3$H$_7$ (35%), C$_2$H$_5$ (12%)

R (yield): t-C$_4$H$_9$ (49%), n-C$_4$H$_9$ (41%), C$_6$H$_5$ (16%), CH$_3$ (7%)

Influence of the Base. A number of examples are cited in the survey of epoxide types in which the nature of the base system plays an important role in determining the course of a reaction. However, relatively little attention is usually given to the choice of an appropriate base in experimental work. Consequently, it is worthwhile summarizing some of the factors that appear to be significant in deciding on a reagent for a specific transformation. Because of a lack of systematic study on representative substrates, there is a fair amount of speculation in the generalizations that follow. Nonetheless, these hypotheses should be useful in devising experimental conditions for controlling isomerizations.

Lithium amides are the most widely used reagents, and in a variety of situations these bases function effectively. A range of amides with different substituents are easily prepared from readily available materials. Lithium diethylamide is often the reagent of choice for promoting β eliminations. This amide generally favors allylic alcohol formation even for the epoxycycloalkanes, which have a tendency to undergo epoxide metalation. The high selectivity for the sequence $CH_3 > CH_2 > CH$ is of substantial importance in determining the regiochemistry of β elimination with lithium diethylamide. There are indications that this strong preference can overcome the propensity for reaction from the more stable complex in some instances. Sterically smaller bases may enhance this tendency, whereas more bulky amides are expected to favor the product derived from the more stable complex.

Nucleophilic addition of the base to the epoxide is occasionally a problem. The use of bulkier lithium amides such as lithium diisopropylamide is indicated in such circumstances.

Increase in the steric hindrance of the reagent also results in a shift toward metalation with epoxycycloalkanes. This leads to ketone formation or transannular insertion depending on the structure. Thus bulky amides should be adopted or avoided according to the desired outcome of the isomerization.

A detailed study of the isomerization of cyclohexene oxide provides some useful insight into the influence of the amide structure.[53] The only important products from this substrate are 2-cyclohexenol, cyclohexanone, the nucleophilic adduct, and occasionally a little 3-cyclohexenol. Monoalkylamides, with the exception of lithium *tert*-butylamide, give large amounts of nucleophilic addition and are, therefore, of limited utility in synthesis. These reactions are slow and usually result in noticeable secondary isomerization of the allylic alcohol to 3-cyclohexenol. The dialkylamides function more rapidly and tend to give less nucleophilic addition. The best conversions to 2-cyclohexenol are achieved with the rarely used di-*n*-propylamide and di-*n*-butylamide that are a little more efficient than lithium diethylamide. Both nucleophilic adducts and cyclohexanone are important products with amides of intermediate steric bulk such as lithium diisopropylamide. With severely hindered amides such as lithium 2,2,6,6-tetramethylpiperidide, nucleophilic addition is minimal, but cyclohexanone is the major product. Thus the less congested amides favor allylic alcohol, whereas ketone formation is promoted by bulky bases. A similar effect is noted for *cis*-epoxycyclooctane, where the competition is between β elimination and transannular insertion.

Hindered amides are also suggested for the isomerization of 1-methylepoxycycloalkanes. These bases should show more selectivity for β elimination into the CH_3 group and limit secondary isomerization with this type of substrate. Bulky amides are also expected to modify the regiochemistry of elimination with other trisubstituted epoxides by directing reaction through the more stable complex.

The amide diethylaluminum 2,2,6,6-tetramethylpiperidide (**38**) shows high regioselectivity with trisubstituted epoxides, indicating that this large, congested reagent displays a very pronounced preference for β elimination by way of the least hindered complex. Consequently, this reagent is suggested when it is important to maximize this regiochemistry, in spite of the inconvenience involved in its preparation. Significantly, neither epoxide metalation nor secondary isomerizations are found with aluminum amide **38**. Nucleophilic additions of diethylaluminum 2,2,6,6-tetramethylpiperidide are also unimportant, although less hindered variants of this species are reported to add to epoxides in a useful synthetic preparation of amino alcohols.[83] Amide **38** does not isomerize normal-ring and medium-ring epoxycycloalkanes presumably because of the severe destabilization of the more hindered complex that is required for β elimination. Diethylaluminum 2,2,6,6-tetramethylpiperidide is generally used in benzene since the reaction is retarded by basic solvents such as ether and tetrahydrofuran that can compete with the epoxide for complex formation.[31,32]

There is a pronounced effect of hexamethylphosphoramide on epoxide rearrangements promoted by lithium amides that strongly favors formation of allylic alcohols as illustrated above.[11,26,42,84] Hexamethylphosphoramide acts to break up base aggregates and undoubtedly renders base–epoxide complexation ineffective.[28] Under these conditions β elimination is the only important pathway for isomerization. The mechanism for this process is probably of the typical E-2 variety.[24,25] This hypothesis lifts the restriction for *syn* elimination and predicts regioselectivities determined by steric approach control of the attacking base and relative product stabilities. As a consequence, the regiochemistry of hexamethylphosphoramide-mediated isomerizations may differ appreciably from the usual lithium amide reactions in nonpolar solvents. This should lead to useful product modifications in some instances. However, the addition of hexamethylphosphoramide is most useful for suppression of the metalation process with epoxycycloalkanes that tend to yield ketones or transannular insertion products.

Isomerizations promoted by potassium *tert*-butoxide in dipolar aprotic solvents such as pyridine, dimethyl sulfoxide, and dimethylformamide have been recorded.[46,59,85] However, *t*-butanol is seldom a suitable solvent with this base.[86] Elevated temperatures are required for these conversions, and it is difficult to drive the reactions to completion since the alcohol generated in the reaction appears to inhibit isomerization. The reaction conditions of these isomerizations are not likely to favor the complexation phenomena proposed for lithium amides. Mechanistic details are probably more akin to those of amide–hexamethylphosphoramide reagents. Thus β elimination is the dominant process even with epoxycyclooctane.[26] The regioselectivity is consistent with base attack at the more accessible proton adjacent to the epoxide ring. For example, the difference in the product distribution from trimethyloxirane with potassium *tert*-butoxide and lithium diethylamide is detailed above. The isomerization of α-pinene oxide, which gives exclusively *trans*-pinocarveol with lithium diethylamide, yields a mixture of this alcohol and the isomeric allylic alcohol **93** with potassium *tert*-butoxide in dimethyl sulfoxide or dimethylformamide.[85] Whereas reasonably clean isomerizations are observed in a few cases with regioselectivities different from or better than those obtained with lithium amides, the potassium *tert*-butoxide isomerizations are not, in general, very useful synthetically. The lithium amide–hexamethylphosphoramide reagent, if carefully controlled to minimize secondary isomerizations, may result in more facile conversions with similar product distributions.

Organolithium reagents are also used to effect epoxide rearrangements. For the most part, these reagents give results that are at best equivalent to those achieved with amide bases. Furthermore, unlike the situation with the latter, the byproducts formed from nucleophilic addition of organolithium reagents are a nuisance to remove from the desired isomerization products. However, *n*-butyllithium in hydrocarbon solvents does show a strong bias for metalation of *cis*-epoxycycloalkanes. Thus normal-ring and large-ring epoxides yield appreciable amounts of the corresponding cycloalkanones in addition to allylic alcohols.[42,51] Medium-ring epoxides are converted almost exclusively to ketones and bicyclic alcohols by transannular insertion under these conditions. There are some obvious synthetic advantages to the use of *n*-butyllithium when the desired products result from the metalated epoxide.

The hindered *tert*-butyllithium reagent displays an enhanced propensity for epoxide metalation that can be used to advantage with substrates that are particularly susceptible to nucleophilic attack or are unreactive to the usual bases. An example of the latter is *trans*-di-*tert*-butylethylene oxide, which is not affected by prolonged treatment with lithium diethylamide at elevated temperatures but does yield isomeric products with *tert*-butyllithium (p. 357).[39]

Finally, limited information regarding the use of aluminum isopropoxide[87–89] and the reagent obtained from isopropylcyclohexylamine and methylmagnesium bromide[90] suggest the need for further evaluation of these bases for epoxide isomerizations.

Unsaturated Epoxides

In many instances the presence of unsaturation in a substrate, particularly if it is remote from the epoxide site, has little effect on the rearrangement reactions. Several examples of this behavior are cited in the discussion on saturated epoxides. However, the activating effect of double bonds, triple bonds, and aryl groups can have a profound influence on the course of rearrangements when these functions are more proximate to the epoxide. Secondary isomerizations are also often promoted by this type of activation. Of course, the position of the activating group relative to the epoxide unit is an important consideration with these epoxides.

α Unsaturation. Rather diverse behavior is found for epoxides that are directly bonded to an activating group, and a consistent pattern of reactivity is not presently discernible. Epoxides **94** and **95**, which do not have acceptable alternative modes of isomerization, display typical β-elimination reactions.[91,92] The high yields of these lithium diisopropylamide rearrangements are noteworthy for terminal epoxides, probably reflecting some assistance of the unsaturated function in the elimination process.

$$\underset{\textbf{94}}{\text{H}_2\text{C}\overset{\displaystyle\text{O}}{\overbrace{\qquad}}\text{C(CH}_3)\text{CH}=\text{CH}_2} \xrightarrow{\text{LDPA}} \text{HOCH}_2\text{C}(=\text{CH}_2)\text{CH}=\text{CH}_2$$

(76%)

$$\underset{\textbf{95}}{\text{H}_2\text{C}\overset{\displaystyle\text{O}}{\overbrace{\qquad}}\text{C(CH}_3)\text{C}_6\text{H}_4\text{CH}_3\text{-}p} \xrightarrow{\text{LDPA}} \text{HOCH}_2\text{C}(=\text{CH}_2)\text{C}_6\text{H}_4\text{CH}_3\text{-}p$$

(85%)

On the other hand, appropriately substituted vinyl epoxides can undergo 1,4 eliminations. Thus epoxide **96** is cleanly and rapidly converted to the *trans* dienol **97** by lithium diethylamide.[44] There is little mechanistic information available about the 1,4-elimination process. Consequently, the involvement of complexes and the geometric requirements for this reaction remain to be fully explored. The preference for 1,4 elimination disappears in the isomerization of epoxide **98**, where the additional methyl substituent must somehow retard this pathway, perhaps by forcing the double bond away from a favorable elimination geometry.[44]

$$\underset{\textbf{96}}{(\text{CH}_3)_2\text{C}\overset{\displaystyle\text{O}}{\overbrace{\qquad}}\text{CHCH}=\text{C(CH}_3)_2}$$

$$\xrightarrow{\text{LiNEt}_2} \underset{\textbf{97}}{(E)\text{-}(\text{CH}_3)_2\text{COHCH}=\text{CHC(CH}_3)=\text{CH}_2} \quad \text{(quant)}$$

$$\underset{\textbf{98}}{(\text{CH}_3)_2\text{C}\overset{\displaystyle\text{O}}{\overbrace{\qquad}}\text{C(CH}_3)\text{CH}=\text{C(CH}_3)_2}$$

$$\xrightarrow{\text{LiNEt}_2} (\text{CH}_3)_2\text{COHC}(=\text{CH}_2)\text{CH}=\text{C(CH}_3)_2$$

$$+ (E)\text{-}(\text{CH}_3)_2\text{COHC(CH}_3)=\text{CHC(CH}_3)=\text{CH}_2$$

$$+ \text{CH}_2=\text{C(CH}_3)\text{COH(CH}_3)\text{CH}=\text{C(CH}_3)_2$$

(76:14:10; 70%)

The *cis* epoxide **99** gives mainly 1,4 elimination with lithium diethylamide.[44] Interestingly, the use of the diisopropylamide diverts most of the isomerization to the β,γ-unsaturated ketone **100**. The *cis* double bond in **100** suggests that it is a primary product derived from epoxide metalation, most probably at the carbon adjacent to the double bond. The *trans* epoxide **101** reacts more slowly with the diethylamide to yield predominantly the nucleophilic adduct. This side reaction can be avoided with lithium diisopropylamide; epoxide **101** gives mainly elimination products, albeit without significant regioselectivity for the 1,4 mode over the 1,2 process.

LiNEt$_2$ 99:1:0 (70%)
LDPA 26:10:64 (70%)

+ aminoalcohols + enones

LiNEt$_2$ 15:6:71:8 (80%)
LDPA 54:30:5:21 (75%)

The reaction of 3,4-epoxycyclohexene (**102**) with lithium diethylamide very rapidly leads to a quantitative yield of benzene.[44] This conversion undoubtedly proceeds by 1,4 elimination to give the lithium salt of benzene hydrate (**103**), followed by metalation and loss of Li$_2$O in a process similar to that that transforms allylic alcohols into dienes. Benzene hydrate can be isolated as the major product, along with some of the nucleophilic adduct, when methyllithium is used as the reagent.[93]

Isomerization of *cis-β*-methylstyrene oxide (**104**) leads to a mixture of propiophenone, a small amount of phenylacetone, and allylic alcohol **105**.[43] Propiophenone results from a facile secondary isomerization of alcohol **105** under the reaction conditions. Phenyl activation for this process is clearly important. The *trans* isomer **106** is more reactive and gives more phenylacetone and alcohol **105**. It is reasonable that the faster reaction results in less rearrangement of **105** to propiophenone, but the more effective isomerization of the *trans* epoxide to phenylacetone is puzzling since metalation leading to ketone formation normally occurs best with *cis* epoxides.

$$C_6H_5 \overset{O}{\triangle} CH_3 \xrightarrow{\text{LiNEt}_2} C_6H_5CH_2COCH_3 + C_6H_5COC_2H_5$$

$$+ \ C_6H_5CHOHCH=CH_2$$
105

cis (**104**) 5 hr 1.5:91:7.5
trans (**106**) 2 hr 15:41:44

Epoxides with α-acetylenic substitution undergo selective 1,4 elimination, as illustrated by the isomerization of **107** by use of potassium *tert*-butoxide in dimethyl sulfoxide under mild conditions. The initial cumulene product **108** is not observed because of its facile tautomerization to vinylacetylene **109** and cyclization to furan derivative **110**.[94]

$$(CH_3)_2C \overset{O}{\diagup\!\!\!\diagdown} CHC\!\equiv\!CCH_3 \xrightarrow[\text{DMSO}]{\text{t-BuOK}} \left[(CH_3)_2\overset{\overset{\displaystyle OLi}{|}}{C}=C=C=CH_2 \right]$$
107 **108**

$$\longrightarrow (E)\text{-}(CH_3)_2COHCH=CHC\!\equiv\!CH +$$
109 (50%)
110 (7%)

The monoepoxide from cyclooctatetraene **111** is transformed rapidly and in good yield to cyclooctatrienone.[95] This unusually efficient formation of a ketone probably benefits from allylic stabilization in the metalation event.

$$\xrightarrow{\text{LiNEt}_2} \quad (71\%)$$

111

Unsaturated epoxides also partake in transannular insertions similar to those found with medium-ring epoxycycloalkanes. An example of this behavior is the conversion of 3,4-epoxycyclooctene (**112**) to the bicyclic alcohol **113**.[26,96] This implies that metalation is directed exclusively to the allylic position of **112** and that the intermediate species interacts very efficiently with a proximate C–H bond. This process can, however, be completely circumvented by the use of hexamethylphosphoramide.[26,27] In this reaction 1,4 elimination gives dienol **114** that is rapidly equilibrated with isomeric dienol **115**. Prolonged base treatment ultimately converts this mixture to 3-cyclooctenone (**116**). Alternatively, a clean, one-step transformation to **116** is effected by the lithium derivative of ethylenediamine.[26]

112   113 (97%)   116 (3%)

112 $\xrightarrow[\text{HMPA}]{\text{LiNEt}_2}$

114   115

112 $\xrightarrow[\text{HMPA}]{\text{LiNH(CH}_2)_2\text{NH}_2}$ **116** (84%)

Medium-ring 3,4-epoxybenzocycloalkenes also show typical carbenoid reactivity.[41] The isomerization of the seven-ring compound to 3,4-benzocycloheptenone is considered in the context of the mechanism of ketone formation (p. 358). The nine-membered ring compound gives mainly bicyclic alcohol **117**. Metalation is again directed by the aryl group and the C–H insertion process is highly selective for the center that is conformationally more accessible to the reactive site.

$\xrightarrow{\text{LDPA}}$ (65%)

117

Finally, the dichotomy in the isomerizations of the stilbene oxides is noteworthy.[47] Whereas the *cis* isomer rearranges by migration of a hydrogen, a more

$C_6H_5\overset{O}{\wedge}C_6H_5 \xrightarrow{\text{LiNEt}_2} C_6H_5COCH_2C_6H_5$
(70%)

$C_6H_5\overset{O}{\wedge}_{C_6H_5} \xrightarrow{\text{LiNEt}_2} (C_6H_5)_2CHCHO$
(66%)

$(C_6H_5)_2C\overset{O}{\overline{\quad\quad}}CHC_6H_5 \xrightarrow{\text{LiNEt}_2} (C_6H_5)_2CHCOC_6H_5$
(80%)

profound change is involved in the transformation of the *trans* isomer to diphenylacetaldehyde. This implies that an efficient phenyl migration to the metalated epoxide center takes place when this group is *cis* to the lithium, but not when it is *trans*. Minor products derived from a similar alkyl group migration are observed with saturated epoxides with the *trans* geometry.[9,15] Interestingly, triphenylethylene oxide apparently isomerizes without phenyl migration.[47]

β Unsaturation. The situation with epoxides that have an adjacent allylic, benzylic, or propargylic substituent is more straightforward. Normally, removal of a proton from the activated center leads to a highly regioselective β elimination, yielding the corresponding conjugated system in a fast, clean transformation.[43,44,94] Interestingly, *trans*-disubstituted epoxides of this type do not exhibit the very high degree of stereoselectivity for the *trans* allylic alcohols that the less reactive saturated analogs display. The more congested *cis* isomers still give exclusively the *trans* product. The use of hexamethylphosphoramide in benzene as the solvent for isomerizations of epoxides of this type results in a nonstereoselective elimination producing roughly equivalent amounts of the isomeric alcohols.[97]

$$C_6H_5CH_2HC\overset{O}{\overbrace{\qquad}}CH_2 \xrightarrow{\text{LiNEt}_2} C_6H_5CH=CHCH_2OH \quad (E:Z = 92:8)$$

$$\textit{trans-}CH_3HC\overset{O}{\overbrace{\qquad}}CHCH_2CH=CH_2$$

$$\xrightarrow{\text{LiNEt}_2} CH_3CHOHCH=CHCH=CH_2$$

$$(E:Z = 88:12; 85\%)$$

$$\textit{cis-}CH_3HC\overset{O}{\overbrace{\qquad}}CHCH_2CH=CH_2$$

$$\xrightarrow{\text{LiNEt}_2} CH_3CHOHCH=CHCH=CH_2$$

$$(E \text{ only}; 100\%)$$

$$C_2H_5HC\overset{O}{\overbrace{\qquad}}CHCH_2C(CH_3)=CH_2$$

$$\xrightarrow[\text{HMPA}]{\text{LiNEt}_2} C_2H_5CHOHCH=CHC(CH_3)=CH_2$$

$$(E:Z \sim 1:1; 77\%)$$

The facility of these β eliminations generally allows for clean conversions without complications from competing or subsequent transformations. However, the treatment of propargylic epoxides with potassium *tert*-butoxide in dimethyl sulfoxide can result in a secondary isomerization to furans as observed with **118**.[94]

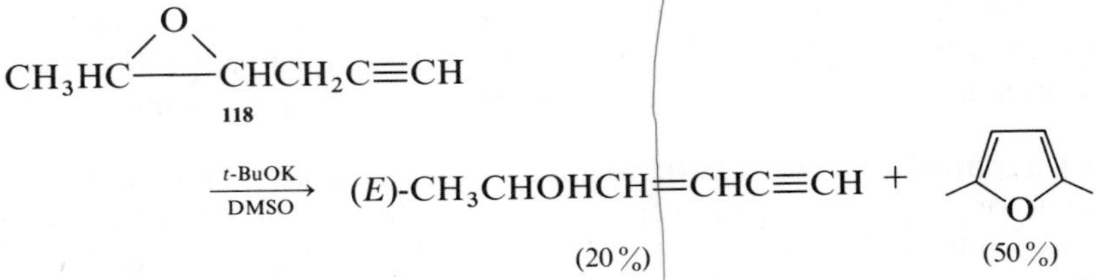
O
CH3HC—CHCH2C≡CH
118
t-BuOK
DMSO
(E)-CH3CHOHCH=CHC≡CH +
O
(20%)
(50%)

The activating effect of a substituent can be negated by other features of a substrate. Thus additional substitution at an activated position can overcome the influence of an activating group. This is demonstrated with epoxide **119**, which shows a competition between the two possible β-elimination modes.[43]

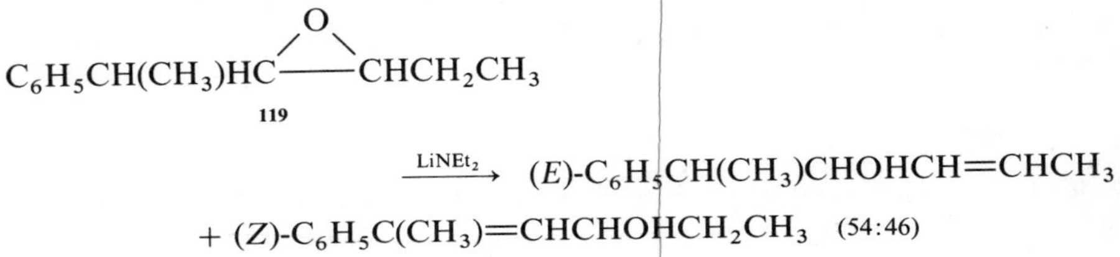
O
C6H5CH(CH3)HC—CHCH2CH3
119
LiNEt2
(E)-C6H5CH(CH3)CHOHCH=CHCH3
+ (Z)-C6H5C(CH3)=CHCHOHCH2CH3
(54:46)

The spiroepoxide **120** prefers proton abstraction from a CH_3 over a benzylic CH_2, especially with the bulky lithium diisopropylamide.[98] In this case the steric congestion on the side of the ring bearing the benzylic substituent favors complex formation on the other face and probably also twists the aryl group away from its most effective geometry.

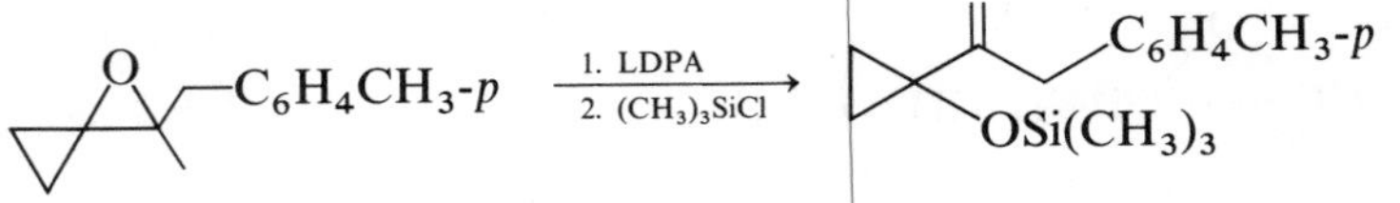
O
C6H4CH3-p
1. LDPA
2. (CH3)3SiCl
C6H4CH3-p
OSi(CH3)3

120

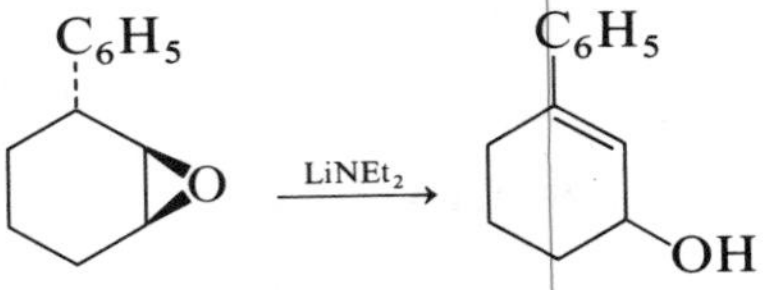
C6H5
O
LiNEt2
C6H5
OH

121

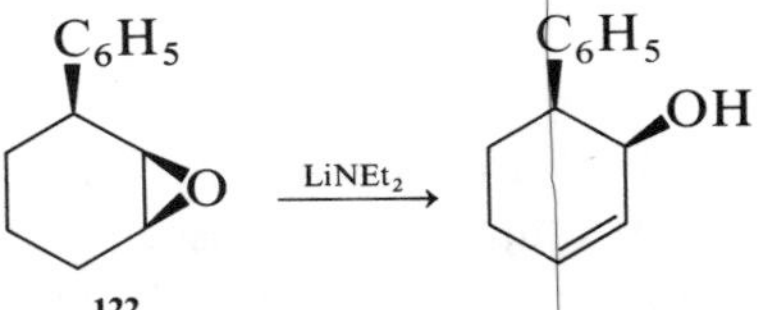
C6H5
O
LiNEt2
C6H5
OH

122

Finally, stereochemical features are clearly important as shown with the isomeric 3-phenylcyclohexene oxides.[43] The *trans* isomer **121** undergoes β elimination by proton abstraction at the benzylic position as anticipated. However, the *cis* isomer **122** slowly yields alcohol by removal of an unactivated proton. These results clearly indicate that *syn* elimination is preferred even with these activated epoxides.

γ Unsaturation. The isomerizations of epoxides in this category depend significantly on the reaction conditions. In nonpolar solvents the rearrangements proceed in a manner analogous to those observed with saturated epoxides. However, the presence of hexamethylphosphoramide drastically affects the reactions in a fashion that is quite different from its effect on the isomerizations of saturated epoxides. This is attributed to the remarkable ability of hexamethylphosphoramide to promote metalation adjacent to sites of unsaturation.

The reaction of medium-ring epoxide **123** with phenyllithium under nonpolar conditions is interesting in that β elimination is the dominant process.[99] Transannular carbenoid reactions are not important with **123**, unlike the situation with medium-ring epoxycycloalkanes. The regioselectivity and stereochemistry of this β elimination are also notable. This unexpected selectivity seems to be derived from conformational features of the substrate.

The isomerization of epoxide **124** with lithium diethylamide in ether–hexane gives the anticipated allylic alcohol **125** as the major product.[11] In hexamethylphosphoramide, on the other hand, the epimeric cyclopropylcarbinols **126** are the predominant products (*cis*:*trans* = 5:95).

$$(CH_3)_2C\overset{O}{\overbrace{}}CH(CH_2)_2CH\!=\!CH_2$$
124

$$\xrightarrow{\text{LiNEt}_2} CH_2\!=\!C(CH_3)CHOH(CH_2)_2CH\!=\!CH_2 + \textbf{126} \quad (97:3)$$
125

$$\textbf{124} \xrightarrow[\text{HMPA}]{\text{LiNEt}_2} (CH_3)_2COH\overset{\triangle}{}CH\!=\!CH_2$$
126

$$+ (E,E)\text{-}(CH_3)_2COHCH\!=\!CHCH\!=\!CHCH_3 \quad (95:5)$$

The unsaturated spiroepoxide **127** is an interesting substrate. Under nonpolar conditions some of the tricyclic alcohol **128** is formed along with the nucleophilic adduct.[11] This reaction is a rare example of an intramolecular capture of a carbenoid center by a double bond. In hexamethylphosphoramide a clean conversion to bicyclic alcohol **129** is observed.

Double-bond isomerizations can take place under the strongly basic conditions of these reactions. This is illustrated with *cis* epoxide **130**, which reacts slowly in ether–hexane to yield a mixture of alcohols **131** and **132**.[11] The *syn* mechanism accounts for the regioselectivity of this β elimination. Alcohol **132** is derived from **131** by base isomerization, presumably after epoxide rearrangement. A much faster reaction occurs in the presence of hexamethylphosphoramide, leading predominantly to alcohols **133** and **134**. It appears likely that double-bond migration precedes epoxide opening in this instance. The isomeric epoxide **135** thus formed as a transient intermediate is activated for elimination in the direction observed. Interestingly, β elimination in the case of **135** probably takes place with *anti* stereochemistry. Cyclopropane formation by intramolecular alkylation of the intermediate allylic anion from **130** is precluded by its stereochemistry, unlike the isomeric *trans* epoxide **34**, which yields the cyclopropyl compound **35** cleanly (p. 360).

Phenyl activation for cyclopropane formation in hexamethylphosphoramide-mediated reactions also works well, as illustrated earlier.[11] However, the ease of metalation of acetylenic compounds leads to complications with epoxides of this type. Thus treatment of epoxide **136** with lithium diethylamide in ether–hexane results in dominant β elimination toward a CH_3 group.[100] Nonetheless, a mixture of products is obtained as a result of concomitant migration of the triple bond. The isomerization of **136** in hexamethylphosphoramide results in cyclopropane formation, although triple-bond migration is a problem.

$(CH_3)_2C\overset{O}{\overbrace{}}CH(CH_2)_2C{\equiv}CCH_3$
136

$$\xrightarrow{\text{LiNEt}_2} CH_2{=}C(CH_3)CHOH(CH_2)_2C{\equiv}CCH_3$$

$$+ CH_2{=}C(CH_3)CHOH(CH_2)_3C{\equiv}CH$$

$$+ CH_2{=}C(CH_3)CHOHCH_2C{\equiv}CC_2H_5$$

$$+ (CH_3)_2COH\text{--}\triangle\text{-}C{\equiv}CCH_3 \quad (47{:}34{:}13{:}6)$$

$$\textbf{136} \xrightarrow[\text{HMPA}]{\text{LiNEt}_2} (CH_3)_2COH\text{--}\triangle\text{-}C{\equiv}CCH_3$$

$$+ (CH_3)_2COH\text{--}\triangle\text{-}CH_2C{\equiv}CH$$
$$(77{:}23)$$

A somewhat different situation is found with the terminal acetylene **137**, which requires two equivalents of base since removal of the acetylenic hydrogen consumes the first equivalent. In this case intramolecular alkylation to give the cyclopropane **138** is the major conversion in ether–hexane.[100] On the other hand, added hexamethylphosphoramide results in the clean production of the vinylacetylene shown in the accompanying formula.[101] Under these conditions rapid isomerization of the triple bond is probably responsible for this conversion. Thus the usual effect of hexamethylphosphoramide is reversed with this particular epoxide.

$(CH_3)_2C\overset{O}{\overbrace{}}CHCH_2CH_2C{\equiv}CH \xrightarrow{\text{LiNEt}_2} (CH_3)_2COH\text{--}\triangle\text{-}C{\equiv}CH$
137   **138**

$$+ CH_2{=}C(CH_3)CHOH(CH_2)_2C{\equiv}CH \quad (85{:}15)$$

$$\textbf{137} \xrightarrow[\text{HMPA}]{\text{LiNEt}_2} (E)\text{-}(CH_3)_2COHCH{=}CHC{\equiv}CCH_3$$

Functional Epoxides

Relatively little work is recorded concerning the strong-base isomerizations of epoxides bearing important additional functionality. This is largely attributable to the anticipated reactivity of many functional groups under the reaction conditions normally utilized. However, epoxides possessing remote functionality in protected form behave in the normal fashion. Several examples of compounds with ether and ketal groups are discussed along with saturated epoxides. A further illustration is provided by the smooth isomerization of the sugar derivative **139** by *n*-butyllithium.[102]

In fact, significant functionality remote from the epoxide group can be tolerated as indicated by the conversions of the carbonyl compounds **140** and **141**.[98,103] Alcohol **142** is also transformed to the indicated diol by the lithium salt of ethylenediamine.[104]

The trimethylsilyl ethers of a number of epoxides of allylic alcohols are isomerized by diethylaluminum 2,2,6,6-tetramethylpiperidide (**38**).[105] These substrates lead to the production of vicinal diols after removal of the protecting group. Thus elimination always involves removal of a proton away from the protected alcohol, even where the possible modes of elimination involve two equivalently situated CH_2 groups, as with epoxide **143**. The stereochemical features of these reactions are identical to those of simple trisubstituted epoxides with diethylaluminum 2,2,6,6-tetramethylpiperidide. This is illustrated for the isomeric epoxides **144** and **145** that yield products derived from the more stable complex. Interestingly, these substrates in their unprotected alcohol forms both react with lithium diisopropylamide to give diol **146** as the predominant product.[106]

$(CH_3)_3SiO(CH_2)_9HC \underset{O}{\overset{\diagup\diagdown}{\text{———}}} CHCH_2OSi(CH_3)_3$

143

$$\xrightarrow[\text{2. KF}]{\text{1. DATMP (38)}} (E)\text{-}HO(CH_2)_8CH{=}CHCHOHCH_2OH$$

(50%)

144 R = Si(CH₃)₃ → **146** (79%)

145 R = Si(CH₃)₃ → **147** (65%)

144 or **145**, R = H $\xrightarrow{\text{LDPA}}$ **146**

144, R = H $\xrightarrow{\text{Ti(OPr-}i)_4}$ **147**

145, R = H $\xrightarrow{\text{Ti(OPr-}i)_4}$ **146**

The above conversions contrast with those of the free alcohols with titanium tetraisopropoxide.[106] This reagent, like diethylaluminum 2,2,6,6-tetramethylpiperidide, results in stereospecific isomerizations, but just the opposite stereochemistry is observed, specifically, **144** → **147** and **145** → **146**. A free hydroxyl group is necessary for the titanium reagent to be effective, implicating a *syn* elimination from an intermediate titanium alkoxide of the substrate. The generality of these results is not entirely clear since complex cationic reactions occur with other substrates. Nonetheless, the concept of using a second functional group to control the site of proton removal is worthy of further exploration.

The isomerization of epoxide **147a** yields allylic alcohol **148** in addition to large amounts of the secondary product **149**.[107] The regiochemistry of this reaction is unexpected and may result from the involvement of the ketal function as an alternative site for base coordination. This complexation could direct proton removal from the adjacent CH_2 group. A second example in which a ketal group is suspected of directing elimination is the steroid epoxide **150**.[108] This compound gives the 1,4-elimination product **151** with potassium *tert*-butoxide, whereas lithium amides promote a 1,2 elimination toward the ketal to yield **152**. However, different stereochemical requirements for the two elimination reagents cannot be discounted with this conformationally rigid molecule.

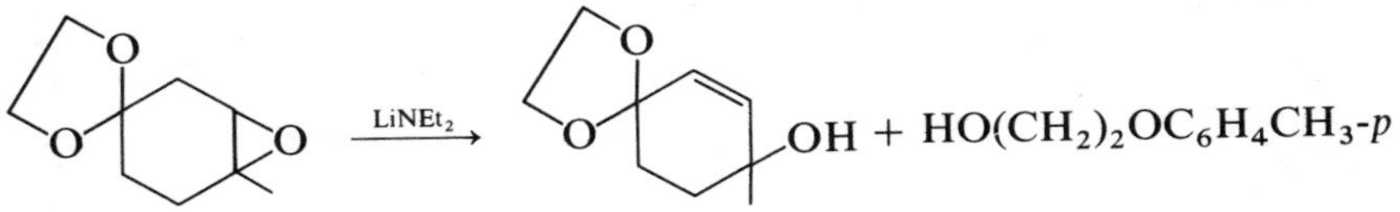

$$\xrightarrow{\text{LiNEt}_2}$$

148 $\quad$ OH $+$ HO(CH$_2$)$_2$OC$_6$H$_4$CH$_3$-p

147a $\qquad\qquad\qquad$ **149**

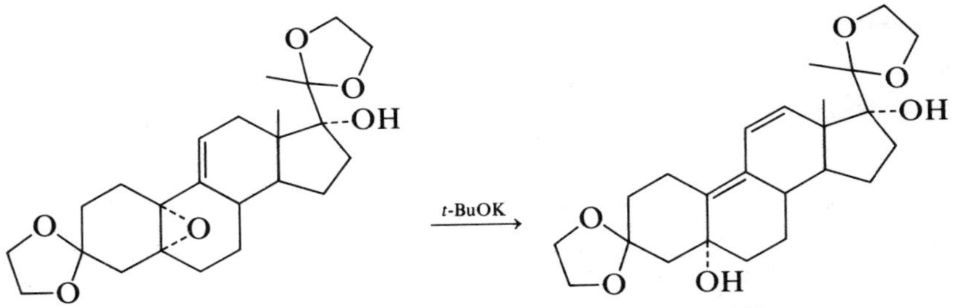

$$\xrightarrow{t\text{-BuOK}}$$

150 $\qquad\qquad\qquad\qquad$ **151**

150 $\quad\xrightarrow{\text{LiNEt}_2}$

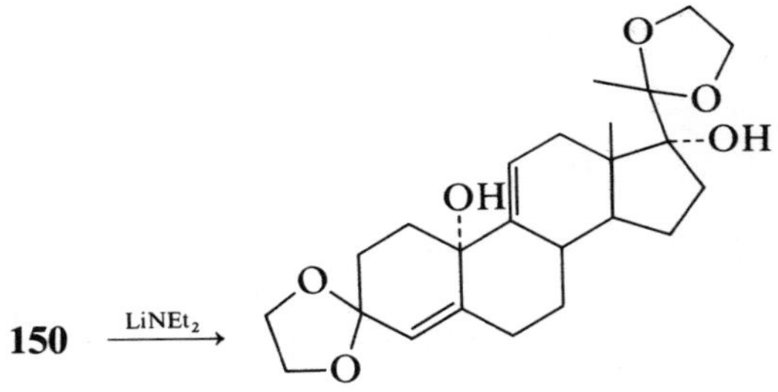

152

Neighboring reactive functions can be more intimately involved in the isomerizations. The diepoxides **153** and **154** demonstrate that rather complicated structural changes occur in some instances.[100,109] The chloroepoxide **155** is a further example of such complex behavior.[110]

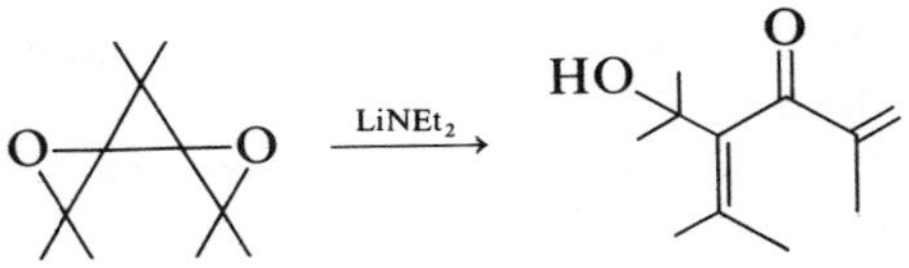

$$\xrightarrow{\text{LiNEt}_2}$$

153

154

$(9:92; 80\%)$

155

A final illustration demonstrating a rather long-range interaction involves the potassium *tert*-butoxide-promoted isomerization of epoxide **156** to conjugate acid **157**.[111] This intriguing conversion is rationalized in terms of a 1,6 elimination to give intermediate **158**, which subsequently tautomerizes to the observed product **157**.

156

158

(78%)

157

It is clear from the examples discussed above that many functionalized epoxides undergo complex transformations involving interplay of the two functions. In some of these cases, useful synthetic conversions take place that can be understood and perhaps anticipated. In other instances the usual reactions occur with or without modification by the second functional group. Surprisingly, reactive functionality (*e.g.*, carbonyl groups) remote from the epoxide site is not detrimental to isomerizations in several cases. These results suggest that much profitable work remains to be done with functionalized epoxides.

Specifically excluded from this survey are the isomerizations of epoxides with neighboring carbonyl groups and other anion-stabilizing substituents that dominate the reactions of such substrates with their own typical chemistry. Nonetheless, some leading references to these rearrangements, eliminations, and intramolecular alkylation reactions are supplied here for completeness.[1,4,5]

EXPERIMENTAL CONDITIONS

The choice of an appropriate base for effecting a desired transformation of a specific epoxide is discussed in detail in the section entitled "Scope and Limitations." The present section is concerned with the experimental aspects of performing an epoxide isomerization in the laboratory, focusing on the other parameters that can exert a significant influence on the outcome of an isomerization.

Lithium Amide Reactions

The lithium amides are the most commonly used bases, especially the diethyl and diisopropyl derivatives. These are usually prepared by the addition of a standardized solution of n-butyllithium in hexane to a solution of the amine in ether. No particular precautions are required in this preparation other than the use of a dry apparatus, anhydrous solvents, an inert atmosphere, and syringe techniques for the transfer of alkyllithium solutions. The quantity of the lithium amide is estimated by assuming complete conversion of the alkyllithium. Methyllithium in ether is also occasionally used to prepare amides. Alternatively, the lithium amides can be prepared directly from the amines and metallic lithium under appropriate conditions.[97,112–114] This approach is not used frequently but provides a more economical source of amides for large-scale preparations. N-Lithioethylenediamine is readily prepared in this fashion and is recommended as a base worthy of more widespread application for epoxide isomerizations.[26,115] It should be noted, however, that direct preparations of lithium amides utilizing hexamethylphosphoramide give a reagent that may behave differently from the lithium amides obtained in the usual fashion because of the special effect of hexamethylphosphoramide on some epoxide rearrangements. Of course, hexamethylphosphoramide can be added to the usual preparations of lithium amides when it is needed to produce a desired effect. Likewise, excess lithium salts, sometimes present in solutions of organolithium reagents prepared in ether, also have a significant effect on the course of certain isomerizations.[58]

Typical reaction conditions for performing an epoxide isomerization call for treatment of the substrate with an excess of the lithium amide, often 2–5 eqs, in an ether–hexane solvent mixture. This is convenient, especially with small-scale reactions, in dealing with impurities that consume base. Furthermore, base decomposition of the ether solvent also occurs.[53] However, a large excess of amide is not, in general, necessary.[6,15,27,33,116] In fact, excess base can be detrimental when subsequent transformations of the primary products are facile (p. 373). Moreover, in the well-studied isomerization of cyclohexene oxide, a decrease in the ratio of base to epoxide results in lowering of the amount of both nucleophilic adduct and cyclohexanone relative to 2-cyclohexenol.[53]

The ether–hexane solvent system is most often used for lithium amide isomerizations, partly as a consequence of the usual method of preparation of this base and partly because this solvent mixture usually results in homogeneous reaction mixtures, which is seldom the case with less polar solvents.

Although the ratio of the solvent components is usually rather arbitrarily determined, it is recommended that the amount of ether be limited to that required to obtain homogeneous solutions. Isomerizations are significantly slower in pure ether solvent, undoubtedly as a result of complexation of the lithium amide with ether. This competition between the substrate and the solvent for the reagent results in lowering of the concentration of the epoxide–base complex. This situation may also affect the selectivity of certain epoxide isomerizations that are controlled by complex formation, but information regarding this point is lacking.

Most isomerizations by lithium amides are run between 0° and reflux temperature of the ether–hexane solvent mixture, usually the latter. Some epoxides react very slowly under these conditions. In this situation benzene can be employed to facilitate reaction by allowing higher reaction temperatures. However, side reactions are also more likely at higher temperatures. Alternatively, tetrahydrofuran is used in some instances, although it should be noted that this polar solvent can significantly affect the product ratios in cases where competing processes are possible.

In general, it is strongly recommended that the course of the reaction be followed closely by a rapid assay technique such as gas chromatography or thin-layer chromatography. This allows the reaction to be stopped when the starting epoxide is consumed or when the desired product is maximized.

The significant influence of hexamethylphosphoramide on epoxide rearrangements is very useful synthetically in a number of instances. Although many of these reactions have been performed in hexamethylphosphoramide as the solvent, its effect appears to be achieved with as little as 2–4 eqs of hexamethylphosphoramide per equivalent of lithium amide.[11,26,27,100] This is recommended in view of the cost and biohazards associated with hexamethylphosphoramide.

Caution!

Hexamethylphosphoramide is an animal carcinogen. Appropriate handling and disposal procedures should be followed.

Organolithium Reactions

Organolithium reagents are only infrequently used as the functioning base in epoxide isomerizations. Nonetheless, these reagents are of central importance because they serve as the source of amide bases in most rearrangement procedures. A recent *Organic Reactions* chapter details the use of organolithium reagents.[117] Solutions of *n*-butyllithium in hexane are generally available from commercial sources, as are methyllithium solutions in ether, *tert*-butyllithium solutions in hydrocarbon solvents, and other organolithium reagents. When these are not available, organolithium reagents can be obtained by literature preparations.[118–120] The quality of fresh commercial solutions is generally good, and the indicated assay is usually sufficiently accurate for most synthetic applications. In reactions where the exact amount of base is critical or when the concentration of the reagent is uncertain (as with aged commercial reagents or those prepared in the laboratory), several methods of analysis are available.[117,121-123] The direct titration with *sec*-butyl alcohol by use of 2,2-biquinoline as an indicator is a simple and reliable method for the assay of most organolithium reagents.[124] The commonly used hexane solutions of *n*-butyllithium can normally be stored for several months in bottles capped with serum stoppers, provided transfers are carefully performed by using dry syringes. The deposit of large amounts of solid material from the clear yellow commercial solutions is usually an indication of reagent deterioration.

Isomerizations using organolithium reagents as the effective base are generally performed with an excess of the reagent in hydrocarbon solvents at temperatures ranging from −78° to reflux temperatures of the solvents employed. Because of possible decomposition of the reagent at higher temperatures, the lowest convenient temperatures consistent with reasonable reaction times are recommended. Ether is sometimes used as a solvent or cosolvent in isomerizations in which organolithium reagents are used as bases, but these reactions appear to give more complicated product mixtures than do those run in hydrocarbon solvents.

Aluminum Amide Reactions

These reagents deserve more widespread application to the isomerization of suitable epoxides. The efficiency of these conversions appears to increase with increasing steric bulk of the substituents on nitrogen. The diethylaluminum derivative of 2,2,6,6-tetramethylpiperidine is most commonly used, although the species derived from diisopropylamine, *N*-methylaniline, and other more readily available amines also effect the reaction, albeit more slowly.[32]

Reactions are normally performed with four equivalents of the reagent at 0° in benzene solution. Hexane is also suitable as a reaction medium, but ether and especially tetrahydrofuran inhibit the reaction. Diethylaluminum *N*-methylanilide is a superior reagent for the analogous isomerization of several oxetanes, which require prolonged reaction times in refluxing benzene.[125] Use of this reagent with certain epoxides may also be advantageous.

The aluminum amides are prepared *in situ* in the desired solvent prior to the addition of the epoxide. The reaction of the appropriate lithium amide, prepared in the usual manner, with diethylaluminum chloride at 0° in benzene for 30 minutes is a convenient procedure.[32] Alternatively, diisobutylaluminum hydride reacts with secondary amines in benzene at slightly higher temperatures with the evolution of hydrogen to give related aluminum amides.[32] Secondary amines also react with trimethylaluminum to give dimethylaluminum amides and methane.[126]

The various organoaluminum reagents required for preparation of the aluminum amides are available commercially in lecture bottles as the pure materials or as standardized solutions in hydrocarbon solvents.[127] The pure materials are highly *pyrophoric* and require special handling.[128] It is much more convenient to use the commercial solutions, and this is recommended whenever possible. These solutions react with air and moisture, but they can be transferred by dry syringe techniques from solutions stored in bottles sealed with rubber septa. Reactions should be run under anhydrous conditions under an inert atmosphere of dry nitrogen or argon.

EXPERIMENTAL PROCEDURES

The following procedures have been chosen to illustrate typical reaction conditions for effecting various types of isomerization with different bases. Reactions are normally performed in an apparatus consisting of a three-necked, round-bottomed flask equipped with a reflux condenser to which a gas inlet tube is attached, with a pressure-regulating dropping funnel and a rubber septum. A stream of dry nitrogen is passed through the assembly while it is flame-dried with a Bunsen burner and subsequently charged with the reactants. At this point the system is closed so as to maintain a slight positive pressure of nitrogen. A magnetic stirrer is usually adequate for agitation. The course of the reaction is followed by withdrawing aliquots by syringe through the rubber septum. These are hydrolyzed and assayed for the disappearance of starting material by gas chromatography or thin-layer chromatography (TLC).

2-p-Tolyl-2-propen-1-ol (Isomerization of a Terminal Epoxide to an Allylic Alcohol by Using Lithium Diisopropylamide in Ether)[92]

To a solution of 3 g (0.03 mol) of diisopropylamine in 25 mL of ether was added 15 mL (0.03 mol) of 2.1 M n-butyllithium in ether. After 30 minutes a solution of 3 g (0.02 mol) of 1,2-epoxy-2-p-tolylpropane in 40 mL of ether was added slowly. The reaction mixture was stirred overnight at room temperature, heated to reflux for 4 hours, and, after cooling, partitioned between ether and water. The organic layer was washed successively with water and brine and dried. Solvent evaporation followed by chromatography on neutral alumina afforded 2.6 g (85%) of 2-p-tolyl-2-propen-1-ol; bp 105–107° (5–7 torr); IR(neat) 3380, 3070, 1650, 1050, 890, and 820 cm^{-1}. This product showed a single spot by TLC analysis.

trans-Pinocarveol [2(10)-Pinen-3α-ol] (Isomerization of a 1-Methylepoxycycloalkane to an Allylic Alcohol with Lithium Diethylamide in Ether–Hexane)[6]

The isomerization of α-pinene oxide to trans-pinocarveol in 90–95% yield is described in Organic Syntheses.[6]

12-Methoxy-19-norpodocarpa-4(18),8,11,13-tetraen-3α-ol (Isomerization of a 1-Methylepoxycycloalkane with Lithium Diethylamide in Ether)[78]

To 12.1 mL (0.02 mol) of a 1.6 M solution of methyllithium in ether at 0° was added 3 mL of dry diethylamine in 5 mL of ether. After 10 minutes the evolution of methane had ceased. A solution of 1.0 g (0.004 mol) of 3α,4α-epoxy-12-methoxy-18-norpodocarpa-8,11,13-triene in 18 mL of ether was added slowly, and the mixture was heated to reflux for 36 hours. The cooled reaction mixture was poured into ice water and extracted with ether. The ether extract was washed with 2 N hydrochloric acid, water, and brine. After drying, the solvent was removed to give 1.0 g (100%) of 12-methoxy-19-norpodocarpa-4(18),8,11,13-tetraen-3α-ol, which crystallized from light petroleum as needles: mp 93–95°, $[\alpha]_D + 163°$ (c, 0.45), IR (CHCl$_3$) 3600, 1660, 1040, and 910 cm^{-1}, NMR δ 1.00 (s, 3), 1.62 (s, 1, exchanged with D$_2$O), 3.88 (s, 3), 4.35 (m, 1), 4.74 (br s, 1), 5.07 (br s, 1), and 6.7–7.0 (m, 3).

2,4,7-Cyclononatrien-1-ol (Isomerization of an Unsaturated Epoxide to an Allylic Alcohol by n-Butyllithium in Ether–Hexane)[129]

To a stirred solution of 40.0 g (0.30 mol) of 1,4,7-cyclononatriene oxide in 650 mL of ether at 0° was added 320 mL (0.51 mol) of a 1.6 M solution of n-butyllithium in hexane over a period of 1 hour. After an additional 4.5 hours, 200 mL of a saturated ammonium chloride solution was added slowly, with stirring and cooling, over a period of approximately 1 hour. The organic layer was separated, washed with 200 mL of water and dried over anhydrous potassium carbonate. The solvent was removed to give a yellow oil that was distilled under reduced pressure to give 6 g of volatile material followed by 26.3 g (66%) of 2,4,7-cyclononatrien-1-ol of bp 80° (0.1 torr) that crystallized on standing, mp 40–42°. The IR spectrum of this material was identical with that of an authentic sample.

endo,cis-Bicyclo[3.3.0]octan-2-ol (Isomerization of an Epoxycycloalkane to a Bicyclic Alcohol by Lithium Diisopropylamide in Ether–Hexane)[58]

To a solution consisting of 90 mL (0.22 mol) of 2.4 M n-butyllithium in hexane and 90 mL of ether in an ice bath was added 35 mL (0.25 mol) of diisopropylamine over a period of 5 minutes. After 30 minutes 12.6 g (0.1 mol) of cis-cyclooctene oxide in 20 mL of ether was added and the reaction mixture was heated to reflux for 2 days. The cooled solution was washed with 100 mL of water, dried, and concentrated under reduced pressure. The residue was distilled under vacuum to afford 10.1 g (80%) of endo,cis-bicyclo[3.3.0]octan-2-ol, bp 103–105° (25 torr). This product contained less than 2% of 2-cyclooctenol as determined by gas chromatographic analysis on a Carbowax 20M column.

exo-Tetracyclo[5.2.1.0.^{3,7}0^{3,8}]decan-9-ol (Transannular Insertion of a Polycyclic Epoxide by Lithium Diethylamide in Benzene–Hexane)[130]

To a solution of 0.45 g (0.006 mol) of dry diethylamine in 2.8 mL of dry benzene at 0° was added 2.8 mL (0.0045 mol) of 1.6 M n-butyllithium in hexane. After stirring for 20 minutes, a solution of 0.40 g (0.0027 mol) of exo-8,9-epoxytricyclo[5.2.1.0^{3,7}]decane in 2.8 mL of benzene was added, and the mixture was heated to reflux for 48 hours. The cooled reaction mixture was poured into ice water and extracted with ether. The extract was washed with saturated aqueous ammonium chloride solution and water, dried, and concentrated. Distillation at reduced pressure gave 0.34 g (86%) of tetracyclic alcohol: bp 50–60° (0.4 torr), mp 56–70°, near IR (CCl$_4$) 1.676 μM (ε 0.38), NMR (CCl$_4$) δ 1.0–2.1 (m, 14), 3.86 (s, 1), and 4.45 (s, 1).

1,2-Benzocyclohepten-4-one (Isomerization of an Epoxide to a Ketone with Lithium Diisopropylamide)[41]

To a solution of 0.77 mL (0.0055 mol) of diisopropylamine in 15 mL of anhydrous ether at $-78°$ was slowly added 2.4 mL (0.0055 mol) of 2.3 M n-butyllithium. After 15 minutes the mixture was allowed to warm to room temperature, and a solution of 0.35 g (0.0022 mol) of 3,4-epoxy-1,2-benzocycloheptene in 5 mL of anhydrous ether was added slowly. After stirring for 1 hour, the reaction was quenched by the addition of 5 mL of water. The aqueous layer was separated and extracted twice with 10 mL of ether. The combined organic layers were washed with 10% aqueous hydrochloric acid, dried, and concentrated. Vacuum transfer gave 0.22 g (63%) of 1,2-benzocyclohepten-4-one: NMR (CDCl$_3$) δ 1.6–2.0 (m, 2), 2.3–2.8 (m, 2), 2.8–3.0 (m, 2), 3.7 (s, 2), and 7.2 (s, 4).

endo-Bicyclo[5.1.0]oct-5-en-2-ol (Isomerization of an Unsaturated Epoxide to a Cyclopropylcarbinol by *N*-Lithioethylenediamine in Benzene–Hexamethylphosphoramide[11]

Preparation of *N*-Lithioethylenediamine in Benzene.[115] Lithium (5 g of a 40% dispersion in mineral oil) was added to the reaction vessel under an inert atmosphere and washed several times with small quantities of pentane, which was removed by syringe. Anhydrous benzene (200 mL) was added, followed by 18 g (0.3 mol) of dry ethylenediamine over a period of 30 minutes. The reaction mixture was heated to reflux; the liberation of gas was followed by attachment of a gas bubbler. After about 3 hours, gas evolution ceased and a white powder formed in the reaction flask.

Isomerization of 5,6-Epoxycyclooctene. The base mixture was cooled to room temperature and 200 mL of hexamethylphosphoramide was added, followed by 10.0 g (0.08 mol) of epoxide. The epoxide disappeared within 15 minutes to give a product containing the desired alcohol and a small amount of 3,5-cyclooctadienol. The mixture was heated to 60° until the latter was completely converted to 3-cyclooctenone. The cooled reaction mixture was poured into a large amount of water and extracted with ether. The extract was washed with saturated ammonium chloride solution, dried, and concentrated. Distillation of the residue under reduced pressure gave, after a small forerun of 3-cyclooctenone, 8.0 g (80%) of *endo*-bicyclo[5.1.0]oct-5-en-2-ol: bp 90° (9 torr); IR 3350, 3080, 3008, 1660, 695 cm^{-1}; NMR (CCl$_4$) δ 0.3–2.1 (m, 8), 3.5 (s, 1), 3.95–4.35 (m, 1), and 5.1–5.9 (m, 2).[84]

(2-Phenylcyclopropyl)methanol (Formation of a Cyclopropylcarbinol by Lithium Diethylamide Prepared in Benzene–Hexamethylphosphoramide[11]

Base Preparation.[114] A mixture of 0.7 g (0.1 mol) of hammered lithium wire, 7.3 g (0.1 mol) of diethylamine, 20 mL of hexamethylphosphoramide and 20 mL of benzene was stirred at temperatures below 25° under an argon atmosphere for several hours until the lithium disappeared, giving a dark red homogeneous solution.

Isomerization of 4-Phenyl-1,2-epoxybutane. To 0.05 mol of the lithium diethylamide solution in benzene–hexamethylphosphoramide was added 2.0 g (0.013 mol) of epoxide. After 30 minutes the reaction mixture was poured into a large quantity of water and extracted with ether. The extract was washed with saturated ammonium chloride solution, dried, and concentrated. The residue was distilled at reduced pressure to give 1.7 g (86 %) of trans-(2-phenylcyclopropyl)methanol: bp 137° (12 torr); IR 3340, 1605, 1595, 740, 695 cm^{-1}; NMR (250 MHz) δ 0.68–0.85 (m, 2), 1.16–1.40 (m, 1), 1.60–1.78 (m, 1), 3.30–3.54 (m, 2), 3.98 (s, 1), and 6.70–7.40 (m, 5). A small doublet at δ 3.8 indicated the presence of 6 % of the cis isomer.

(E)-2-Cyclododecenol (Isomerization of an Epoxide to an Allylic Alcohol by Diethylaluminum 2,2,6,6-Tetramethylpiperidide).[32]

Preparation of Diethylaluminum 2,2,6,6-Tetramethylpiperidide. A benzene solution of 1 eq of diethylaluminum chloride was added dropwise at 0° to a solution of 1 eq of lithium 2,2,6,6-tetramethylpiperidide prepared in the usual fashion in benzene.[131] The resulting slurry was stirred for 30 minutes and used immediately.

Isomerization of trans-Epoxycyclododecane. To a stirred mixture of 0.004 mol of diethylaluminum 2,2,6,6-tetramethylpiperidide in 10 mL of benzene at 0° was added dropwise over 5 minutes a solution of 0.18 g (0.001 mol) of epoxide in 3 mL of benzene. The mixture was stirred at 0° until analysis indicated the absence of starting material. The reaction was quenched by the addition of ice-cold 1 N hydrochloric acid. The organic layer was separated, and the aqueous layer was extracted with ether. The organic layers were combined, washed with brine, dried, and concentrated. The residue was purified by preparative TLC (R_f 0.22 in 1:2 ether–hexane) to give 99 % of (E)-2-cyclododecenol: IR (neat) 3330–3370, 1465, 1450, 970 cm^{-1}; NMR (CCl$_4$) δ 3.73–4.20 (1, m), 4.97–5.82 (2, m); mass spectrum (m/z) 182 (16), 164 (13), 139 (32), 125 (46), and 98 (100).

TABULAR SURVEY

Examples of base-promoted isomerizations of unactivated epoxides falling within the limitations described in the text are summarized in the following tables. An attempt has been made to include all reports of such reactions appearing in the literature through late 1982. The tables are organized to parallel the discussion in the section entitled "Scope and Limitations." Table I lists the saturated epoxides that have been exposed to base isomerization. This table is further divided into acyclic, spiro, and cyclic epoxides. Table II gives epoxides that have unsaturation present in the molecule as olefinic, acetylenic, or aromatic functions and is subdivided according to the type of unsaturation present. Table III groups those epoxides that have additional functionality in the molecule. Within each subdivision of these tables the epoxides utilized as substrates are listed in order of increasing number of carbon atoms and then increasing number of hydrogen atoms in the molecular formula. A dash in the yield or reaction conditions column indicates that the appropriate information was not reported. In many instances products that are not isomeric with the starting epoxide are given in order to help define the limitations on the title reaction.

The following abbreviations are used in the tables:

DATMP	Diethylaluminum 2,2,6,6-tetramethylpiperidide
Diglyme	Diethylene glycol dimethyl ether
DMF	Dimethylformamide
DMSO	Dimethyl sulfoxide
Ether	Diethyl ether
HMPA	Hexamethylphosphoramide
LDPA	Lithium diisopropylamide
THF	Tetrahydrofuran
THP	Tetrahydropyranyl ether

TABLE I. SATURATED EPOXIDES

A. Acyclic Epoxides

	Epoxide	Base	Reaction Conditions	Products and Yields (%)	Refs.
C_5	CH$_3$HC$\overset{O}{\diagup\diagdown}CHC_2H_5$ cis:trans (44:56)	LiNEt$_2$, 2.5 eq	Ether–hexane, reflux	CH$_2$=CHCHOHC$_2$H$_5$ (68)	21
	CH$_3$HC$\overset{O}{\diagup\diagdown}$C(CH$_3$)$_2$	LiNEt$_2$, 2.5 eq	Ether–hexane, reflux	CH$_2$=CHCOH(CH$_3$)$_2$ (I), CH$_3$CHOHC(CH$_3$)=CH$_2$ (II) 41:59 (—)	21
		t-BuOK, 1 eq	DMSO, 45°, 2 weeks	I (80) + II (15)	46
C_6	(CH$_3$)$_2$C$\overset{O}{\diagup\diagdown}$C(CH$_3$)$_2$	t-BuOK, 1 eq	DMSO, 45°, 2 weeks	CH$_2$=C(CH$_3$)COH(CH$_3$)$_2$ (98)	46
	t-C$_4$H$_9$HC$\overset{O}{\diagup\diagdown}CH_2$	t-BuOK, 1 eq	DMSO, room temp. 2 weeks	No reaction	46
C_8	n-C$_3$H$_7$HC$\overset{O}{\diagup\diagdown}CHC_3H_{7}$-n cis	LiNEt$_2$, 1.2 eq	Ether, room temp, 3 days	(E)-n-C$_3$H$_7$CHOHCH=CHC$_2$H$_5$ (55), threo-n-C$_3$H$_7$CHOHCH(C$_3$H$_{7}$-n)N(C$_2$H$_5$)$_2$ (14)	15
	n-C$_3$H$_7$HC$\overset{O}{\diagup\diagdown}CHC_3H_{7}$-n trans	LiNEt$_2$, 1.2 eq	Ether, room temp, 3 days	(E)-n-C$_3$H$_7$CHOHCH=CHC$_2$H$_5$ (79), (n-C$_3$H$_7$)$_2$CHCHO (3)	15
	i-C$_3$H$_7$HC$\overset{O}{\diagup\diagdown}CHC_3H_{7}$-n cis:trans (88:12)	LiNEt$_2$, 2.5 eq	Ether–hexane, reflux	(E)-i-C$_3$H$_7$CHOHCH=CHC$_2$H$_5$ (—)	21
	i-C$_3$H$_7$HC$\overset{O}{\diagup\diagdown}CHC_3H_{7}$-i	LiNEt$_2$	Ether–hexane, reflux, 2 days	No reaction	29
	t-C$_4$H$_9$HC$\overset{O}{\diagup\diagdown}$C(CH$_3$)$_2$	LiNEt$_2$, 2.5 eq t-BuLi, 3 eq	Ether–hexane, reflux Pentane, reflux, 1 day	t-C$_4$H$_9$CHOHC(CH$_3$)=CH$_2$ (87), " (74)	52 50

C_{10} 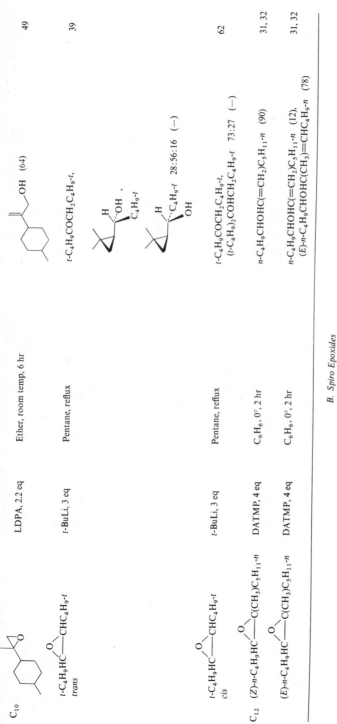	LDPA, 2.2 eq	Ether, room temp, 6 hr	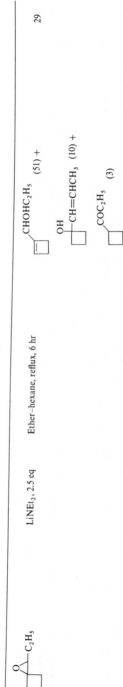 OH (64)	49
$t-C_4H_9HC\!-\!CHC_4H_9\text{-}t$ (O) trans	t-BuLi, 3 eq	Pentane, reflux	$t-C_4H_9COCH_2C_4H_9\text{-}t$,	39
$t-C_4H_9HC\!-\!CHC_4H_9\text{-}t$ (O) cis	t-BuLi, 3 eq	Pentane, reflux	$t-C_4H_9COCH_2C_4H_9\text{-}t$, $(t-C_4H_9)_2COHCH_2C_4H_9\text{-}t$ 73:27 (—)	62
C_{12} (Z)-$n-C_4H_9HC\!-\!C(CH_3)C_5H_{11}\text{-}n$ (O)	DATMP, 4 eq	C_6H_6, 0°, 2 hr	$n-C_4H_9CHOHC(\!=\!CH_2)C_5H_{11}\text{-}n$ (90)	31, 32
(E)-$n-C_4H_9HC\!-\!C(CH_3)C_5H_{11}\text{-}n$ (O)	DATMP, 4 eq	C_6H_6, 0°, 2 hr	$n-C_4H_9CHOHC(\!=\!CH_2)C_5H_{11}\text{-}n$ (12), $(E)-n-C_4H_9CHOHC(CH_3)\!=\!CHC_4H_9\text{-}n$ (78)	31, 32

B. Spiro Epoxides

C_7	LiNEt$_2$, 2.5 eq	Ether–hexane, reflux, 6 hr	CHOHC$_2$H$_5$ (51) + OH CH=CHCH$_3$ (10) + COC$_2$H$_5$ (3)	29

TABLE I. SATURATED EPOXIDES (*Continued*)

B. *Spiro Epoxides* (*Continued*)

Epoxide	Base	Reaction Conditions	Products and Yields (%)	Refs.
C₇ (*Contd.*)	LiNEt₂, 2.5 eq	Ether–hexane, reflux, 1.5 hr	CHOHCH₃ (57) + (25)	29
C₈	LiNEt₂, 2 eq	Hexane, room temp	(94)	79
	LiNEt₂, 2.5 eq	Ether–hexane, reflux, 1 hr	CHOHC₂H₅ (84)	29
	LiNEt₂, 2.5 eq	Ether–hexane, reflux, 2 hr	CH=CH₂ (66)	29
C₉	LiNEt₂, 1.1 eq	Ether, room temp, 2 hr	(I) + (II) 60:40 (—)	79
	LiNMe₂, 1.1 eq	Ether, room temp, 2 hr	I + II 14:86 (—)	79
	LiNEt₂, 1.1 eq	Pentane, room temp, 2 hr	I + II 90:10 (—)	79
	LDPA, 1.1 eq	Ether, room temp, 2 hr	I + II 12:88 (—)	79
	n-BuLi, 1.1 eq	Pentane, room temp, 2 hr	I + II 78:22 (—)	79
	LiNEt₂, 2.5 eq	Ether–hexane, reflux, 49 hr	CHOHC₂H₅ (3.5) +	29

402

Substrate	Reagent	Conditions	Product(s) (%)	Ref.
C_{10} — cycloheptane epoxide (CH₃)	LiNEt₂, 2.5 eq	Ether–hexane, reflux, 2.5 hr	OH, CH=CHCH₃ (66) (E) + CHOHCH₃ (28)	29
cycloheptane epoxide (cyclopropyl)	LiNEt₂, 2 eq	Hexane, room temp	OH, CH=CH₂ (46); OH^a (91)	79
cyclohexane epoxide (cyclopropyl)	LiNEt₂	—	No reaction	79
bicyclic epoxide	LiNEt₂, 2.5 eq	Ether, reflux, 2 days	OH (81)	48
cyclohexane epoxide (C_3H_{7}-i)	LiNEt₂, 2.5 eq	Ether–hexane, reflux, 72 hr	CHOHC₃H₇-i (—) + Starting epoxide (76)	29
cycloheptane epoxide (C_2H_5)	LiNEt₂, 2.5 eq	Ether–hexane, reflux, 5 hr	CHOHC₂H₅ (74.5) + OH, CH=CHCH₃ (1.5)	29

TABLE I. SATURATED EPOXIDES (Continued)

B. Spiro Epoxides (Continued)

Epoxide	Base	Reaction Conditions	Products and Yields (%)	Refs.
C_{10} (Contd.) 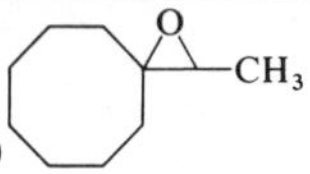 [structure: CH₃ spiro epoxide cyclooctane]	LiNEt₂, 2.5 eq	Ether–hexane, reflux, 2 hr	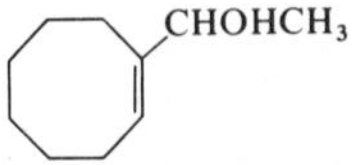 [structure: cyclooctene-CHOHCH₃] (48) 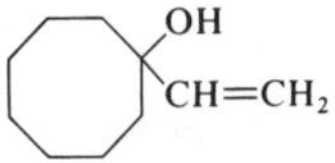 +	29
C_{11} 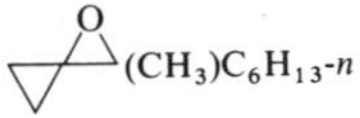 [structure: (CH₃)C₆H₁₃-n, C₂H₅ spiro epoxide cyclooctane]	LiNEt₂, 2 eq	Hexane, room temp	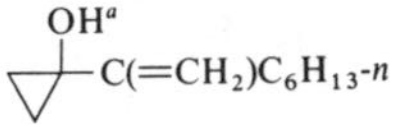 [structure: OH CH=CH₂] (25); [structure: cyclopropyl OHª C(=CH₂)C₆H₁₃-n] (96)	79
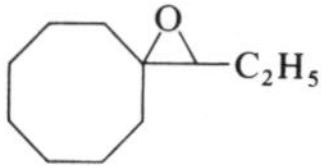	LiNEt₂, 2.5 eq	Ether–hexane, reflux, 2 hr	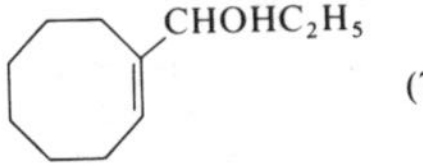 [structure: cyclooctene-CHOHC₂H₅] (74)	29
C_{13} 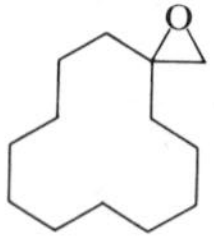 [structure: C₂H₅ spiro epoxide cyclodecane]	DATMP, 4 eq	C₆H₆, 0°, 2 hr	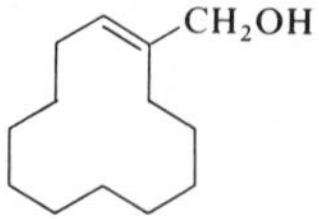 [structure: CH₂OH] (63) 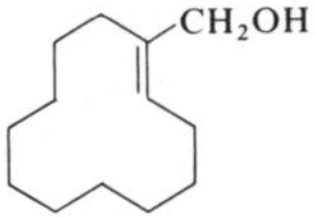 + [structure: CH₂OH] (29)	31, 32
C_{15} 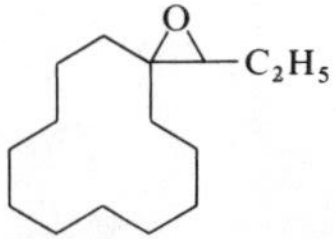 [structure: C₂H₅ spiro epoxide cyclododecane]	LiNEt₂, 2.5 eq	Ether–hexane, reflux, 22 hr	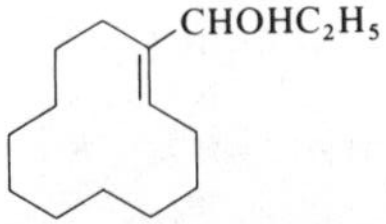 [structure: CHOHC₂H₅] (55)	29

404

C. Alicyclic Epoxides

Substrate	Reagent	Solvent / Conditions	Product (yield)	Refs.
C_{20} epoxide	n-BuLi	Ether	t-C_4H_9 ···C_4H_9-t + (—) ···C_4H_9-t (OH)	133
C_5 epoxide	LiNEt$_2$, 2.5 eq	Ether–hexane, reflux, 2 days	(I) + (II) (I + II, 12) + N(C$_2$H$_5$)$_2$ (36)	52
	LDPA / DATMP, 4 eq	Ether–hexane, reflux, 7 hr / C$_6$H$_6$, 0°, 2 hr	I + II (50) / No reaction	52 / 32
C_6 bicyclic epoxide	LiNEt$_2$, 3 eq	C$_6$H$_6$–hexane, room temp, 65 hr	(46)	10
methyl cyclopentene epoxide	LiNEt$_2$, 2.5 eq	Ether–hexane, reflux, 5 hr	(7.5) + OH (61)	67
	LiNEt$_2$, 2.5 eq	Ether–hexane, reflux, 2 days	OH (I, 67)	52, 53
cyclohexene epoxide	LDPA, 2.5 eq	Ether–hexane, HMPA (5 eq), room temp, 1.5 hr	I (—)	26
	n-BuLi, 3 eq	Ether–hexane, −78°, 3 hr; 25°, 15 hr	I + 34:66 (—)	42

TABLE I. SATURATED EPOXIDES (Continued)

C. Alicyclic Epoxides (Continued)

Epoxide	Base	Reaction Conditions	Products and Yields (%)	Refs.
C₆ (Contd.)	t-BuOK, 2.8 eq DATMP	Pyridine, reflux, 1 hr C₆H₆, 0°, 3 hr	I (—) No reaction	134 32
C₇ 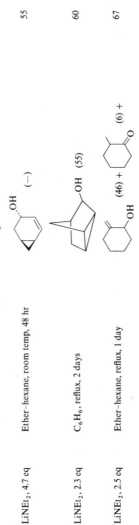	LiNEt₂, 4.7 eq	Ether–hexane, room temp, 48 hr	(40)	55
	LiNEt₂, 4.7 eq	Ether–hexane, room temp, 48 hr	(—)	55
	LiNEt₂, 2.3 eq	C₆H₆, reflux, 2 days	(55)	60
	LiNEt₂, 2.5 eq	Ether–hexane, reflux, 1 day	(46) + (6) +	67
cis:trans ca. 50:50 	LiNEt₂, 2.5 eq	Ether–hexane, reflux	(66) + (6) (2) cis:trans ca. 50:50	21
	LiNEt₂, 2.5 eq	C₆H₆, reflux, 2 days	(13) + (1, 16) +	26, 39

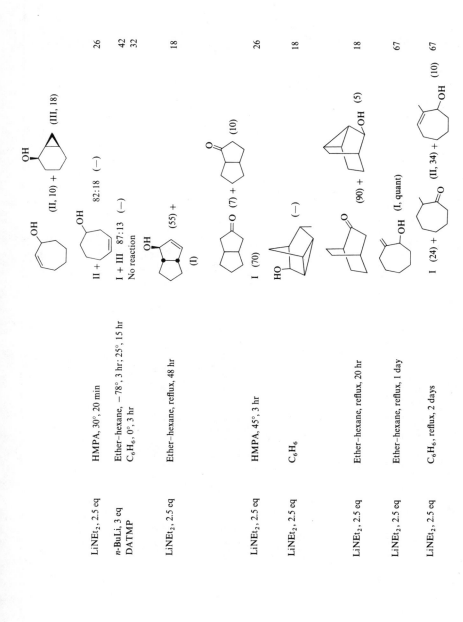

LiNEt₂, 2.5 eq	HMPA, 30°, 20 min	(II, 10) + (III, 18)	26
n-BuLi, 3 eq DATMP	Ether–hexane, −78°, 3 hr; 25°, 15 hr	82:18 (—)	42
	C₆H₆, 0°, 3 hr	II + III 87:13 (—)	32
LiNEt₂, 2.5 eq	Ether–hexane, reflux, 48 hr	No reaction; (55) + (I)	18
LiNEt₂, 2.5 eq	HMPA, 45°, 3 hr	(7) + (10); I (70)	26
LiNEt₂, 2.5 eq	C₆H₆	(—)	18
LiNEt₂, 2.5 eq	Ether–hexane, reflux, 20 hr	(90) + (5)	18
LiNEt₂, 2.5 eq	Ether–hexane, reflux, 1 day	(I, quant)	67
LiNEt₂, 2.5 eq	C₆H₆, reflux, 2 days	I (24) + (II, 34) + (10)	67

C₈

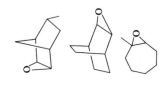

TABLE I. Saturated Epoxides (Continued)

C. Alicyclic Epoxides (Continued)

Epoxide	Base	Reaction Conditions	Products and Yields (%)	Refs.
C_8 (Contd.) [cyclooctene oxide]	t-BuLi, 3 eq	Pentane, reflux 2 days	I (52) + II (6) + [1-(t-butyl)cycloheptanol] (9)	67
	LiNEt$_2$	Ether–hexane, (1:1), reflux, 2 days	[bicyclo[3.3.0]octanol (I)] + [cyclooct-2-enol (II)] 8:92 (≥80)	42
	LiNEt$_2$	Ether–hexane	I + II 20:80 (—)	58
	LiNEt$_2$	Ether–hexane, LiBr	I + II 80:20 (—)	58
	LiNEt$_2$	Ether–LiBr, reflux, 2 days	I (70) + II (16)	9
	LiNEt$_2$	Ether, reflux, 2 days	I + II 65:35 (≥80)	42
	LiNEt$_2$	Ether–HMPA, room temp	I^b + II^b 4:96 (≥80)	42
	LiNEt$_2$	THF, reflux, 2 days	I^b + II^b 7:93 (≥80)	42
	LiNEt$_2$	THF, HMPA, 1 eq	I^b + II^b 9:91 (≥80)	42
	LDPA, 2.2 eq	Ether–hexane, reflux, 2 days	I + II 98:2 (80)	26, 58
	LDPA, 2.5 eq	HMPA, room temp, 72 hr	II (70)	26
	n-BuLi, 3 eq	Ether–hexane, −78°, 3 hr; 25°, 15 hr	I (Good)	42
	C_6H_5Li, 1.8 eq	C_6H_6, LiBr, reflux, 18 hr	I (50) + II (27)	9
	t-BuOK, 1 eq	DMSO, 80°, 22 hr	II (40)	59
[cyclooctene oxide, dashed]	LiNEt$_2$, 2.4 eq	C_6H_6, reflux, 72 hr	[cycloheptanecarbaldehyde] (27) + [cyclooct-2-enol] (10) + [bicyclo[3.3.0]octanol] (47)	9

408

	Reagent	Conditions	Products (yield %)	
C₉	LiNEt₂, 2.5 eq	C₆H₆, reflux, 2 days	(17) + (20) + O (40) +	52
	LiNEt₂, 2.5 eq	Ether–hexane	(86)	18
	LiNEt₂, 4.7 eq	Ether–hexane, room temp, 48 hr	OH (—)	55
	LiNEt₂, 4.7 eq	Ether–hexane, room temp, 48 hr	OH (—)	55
	LiNEt₂, 2.5 eq	Ether–hexane, room temp, 1 day	OH (I, 88) + II (6)	67
	LiNEt₂, 2.5 eq	Ether–hexane, reflux, 1 day	I (37) + II (41) + OH (II, 6)	67
	LiNEt₂, 2.5 eq	C₆H₆, reflux, 2 days	I (5) + II (68) + III (22) O (III, 9)	67
	LiN(C₆H₁₁)Pr-i, 2 eq	10 min	OH (56)	135
C₁₀ 	LiNEt₂, 1.7 eq	C₆H₆, reflux, 2 days	OH (86)	130

TABLE I. SATURATED EPOXIDES (Continued)

C. Alicyclic Epoxides (Continued)

Epoxide	Base	Reaction Conditions	Products and Yields (%)	Refs.
C_{10} (Contd.)	LiNEt$_2$, 4.7 eq	Ether–hexane, room temp, 2 days	(—)	55
	LiNEt$_2$, 4.7 eq	Ether–hexane, room temp, 2 days	(—)	55
	LiNEt$_2$, 2.5 eq	Ether–hexane, reflux	2:98 (—)	21
	LiNEt$_2$, 2.5 eq	C$_6$H$_6$, reflux, 12 hr	(I, 32) + (II, 26) + (III, 9) + (IV, 11)	18
	LiNEt$_2$, 2.5 eq	C$_6$H$_6$, room temp, 12 hr	I + II + III + IV 35:26:24:15 (—)	18

Substrate	Reagent	Conditions	Product(s)	Refs.
	LiNEt₂, 1 eq	Ether–hexane, reflux, 6 hr	(I, 90–95)	6, 52
	n-BuLi	Ether, reflux, 6 hr	I (51)	132
	t-BuOK, 4.5 eq	t-BuOH, reflux, 3 hr	No reaction	86
	t-BuOK, 4.4 eq	DMF, 115°, 1 hr	I + (II) 71:29 (—)	85
	t-BuOK, 4.5 eq	Pyridine, reflux, 4.5 hr	I + II (91)	86
	t-BuOK, 1.1 eq	DMSO, 90°, 48 hr	I + II 62:38 (—)	85
	n-PrLi, 5 eq	Ether, room temp, 24 hr; reflux, 16 hr	(26) + (27) + (23)	80
	n-PrLi, 5 eq	Ether, room temp, 24 hr; reflux, 3 hr	(46) + (I, 26)	80
	t-BuOK, 4.4 eq	Pyridine, reflux 2.5 hr	I (75) + (6)	134

TABLE I. SATURATED EPOXIDES (*Continued*)

C. Alicyclic Epoxides (*Continued*)

Epoxide	Base	Reaction Conditions	Products and Yields (%)	Refs.
C_{10} (*Contd.*) epoxide, $t\text{-}C_4H_9$	$LiNEt_2$, 2.5 eq	Ether–hexane, reflux	(products, $t\text{-}C_4H_9$, OH) 50:24:22:4 (—)	21
epoxide, $t\text{-}C_4H_9$	$LiNEt_2$, 2.5 eq	Ether–hexane, reflux	(products, $t\text{-}C_4H_9$, OH) 97:3 (—)	21
epoxide, $i\text{-}C_3H_7$	n-PrLi, 6 eq	Ether, reflux, 30 hr	(I, 58) + (II, 31)	138
	t-BuOK, 3 eq	Pyridine, reflux, 3 hr	II (—)	134
	t-BuOK, 3 eq	DMF, 120°, 1 hr	II (—)	134
	t-BuOK, 3 eq	DMSO, 115°, 3 hr	II (—)	134
epoxide, $i\text{-}C_3H_7$	n-PrLi, 5 eq	Ether, room temp, 24 hr; reflux, 6 hr	(82) + (13)	138
epoxide	$LiNEt_2$, 1 eq	C_6H_6, reflux, 72 hr	(I) (64) + (II) (7) +	33

412

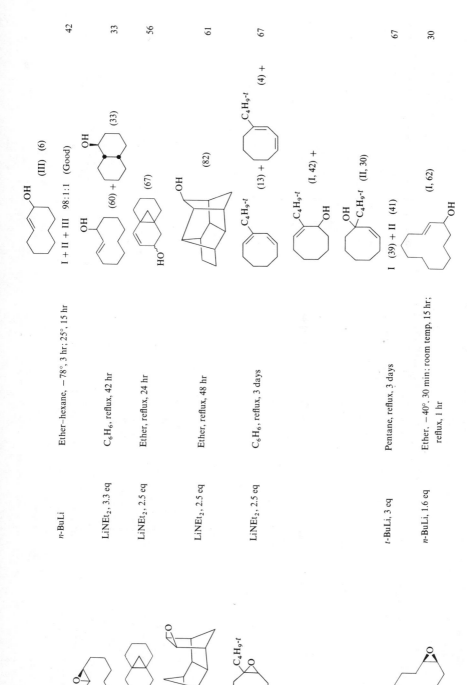

C_{11}	*n*-BuLi	Ether–hexane, −78°, 3 hr; 25°, 15 hr	(III) (6)
			I + II + III 98:1:1 (Good) 42
	LiNEt$_2$, 3.3 eq	C$_6$H$_6$, reflux, 42 hr	(60) + (33) 33
	LiNEt$_2$, 2.5 eq	Ether, reflux, 24 hr	(67) 56
C_{12}	LiNEt$_2$, 2.5 eq	Ether, reflux, 48 hr	(82) 61
	LiNEt$_2$, 2.5 eq	C$_6$H$_6$, reflux, 3 days	(13) + (4) + 67
			(I, 42) +
	t-BuLi, 3 eq	Pentane, reflux, 3 days	(II, 30)
			I (39) + II (41) 67
	n-BuLi, 1.6 eq	Ether, −40°, 30 min: room temp, 15 hr; reflux, 1 hr	(I, 62) 30

413

TABLE I. SATURATED EPOXIDES (Continued)

C. Alicyclic Epoxides (Continued)

Epoxide	Base	Reaction Conditions	Products and Yields (%)	Refs.
C_{12} (Contd.) 	n-BuLi, 3 eq	Ether–hexane, −78°, 3 hr; 25°, 15 hr	I + 89:11 (Good)	42
	DATMP, 4 eq	C_6H_6, 0°, 3 hr	I (20)	31
	n-BuLi, 1.6 eq	Ether, reflux	(I, 72–81)	30
	NLi	C_6H_6, 0°, 1 hr	I (<5)	31
	Et_2NAlEt_2	C_6H_6, 0°, 1 hr	I (<5)	32
	$(i\text{-}Pr)_2NAlEt_2$	C_6H_6, 0°, 1 hr	I (45)	31
	$(C_6H_{11})_2NAlEt_2$	C_6H_6, 0°, 1 hr	I (36)	32
	DATMP, 4 eq	C_6H_6, 0°, 3 hr	I (90)	31
C_{14} α, β epoxides (ca. 50:50)	LDPA, 3 eq	THF, reflux, 6 hr	(22) + OH[b] (47)	73

414

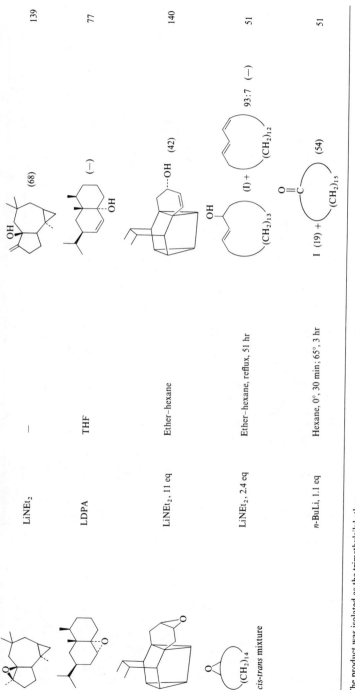

	Reagent	Solvent/Conditions	Product (yield)	Ref.
C_{15}	LiNEt$_2$	—	(68)	139
	LDPA	THF	(—)	77
C_{16}	LiNEt$_2$, 11 eq	Ether–hexane	⋯OH (42)	140
$(CH_2)_{14}$ cis-trans mixture	LiNEt$_2$, 2.4 eq	Ether–hexane, reflux, 51 hr	(I) + 93:7 (—)	51
	n-BuLi, 1.1 eq	Hexane, 0°, 30 min; 65°, 3 hr	I (19) + (54)	51

[a] The product was isolated as the trimethylsilyl ether.

[b] The product was analyzed after oxidation to the corresponding ketone with CrO_3.

415

TABLE II. UNSATURATED EPOXIDES

A. Olefinic Epoxides

	Epoxide	Base	Reaction Conditions	Products and Yields (%)	Refs.
C_5	H_2C—$C(CH_3)CH{=}CH_2$ (epoxide)	LDPA, 1.3 eq	Ether	$HOCH_2C({=}CH_2)CH{=}CH_2$ (76)	91
C_6	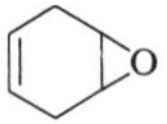 (cyclohexadiene epoxide)	$LiNEt_2$, 2.5 eq	Ether–hexane, room temp	Benzene (Quant)	44
		CH_3Li		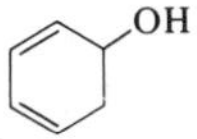 (phenol) OH + benzene 75:25 (—)	93
		$CH_2{=}CHCH_2Li$	Ether–THF, HMPA	" 65:35 (—)	84
	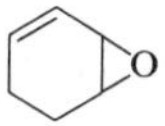 (cyclohexene epoxide)	$LiNEt_2$, 2.5 eq	Ether–hexane, room temp	Benzene (Quant)	44
		CH_3Li	—	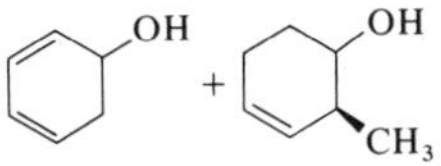 (phenol) OH + OH···CH_3 63:37 (—)	93
	CH_3HC—O—$CHCH_2CH{=}CH_2$ *cis*	$LiNEt_2$, 2.5 eq	Ether–hexane, several minutes	(E)-$CH_3CHOHCH{=}CHCH{=}CH_2$ (Quant)	44
	CH_3HC—O—$CHCH_2CH{=}CH_2$ *trans*	$LiNEt_2$, 2.5 eq	Ether–hexane, 4 min	$CH_3CHOHCH{=}CHCH{=}CH_2$ (85) ($E{:}Z = 88{:}12$)	44
	H_2C—O—$CH(CH_2)_2CH{=}CH_2$	$LiNEt_2$, 2.5 eq	Ether–HMPA, room temp, 15 min	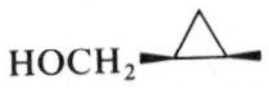 $HOCH_2$–△–$CH{=}CH_2$ (37) 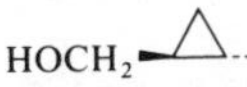 + $HOCH_2$–△···$CH{=}CH_2$ (63)	11
		$LiNH(CH_2)_2NH_2$	Ether–hexane, reflux, 72 hr	$H_2N(CH_2)_2NHCH_2CHOH(CH_2)_2CH{=}CH_2$ (—) +	11

416

Substrate	Reagent	Conditions	Products	Ref.
(bicyclic structure with OH) (I, –)	t-BuLi	Hydrocarbon	t-C$_4$H$_9$CH=CH(CH$_2$)$_2$CH=CH$_2$ (34) + (E, Z mixture) t-C$_4$H$_9$CH$_2$CHOH(CH$_2$)$_2$CH=CH$_2$ (11) + I (9)	39, 50
CH$_3$HC—O—CHCH=CHCH$_3$ cis, cis	LiNEt$_2$, 2.5 eq	Ether–hexane, 20 min	(E)-CH$_3$CHOHCH=CHCH=CH$_2$ (I, 65) + (E)-CH$_3$COCH=CHC$_2$H$_5$ (II, 1)	44
	LDPA, 2.5 eq	Ether–hexane, 6 hr	I (18) + II (7) + (Z)-CH$_3$COCH$_2$CH=CHCH$_3$ (45)	44
CH$_3$HC—O—CHCH=CHCH$_3$ trans, trans	LiNEt$_2$, 2.5 eq	Ether–hexane, 1 hr	(E)-CH$_3$CHOHCH=CHCH=CH$_2$ (I, 11) + (E)-CH$_2$=CHCHOHCH=CHCH$_3$ (II, 5) + aminoalcohols (54)	44
	LDPA, 2.5 eq	Ether–hexane, 6 hr	I (38) + II (22) + (E)-CH$_3$COCH$_2$CH=CHCH$_3$ (12) + (E)-CH$_3$COCH=CHC$_2$H$_5$ (3)	44
C$_7$ (cyclohexene oxide structure)	LiNEt$_2$, 2.5 eq	Ether–hexane, room temp	Toluene (Quant)	44
C$_2$H$_5$HC—O—CHCH$_2$CH=CH$_2$	LiNEt$_2$, 1 eq	C$_6$H$_6$–HMPA, room temp, 1 hr	C$_2$H$_5$CHOHCH=CHCH=CH$_2$ (52) (E:Z ~50:50)	97
C$_8$ (cyclooctatriene oxide structure)	LiNEt$_2$, 2 eq	Ether, −10°, 15 min	(structure) (71)	95
	(CH$_3$)$_3$C$_6$H$_2$Li	Ether, room temp, 40 hr	" (39)	95
(vinylcyclohexene oxide structure)	LiNEt$_2$	HMPA, room temp, 30 min	(structure) (52) + (structure) (28)	11
(cyclooctadiene oxide structure)	LiNEt$_2$, 2.5 eq	Ether–hexane, reflux, 22 hr	(structure) (I, 6) + (structure) (II, 65)	96
	LiNEt$_2$, 2.5 eq	C$_6$H$_6$, 50°, 1 hr	(3) + " (97)	26

417

TABLE II. UNSATURATED EPOXIDES (Continued)

A. Olefinic Epoxides (Continued)

Epoxide	Base	Reaction Conditions	Products and Yields (%)	Refs.
C_8 (Contd.)	$LiNEt_2$, 2.5 eq	HMPA, room temp, 10 min	I + (structures) 20:80 (—)	26
	$LiNH(CH_2)_2NH_2$	HMPA, 60°	I (84)	26
(epoxide structure)	$LiNEt_2$, 2.5 eq	Ether–hexane, room temp, 7 hr	(structures) (I, 32)	96
	$LiNEt_2$, 2.5 eq	Ether–hexane, room temp, 2 days	I (15) + (30) + (22) (structures)	96
	$LiNEt_2$, 2.5 eq	HMPA, 35°, 15 min	I (8) + (II, 92) (structures)	11
	$LDPA$, 2.5 eq	HMPA	I + II 21:79 (—)	11
	$LiNHC_6H_{11}$, 2.5 eq	HMPA	I + II 8:92 (—)	11
	$LiNH(CH_2)_2NH_2$, 3.7 eq	C_6H_6–HMPA, room temp, 15 min	I (4) + II (76)	11
$C_2H_5HC\overset{O}{-}CHCH_2C(CH_3)=CH_2$	$LiNEt_2$, 1 eq	C_6H_6–HMPA, room temp, 1 hr	$C_2H_5CHOHCH=CHC(CH_3)=CH_2$ (77) $(E:Z \sim 50:50)$	97
$C_2H_5HC\overset{O}{-}CHCH(CH_3)CH=CH_2$	$LiNEt_2$, 1 eq	C_6H_6–HMPA, room temp, 1 hr	$C_2H_5CHOHCH=C(CH_3)CH=CH_2$ (79) $(E:Z \sim 50:50)$	97
$(CH_3)_2C\overset{O}{-}CH(CH_2)_2CH=CH_2$	$LiNEt_2$, 2.5 eq	Ether–hexane, reflux, 2 hr	$CH_2=C(CH_3)CHOH(CH_2)_2CH=CH_2$ + $(CH_3)_2COH$—(structure)$\cdots CH=CH_2$ (I) 97:3 (—)	11

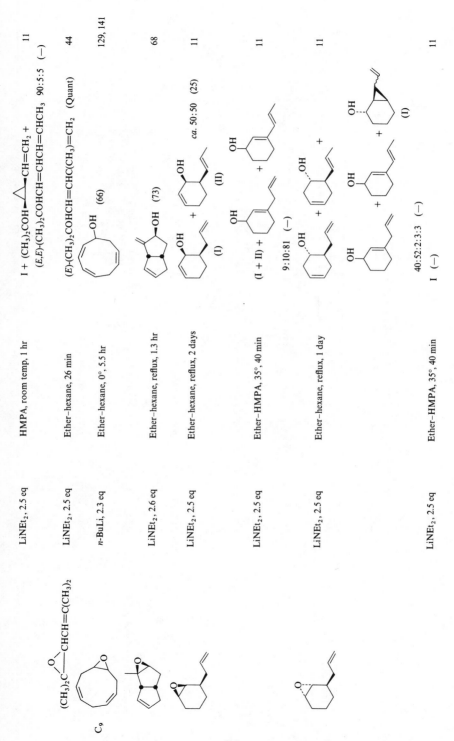

TABLE II. UNSATURATED EPOXIDES (*Continued*)

A. *Olefinic Epoxides* (*Continued*)

Epoxide	Base	Reaction Conditions	Products and Yields (%)	Refs.
C₉ (*Contd.*)	LiNEt₂, 2.5 eq	Ether–hexane, reflux, 4 days	[structure] OH + [structure] OH CH₂N(C₂H₅)₂ 20:80 (—)	11
	LiNEt₂, 2.5 eq	HMPA, room temp, 3 hr	[structure] CH₂OH (—)	11
	LDPA, 1.1 eq	Ether–hexane, 0°, 6 hr	[structure] OH (30) + [structure] OH (20)	136
(CH₃)₂C—C(CH₃)CH=C(CH₃)₂ (epoxide)	LiNEt₂, 2.5 eq	Ether–hexane, reflux, 2 hr	(CH₃)₂COHC(=CH₂)CH=C(CH₃)₂ (53) + (E)-(CH₃)₂COHC(CH₃)=CHC(CH₃)=CH₂ (10) + CH₂=C(CH₃)C(CH₃)OHCH=C(CH₃)₂ (7)	44
C₁₀	LiNEt₂, 2.5 eq	Ether–hexane, 1 min	[structure] OH (I, 86) + [structure] OH (II, 8) + [structure] (III, 6)	44
	LiNEt₂, 2.5 eq	Ether–hexane, 45 min	I + II + III 13:45:42 (—)	44

Substrate	Reagent	Conditions	Products	Ref.
α:β = 55:45	t-BuOK, 3 eq	Pyridine, reflux, 3.5 hr	(I) 10:77 (—)	134
	t-BuOK, 2.4 eq	DMF, 120°, 5 hr	I (—)	134
	t-BuOK, 2.2 eq	DMSO, 120°, 5 hr	I (—)	134
	LiNEt$_2$, 2.5 eq	Ether, reflux, 6 hr	(I) + (II) + (III) + (IV) 18:17:17:10 (—) I + II + III + IV 32:43:22:3 (—)	70
	LiNEt$_2$ LDPA	THF, 50°, 1 hr Hexane, 37°, 3 hr	I + II + III + IV + 32:35:18:10:1 (—)	70 70
	C$_6$H$_5$Li, 1.3 eq	Ether, reflux, 8–12 hr	(72)	99
C$_{11}$	LiNEt$_2$, 1.6 eq	Ether–hexane	(I, 25) +	142
	t-BuOK, 3 eq	C$_6$H$_6$, reflux, 6 hr	(II, 50) I (76) + II (9)	142

421

TABLE II. Unsaturated Epoxides (*Continued*)

A. *Olefinic Epoxides* (*Continued*)

Epoxide	Base	Reaction Conditions	Products and Yields (%)	Refs.
C$_{11}$ (*Contd.*)	LiNEt$_2$, 2.5 eq	Ether–hexane, reflux, 3 hr	(65) + (3) +	27
	LiNEt$_2$, 2.5 eq	Ether–hexane, 5 eq HMPA, room temp, several hours	(19) + (I, 2) I + II 37:24:39 (—)	27
	LiNEt$_2$, 2.5 eq	Ether–hexane, 10 eq, HMPA, room temp, several hours	I + II 25:75 (—)	27
C$_{12}$	*n*-BuLi, 2.7 eq	Ether	(66)	143
C$_{14}$	LiN , 2 eq	THF–hexane, room temp, 12 hr	(54)	144

422

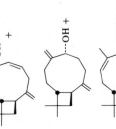

OH +

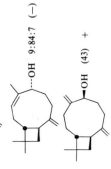

--OH +

--OH 9:84:7 (−)

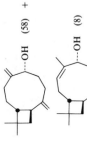

OH (43) +

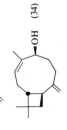

OH (34)

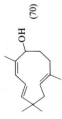

--OH (58) +

--OH (8)

OH (70)

71

71

71

145

Ether–C₆H₆, 35°, 3 hr

Ether–C₆H₆, 35°, 3 hr

Ether–C₆H₆, 35°, 3 hr

THF, room temp, 2 hr

n-BuLi, 1.4 eq

n-BuLi, 1.4 eq

n-BuLi, 1.4 eq

LDPA, 3 eq

C₁₅

TABLE II. UNSATURATED EPOXIDES (*Continued*)

A. Olefinic Epoxides (Continued)

Epoxide	Base	Reaction Conditions	Products and Yields (%)	Refs.
C$_{15}$ (*Contd.*) [structure]	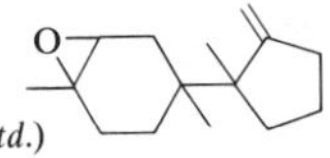 LiNEt$_2$	C$_6$H$_6$, Reflux	[structure] (—)	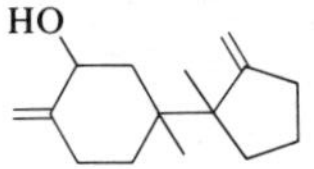 146
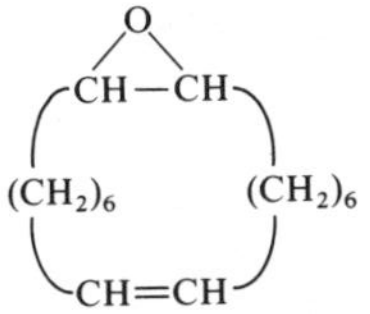 C$_{16}$ [structure] CH–CH (CH$_2$)$_6$ / (CH$_2$)$_6$ CH=CH (Mixture of four isomers)	n-BuLi, 1.3 eq	Hexane, 65°, 3 hr	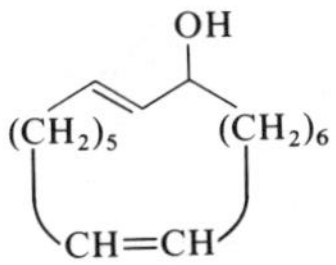 OH (CH$_2$)$_5$ CH=CH (CH$_2$)$_5$ (51) + [structure] (Mixture of *E,E* and *E,Z* isomers) 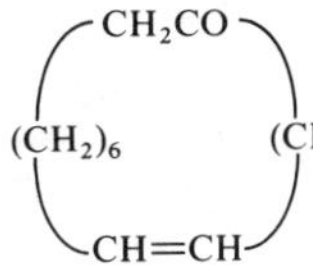	51
C$_{30}$ [structure]	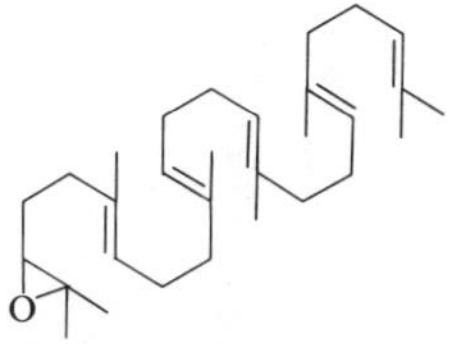 DATMP, 4 eq	C$_6$H$_6$, 0°, 2 hr	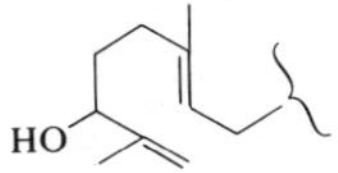 CH$_2$CO (CH$_2$)$_6$ / (CH$_2$)$_6$ CH=CH (—) [structure] (84)	31, 32

B. Acetylenic Epoxides

| C$_6$ CH$_3$HC(O)CHCH$_2$C≡CH *trans* | t-BuOK, 1.1 eq | DMSO, 25–35°, 45 min | (E)-CH$_3$CHOHCH=CHC≡CH (20) + | 94 |

424

Substrate	Base	Conditions	Product(s)	Refs.
CH₃HC—O—CHC≡CCH₃ (epoxide), trans	t-BuOK, 1.1 eq	DMSO, 25–35°, 45 min	![2,5-dimethylfuran] (50) + (E)-CH₃CHOHCH=CHC≡CH (9) +	94
H₂C—O—C(CH₃)C≡CCH₃ (epoxide)	LDPA, 2.2 eq	Hexane, 0°, 2 hr: room temp, 1 hr	![2,5-dimethylfuran] (40), HOCH₂C(CH₃)=CHC≡CH (40)	147
C₇ C₂H₅HC—O—CHCH₂C≡CH (epoxide)	LiNEt₂, 2 eq	C₆H₆–HMPA, room temp, 1 hr	C₂H₅CHOHCH=CHC≡CH (58), *E:Z = ca.* 50:50	97
(CH₃)₂C—O—CHCH₂C≡CH (epoxide)	t-BuOK, 1.1 eq	DMSO, 25–35°, 45 min	(E)-(CH₃)₂COHCH=CHC≡CH (70)	94
(CH₃)₂C—O—CHC≡CCH₃ (epoxide)	t-BuOK, 1.1 eq	DMSO, 25–35°, 45 min	(E)-(CH₃)₂COHCH=CHC≡CH (50) + ![dimethyl furanone] (7)	94
C₈ (CH₃)₂C—O—CH(CH₂)₂C≡CH (epoxide)	LiNEt₂, 2 eq	Ether–HMPA, 45°, 2.3 hr	(E)-(CH₃)₂COHCH=CHC≡CCH₃ (—)	100, 101
	LiNEt₂, 2.5 eq	Ether–hexane, reflux, 1 hr	(CH₃)₂COH—△—C≡CH + CH₂=C(CH₃)CHOH(CH₂)₂C≡CH 85:15 (—)	100, 101
C₉ (CH₃)₂C—O—CH(CH₂)₂C≡CCH₃ (epoxide)	LiNEt₂, 2.5 eq	Ether–hexane, reflux, 2 hr	CH₂=C(CH₃)CHOHCH₂CH₂C≡CCH₃ + CH₂=C(CH₃)CHOHCH₂C≡CC₂H₅ + CH₂=C(CH₃)CHOH(CH₂)₃C≡CH + (CH₃)₂COH—△—C≡CCH₃ (I) 47:13:34:6 (—)	100
	LiNEt₂, 2 eq	HMPA, 30°, 1 hr	I + (CH₃)₂COH—△—CH₂C≡CH 77:23 (—)	100
H₂C—O—C(CH₃)C≡CC₄H₉-t (epoxide)	LDPA, 2.2 eq	Hexane, reflux, 1 hr	HOCH₂C(=CH₂)C≡CC₄H₉-t (15)	147

TABLE II. UNSATURATED EPOXIDES (*Continued*)

C. *Aryl Epoxides*

Epoxide	Base	Reaction Conditions	Products and Yields (%)	Refs.
C$_9$	LiNEt$_2$	Ether–hexane, 1 min	(10)	43
C$_6$H$_5$CH$_2$HC—CH$_2$ (epoxide)	LiNEt$_2$	Ether–hexane, 7 min	C$_6$H$_5$CH=CHCH$_2$OH (High) (*E:Z* = 92:8)	43
C$_6$H$_5$HC—CHCH$_3$ (epoxide) *cis*	LiNEt$_2$	Ether–hexane, 5 hr	C$_6$H$_5$COC$_2$H$_5$ + C$_6$H$_5$CH$_2$COCH$_3$ + C$_6$H$_5$CHOHCH=CH$_2$ 91:2:7 (—)	43
C$_6$H$_5$HC—CHCH$_3$ (epoxide) *trans:cis* 94:6	LiNEt$_2$	Ether–hexane, reflux, 2 hr	C$_6$H$_5$COC$_2$H$_5$ + C$_6$H$_5$CH$_2$COCH$_3$ + C$_6$H$_5$CHOHCH=CH$_2$ + C$_6$H$_5$COCH$_3$ 33:13:35:19 (—)	43
C$_{10}$ C$_6$H$_5$CH$_2$CH$_2$HC—CH$_2$ (epoxide)	LiNEt$_2$	HMPA, room temp, 30 min	C$_6$H$_5$—△—CH$_2$OH (81) + C$_6$H$_5$—△—CH$_2$OH (5)	11
C$_6$H$_5$CH$_2$HC—CHCH$_3$ (epoxide) *cis*	LiNEt$_2$	Ether–hexane, 9 min	(*E*)-C$_6$H$_5$CH=CHCHOHCH$_3$ (Quant)	43
C$_6$H$_5$CH$_2$HC—CHCH$_3$ (epoxide) *trans*	LiNEt$_2$	Ether–hexane, 5 min	C$_6$H$_5$CH=CHCHOHCH$_3$ (High) (*E:Z* = 95:5)	43
H$_2$C—C(CH$_3$)C$_6$H$_4$CH$_3$-*p* (epoxide)	LDPA, 1.5 eq	Ether, reflux, 4 hr	HOCH$_2$C(=CH$_2$)C$_6$H$_4$CH$_3$-*p* (85)	92
C$_{11}$	LDPA, 2.5 eq	Ether, room temp, 1 hr	(64)	41

426

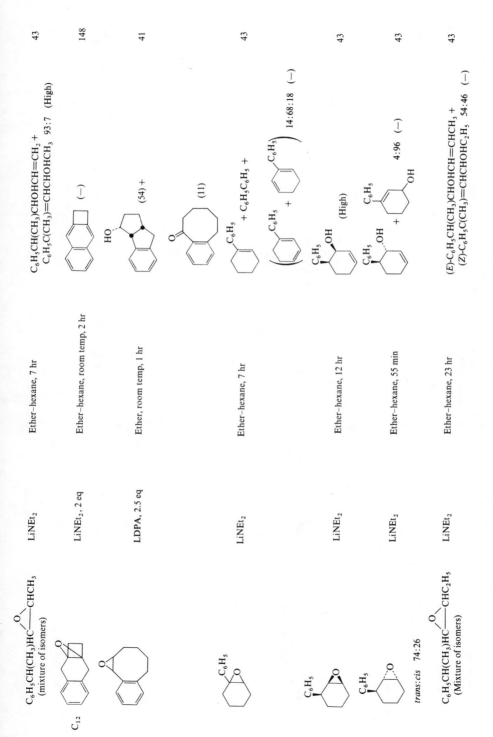

Substrate	Reagent	Conditions	Product(s)	Ref.
C₆H₅CH(CH₃)HC—O—CHCH₃ (mixture of isomers)	LiNEt₂	Ether–hexane, 7 hr	C₆H₅CH(CH₃)CHOHCH=CH₂ + C₆H₅C(CH₃)=CHCHOHCH₃ 93:7 (High)	43
C₁₂ (fused bicyclic epoxide)	LiNEt₂, 2 eq	Ether–hexane, room temp, 2 hr	(—)	148
(fused epoxycyclooctane)	LDPA, 2.5 eq	Ether, room temp, 1 hr	(54) + (11)	41
C₆H₅ epoxycyclohexane	LiNEt₂	Ether–hexane, 7 hr	+ C₆H₅C₆H₅ + 14:68:18 (—)	43
C₆H₅ epoxycyclohexane	LiNEt₂	Ether–hexane, 12 hr	C₆H₅ OH (High)	43
C₆H₅ epoxycyclohexane	LiNEt₂	Ether–hexane, 55 min	C₆H₅ OH + C₆H₅ OH 4:96 (—)	43
C₆H₅CH(CH₃)HC—O—CHC₂H₅ (Mixture of isomers)	LiNEt₂	Ether–hexane, 23 hr	(E)-C₆H₅CH(CH₃)CHOHCH=CHCH₃ + (Z)-C₆H₅C(CH₃)=CHCHOHC₂H₅ 54:46 (—)	43

trans:cis 74:26

TABLE II. UNSATURATED EPOXIDES (*Continued*)

C. Aryl Epoxides (*Continued*)

Epoxide	Base	Reaction Conditions	Products and Yields (%)	Refs.
C_{13} (bicyclic epoxide)	LDPA, 2.5 eq	Ether, room temp, 1 hr	HO— (structure) (65)	41
(cyclopropane epoxide) $CH_2C_6H_4CH_3$-p	$LiNEt_2$	Hexane, $-78°$, 5 min ; room temp, 30 min	$CH_2C_6H_4CH_3$-p^a (I) + $\quad$ $C_6H_4CH_3$-p^a (II) 81:19 (—)	98
	LDPA	Hexane, $-78°$, 5 min ; room temp, 30 min	I + II 99:1 (—)	98
	LiN (pyrrolidine)	Hexane, $-78°$, 5 min ; room temp, 15 min	I + II+ $\quad$ OH $\;$ $C_6H_4CH_3$-p^a (III) 34:56:10 (—)	98
	LiN (pyrrolidine)	Hexane, $-78°$, 5 min ; room temp, 120 min	II + III 98:2 (—)	98
	LiN (pyrrolidine)	Ether, $-78°$, 5 min ; room temp, 30 min	II + III 52:48 (—)	98
C_{14} $\quad$ $(C_6H_5)_2C$—CH_2 (epoxide)	$LiNEt_2$	Ether, reflux	$(C_6H_5)_2COCHCH_2N(C_2H_5)_2$ (26)	47
C_6H_5HC—CHC_6H_5 (epoxide) *cis*	$LiNEt_2$, 2 eq	Ether, reflux, 1 hr	$C_6H_5COCH_2C_6H_5$ (70)	47

428

C_6H_5HC——CHC_6H_5 with O, *trans*	$LiNEt_2$, 2 eq	Ether, reflux, 1 hr	$(C_6H_5)_2CHCHO^b$ (66)	47
C_{18}	$LiNEt_2$, 7.5 eq	Ether–hexane, room temp, 72 hr	(98)	140
C_{19}	$LiNEt_2$, 4 eq	Ether, reflux, 36 hr	(100)	81
	$LiNEt_2$, 2.5 eq	Ether, reflux, 2 days	(60)	78
C_{20} $(C_6H_5)_2C$——CHC_6H_5 with O	$LiNEt_2$	C_6H_6, 63°, 5 hr	$(C_6H_5)_2CHCOC_6H_5$ (80)	47
C_{21} $(C_6H_5)_2C$——$CHC_6H_4CH_3\text{-}p$ with O	$LiNEt_2$	C_6H_6, reflux, 18 hr	$(C_6H_5)_2CHCOC_6H_4CH_3\text{-}p$ (31) + $(C_6H_5)_2CHCHOHC_6H_4CH_3\text{-}p$ (18)	47
C_{26} $(C_6H_5)_2C$——$C(C_6H_5)_2$ with O	$LiNEt_2$	C_6H_6, reflux	No reaction	47

[a] Isolated as the trimethylsilyl ether.
[b] Isolated as the dimedone derivative.

TABLE III. FUNCTIONAL EPOXIDES

Epoxide	Base	Reaction Conditions	Products and Yields (%)	Refs.
C3 ClCH$_2$HC—CH$_2$ (epoxide)	n-BuLi	THF, −80°	(E)-ClCH=CHCH$_2$OH (I, 40–60)	149
	n-BuLi	THF, room temp	I + (Z)-ClCH=CHCH$_2$OH (II) 70:30 (—)	149
	n-BuLi	Ether–heptane, −50°, 1 hr; warm	I (15) + II (60) + HC≡CCH$_2$OH (15)	150
C6 (epoxide structure)	n-BuLi, 2 eq	Ether–hexane, room temp, 12 hr	(82)	102
(epoxide structure)	n-BuLi, 2 eq	Ether–hexane, room temp, 12 hr	(65)	151
(epoxide structure)	LiNEt$_2$, 3 eq	C$_6$H$_6$–hexane, room temp, 18 hr; reflux, 1 hr	(20)	151
(epoxide structure)	LiNEt$_2$, 3 eq	C$_6$H$_6$–hexane, room temp, 18 hr; reflux, 1 hr	(50)	151
C7 (epoxide structure)	LiNEt$_2$, 1.7 eq	C$_6$H$_6$–ether, reflux, 50 hr	(59) + (6)	110
C8 (epoxide structure)	LiNEt$_2$, 2.1 eq	Ether, −15°, 8 min, 0°, 1 hr	(44) + " (20)	110
	LiNEt$_2$, 1.5 eq	C$_6$H$_6$–ether, reflux, 4 days	CON(C$_2$H$_5$)$_2$ 10:1:62:27 (—)	110

430

Substrate	Reagent	Conditions	Product(s) and Yield(s) (%)	Refs.
	LiNH(CH₂)₂NH₂	C₆H₆, 35°, 1 hr	(7) + (I, 73)	100
	LiNH(CH₂)₂NH₂	HMPA, 30°, 1 hr	I (—), (90)	100
	LiNH(CH₂)₂NH₂	HMPA, 50°, 12 hr	(85)	100
	LiNH(CH₂)₂NH₂, 4.5 eq	C₆H₆, 50°, 30 min	HO(CH₂)₂OC₆H₄CH₃-p (—)	100
	LiNH(CH₂)₂NH₂, 4.5 eq	HMPA, 50°, 30 min	" (—)	100
C₉	LiNEt₂	Ether	(61)	107
C₁₀	LiNEt₂, 1.5 eq	THF–hexane, 0°, 5 min	(88) + (12)	54
	LiNEt₂, 2.5 eq	Ether, reflux, 12 hr		72
	LDPA, 2.2 eq	THF, 0°, 4 hr	(65)	106

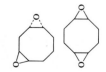

TABLE III. Functional Epoxides (*Continued*)

Epoxide	Base	Reaction Conditions	Products and Yields (%)	Refs.
C$_{10}$ (*Contd.*)	Ti(OPr-*i*)$_4$, 1.2 eq	CH$_2$Cl$_2$, room temp, 12 hr	(70)	106
	LDPA, 2.2 eq	THF, 0°, 4 hr	(65)	106
	Ti(OPr-*i*)$_4$, 1.2 eq	CH$_2$Cl$_2$, room temp, 12 hr	" (70)	106
	LiNH(CH$_2$)$_2$NH$_2$	H$_2$N(CH$_2$)$_2$NH$_2$, 110°, 30 min	(55)	104
	DATMP, 4 eq	C$_6$H$_6$, 0°, 30 min	(88)	31, 32
	Al(OPr-*i*)$_3$	No solvent, 120°	(−)	87
	Al(OPr-*i*)$_3$	No solvent, 120°	(−)	87
C$_{11}$	LiNEt$_2$, 3.5 eq	Ether–hexane, reflux, 12 hr	(76)	109

432

Substrate	Reagent	Conditions	Product(s) (%)	Refs.
C₁₂ $C_4H_9\text{-}n$ epoxide, $OSi(CH_3)_3$	DATMP	C_6H_6, 0°, 2 hr	HO— $\overset{OH}{\underset{}{}}$ $C_4H_9\text{-}n^a$ (75)	105
C₁₃ $C_6H_5CH_2OCH_2CH_2HC\!-\!\!\overset{O}{\diagup}\!\!-\!C(CH_3)_2$	$Al(OPr\text{-}i)_3$, 1 eq	Toluene, reflux	$C_6H_5CH_2OCH_2CHOHC(CH_3)=CH_2$ (99)	89
aryl epoxide, CO_2H	$t\text{-BuOK}$, 3 eq	THF, 15°	aryl CO_2H, $CH_2COH(CH_3)_2$ (78)	111
epoxide chain, $OSi(CH_3)_3$	DATMP	C_6H_6, 0°, 2 hr	OH^a / OH (79)	105
C₁₄ epoxide chain, $OSi(CH_3)_3$	DATMP	C_6H_6, 0°, 2 hr	OH^a / HO (65)	105
steroidal dioxolane epoxide	LDPA, 2.5 eq	Ether, room temp, 24 hr	(82)	152
$C_6H_5CHHC\!-\!\!\overset{O}{\diagup}\!\!-\!CHC_2H_5$, $OSi(CH_3)_3$	DATMP	C_6H_6, 0°, 2 hr	$(E)\text{-}C_6H_5CHOHCHOHCH=CHCH_3{}^a$ (60)	105
C₁₅ *trans* macrocyclic diepoxide, OH	$LiNEt_2$, 5 eq	C_6H_6, reflux, 2 hr	(25)	153

433

TABLE III. FUNCTIONAL EPOXIDES (*Continued*)

Epoxide	Base	Reaction Conditions	Products and Yields (%)	Refs.
C$_{15}$ (*Contd.*)	DATMP, 5 eq	C$_6$H$_6$, 0°, 30 min	(90)	31
	LiNEt$_2$, 4 eq	THF–hexane, reflux, 3 hr	(95)	76
	LiNEt$_2$, 5 eq	C$_6$H$_6$, reflux, 1 hr	OH (>50)	65
	DATMP, 10 eq	C$_6$H$_6$, 0°, 3 hr	OH (41)	32
(CH$_3$)$_2$C—O—CH(CH$_2$)$_2$CH(CH$_3$)(CH$_2$)$_2$OTHP	LDPA, 1.8 eq	Ether, reflux	CH$_2$=C(CH$_3$)CHOH(CH$_2$)$_2$CH(CH$_3$)(CH$_2$)$_2$OTHP + (major) (E)-(CH$_3$)$_2$COHCH=CHCH$_2$CH(CH$_3$)(CH$_2$)$_2$OTHP (minor) (60, total)	64
(structure) OSi(CH$_3$)$_3$	DATMP	C$_6$H$_6$, 0°, 2 hr	OH[a] OH (63)	105
(CH$_2$)$_6$CO$_2$C$_4$H$_9$-t	LDPA	Hexane	(CH$_2$)$_6$CO$_2$C$_4$H$_9$-t[b] (High)	98
C$_{16}$ (structure) CH(OCH$_3$)$_2$	MgBrN(Pr-i)C$_6$H$_{11}$, 27 eq	THF, 45°, 30 hr	CH(OCH$_3$)$_2$ (62) +	137

Substrate	Reagent	Conditions	Product(s) (%)	Refs.
C_{17} (aromatic OCH_3 decalin epoxide structure)	$LiNEt_2$, 5 eq	Ether, reflux, 36 hr	$CH(OCH_3)_2$ structure (10); (100) HO structure	78
$(CH_3)_2C$—O—$CH(CH_2)_2C(CH_3)=CHCH_2OCH_2C_6H_5$	$Al(OPr\text{-}i)_3$, 1 eq	Toluene, reflux	$CH_2=C(CH_3)CHOH-\}$ (93)	89
$(CH_3)_2C$—O—$CH(CH_2)_2C(CH_3)=CHCH_2SO_2C_6H_4CH_3\text{-}p$	$Al(OPr\text{-}i)_3$, 1 eq	Toluene, reflux	$CH_2=C(CH_3)CHOH-\}$ (94)	89
C_{18} (epoxide—$OSi(CH_3)_3$) *trans*	DATMP	C_6H_6, 0°, 2 hr	$\}$—CH_2OH, OH^a structure (71)	105
(epoxide—$OSi(CH_3)_3$) *cis*	DATMP	C_6H_6, 0°, 2 hr	$\}$—CH_2OH, OH^a structure (70)	105
$(CH_3)_3SiO(CH_2)_9HC$—O—$CHCH_2OSi(CH_3)_3$	DATMP	C_6H_6, 0°, 2 hr	$(E)\text{-}HO(CH_2)_8CH=CHCHOHCH_2OH^a$ (50)	105
C_{19} $n\text{-}C_5H_{11}C{\equiv}CCH_2HC$—O—$CH(CH_2)_7CO_2CH_3$	$LiNEt_2$	Ether	$(E)\text{-}n\text{-}C_5H_{11}C{\equiv}CCH=CHCHOH(CH_2)_7CO_2CH_3$ (50) (+ diethylamide)	154
$n\text{-}C_5H_{11}HC$—O—$CHCH_2CH=CH(CH_2)_7CO_2CH_3$	$LiNEt_2$	Ether, 0°, 1 hr	$n\text{-}C_5H_{11}CHOHCH=CHCH=CH(CH_2)_7CO_2CH_3$ (~60) (+ diethylamide)	154
(steroid epoxide structure)	$LiN(Pr\text{-}n)_2$, 1.5 eq	THF–hexane, reflux, 2 hr	(56) + (38) structures	69

435

TABLE III. FUNCTIONAL EPOXIDES (*Continued*)

Epoxide	Base	Reaction Conditions	Products and Yields (%)	Refs.
C₁₉ (*Contd.*)	LiN(Pr-*n*)₂, 1.5 eq	THF–hexane, reflux, 6 hr	(89)	69
C₂₀	LiNEt₂, 8 eq	Ether–hexane, room temp, 60 hr	(44)	140
	t-BuOK	THF, reflux	(>80)	108
	LiNEt₂, 8 eq	THF–hexane, 25°, 2 hr	(69)	103
	LiNEt₂, 8 eq	THF–hexane, 25°, 2 hr	(—)	103
	MgBrN(Pr-*i*)C₆H₁₁, 5 eq	THF, 0°–23°, 2 hr; 23°, 3.5 hr	(70)	90

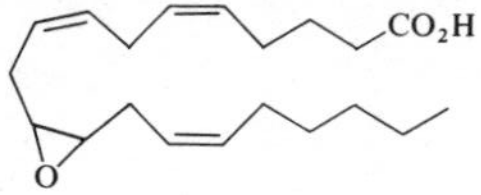

90

90

66

108

74

(28) +

(42)

(—)

(74)

(15)

(I, 71) +

(II, 18)

MgBrN(Pr-i)C$_6$H$_{11}$, 5 eq   THF, 0°, 0.5 hr; 23°, 3.5 hr

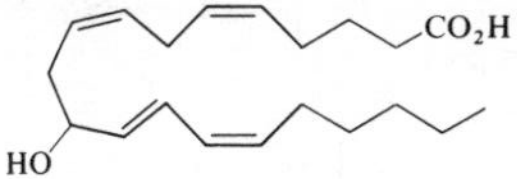

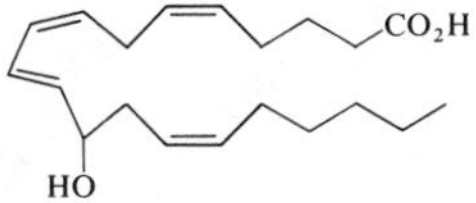

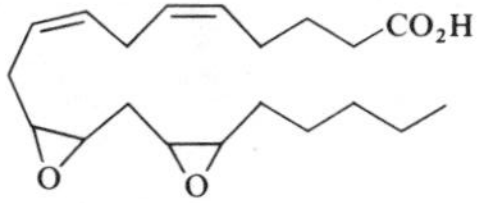

MgBrN(Pr-i)C$_6$H$_{11}$

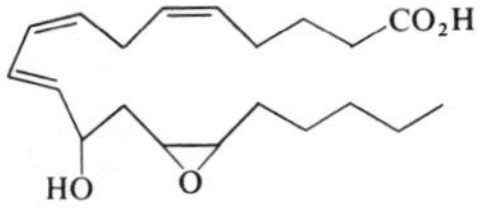

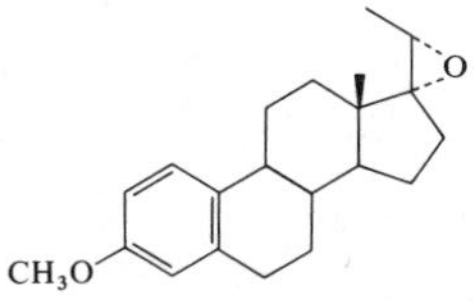

LiNEt$_2$, 5 eq   Hexane, HMPA, (2 eq), room temp, 10 hr

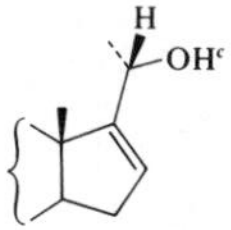

t-BuOK   THF, reflux

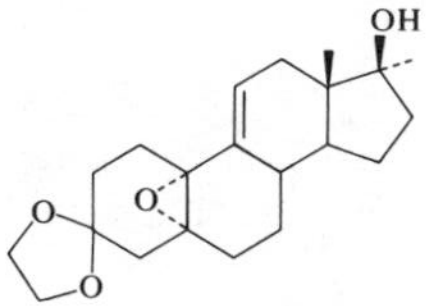

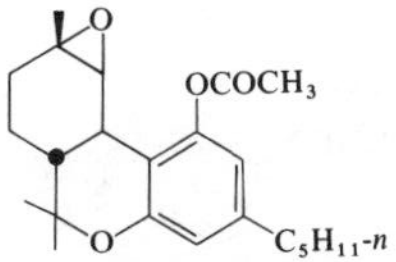

LiN(Bu-t)Si(CH$_3$)$_3$, 4.5 eq   Ether–hexane, reflux, 4 days

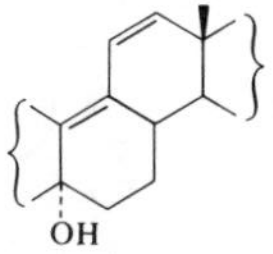

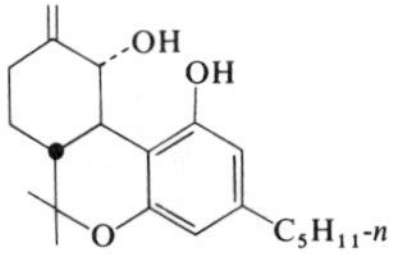

C$_{21}$

C$_{23}$

CO$_2$H

CO$_2$H

CH$_3$O

OCOCH$_3$

C$_5$H$_{11}$-n

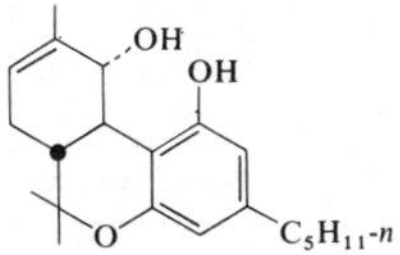

437

TABLE III. FUNCTIONAL EPOXIDES (Continued)

Epoxide	Base	Reaction Conditions	Products and Yields (%)	Refs.
C_{23} (Contd.) [structure: OCOCH₃, C_5H_{11}-n]	LiNEt₂ or LiN(Pr-i)C₆H₅	Ether–hexane	I + II 17:83 (—)	74
	LiN[Si(CH₃)₃]₂, or LDPA, LiN [piperidide]	Ether–hexane	I + II 50:50 (—)	74
[vinyl epoxide structure]	LiNEt₂, 6 eq	Hexane, room temp, 24 hr	[structure] (58) + [structure] (18)	66
C_{24} [structure]	LiNEt₂ or LDPA	Room temp, 10 min	[structure] (—)	108
	t-BuOK	THF, reflux	" (—)	108
[structure]	LiNEt₂ or LDPA	Room temp, 10 min	[structure] (—)	108
	t-BuOK	THF, reflux	[structure] (>80)	108

438

Substrate	Reagent	Conditions	Product (Yield, %)	Refs.
C₂₈ (steroid epoxide)	NaNH₂	H₂N(CH₂)₂NH₂	(steroid) (51)	108
	"	"	(–)	155
C₃₁ (aryl ether)	n-BuLi, 15 eq, H₂O (1 eq)	Diglyme–hexane, 200°, 18 hr		
	Al(OPr-i)₃	Toluene, reflux, 7 hr	{–CH₂CHOHCH(CH₃)}=CH₂ (91)	156
C₃₁ (silyl ether)	DATMP	C₆H₆, 0°, 3 hr	OSi(C₆H₅)₂C₄H₉-t (92)	75
C₃₄	Al(OPr-i)₃, 1 eq	Toluene, reflux	OH (97)	89
C₅₂	Al(OPr-i)₃, 1 eq	Toluene, reflux	OH (74)	89
C₅₈	LDPA, 5 eq	THF, 5°, 10 hr	{–[CH₂CH=C(CH₃)CH₂]₆CH₂CHOHC(CH₃)}=CH₂ (86)	62

439

a The product was obtained after removal of the trimethylsilyl group with potassium fluoride in methanol. The indicated yield is for the overall process starting with the allylic alcohol, which was epoxidized with V(acac)₃/t-BuOOH and silylated prior to treatment with base.
b The product was isolated as the trimethylsilyl ether.
c The product was isolated as the acetate.

REFERENCES

[1] A. Rosowsky, in *Heterocyclic Compounds with Three- and Four-Membered Rings*, Part I, A. Weissberger, Ed., Wiley-Interscience, New York, 1964, p. 1.

[2] G. Berti, in *Topics in Stereochemistry*, Vol. 7, N. L. Allinger and E. L. Eliel, Eds., Wiley-Interscience, New York, 1973, p. 93.

[3] J. G. Buchanan and H. Z. Sable, in *Selective Organic Transformations*, Vol. 2, B. S. Thyagarajan, Ed., Wiley-Interscience, New York, 1972, p. 1.

[4] V. N. Yandovskii and B. A. Ershov, *Russ. Chem. Rev.*, **41**, 403 (1972).

[5] C. J. M. Stirling, *Chem. Rev.*, **78**, 517 (1978).

[6] J. K. Crandall and L. C. Crawley, *Org. Synth.*, **53**, 17 (1973).

[7] R. L. Letsinger, J. G. Traynham, and E. Bobko, *J. Am. Chem. Soc.*, **74**, 399 (1952).

[8] L. J. Haynes, I. Heilbron, E. H. R. Jones, and F. Sondheimer, *J. Chem. Soc.*, 1583 (1947).

[9] A. C. Cope, H. H. Lee, and H. E. Petree, *J. Am. Chem. Soc.*, **80**, 2849 (1958).

[10] F. T. Bond and C. Y. Ho, *J. Org. Chem.*, **41**, 1421 (1976).

[11] M. Apparu and M. Barrelle, *Tetrahedron*, **34**, 1691 (1978).

[12] H. Yamamoto and H. Nozaki, *Angew. Chem., Int. Ed. Engl.*, **17**, 169 (1978).

[13] M. F. Lappert, P. P. Power, A. R. Sanger, and R. C. Srivastave, *Metal and Metalloid Amides*, Wiley, New York, 1980, Chapter 2.

[14] T. L. Brown, *Adv. Organomet. Chem.*, **3**, 365 (1965).

[15] A. C. Cope and J. K. Heeren, *J. Am. Chem. Soc.*, **87**, 3125 (1965).

[16] A. C. Cope, G. A. Berchtold, P. E. Peterson, and S. H. Sharman, *J. Am. Chem. Soc.*, **82**, 6370 (1960).

[17] R. P. Thummel and B. Rickborn, *J. Am. Chem. Soc.*, **92**, 2064 (1970).

[18] J. K. Crandall, L. C. Crawley, D. B. Banks, and L. C. Lin, *J. Org. Chem.*, **36**, 510 (1971).

[19] J. J. Eisch and J. E. Galle, *J. Organomet. Chem.*, **121**, C10 (1976).

[20] J. J. Eisch and J. E. Galle, *J. Am. Chem. Soc.*, **98**, 4646 (1976).

[21] B. Rickborn and R. P. Thummel, *J. Org. Chem.*, **34**, 3583 (1969).

[22] M. Hassan, A. R. O. Abdel Nour, and A. M. Satti, *Revue Roumaine de Chimie*, **23**, 747 (1978).

[23] J. Poláková, M. Palecek, and M. Procházka, *Collect. Czech. Chem. Commun.*, **44**, 3705 (1979).

[24] W. H. Saunders and A. F. Cockerill, *Mechanisms of Elimination Reactions*, Wiley-Interscience, New York, 1973.

[25] R. A. Bartsch and J. Závada, *Chem. Rev.*, **80**, 453 (1980).

[26] M. Apparu and M. Barrelle, *Tetrahedron*, **34**, 1541 (1978).

[27] M. Apparu and M. Barrelle, *Tetrahedron*, **34**, 1817 (1978).

[28] J. Smid, *Angew. Chem., Int. Ed., Engl.*, **11**, 112 (1972).

[29] R. P. Thummel and B. Rickborn, *J. Org. Chem.*, **36**, 1365 (1971).

[30] H. Nozaki, T. Mori, and R. Noyori, *Tetrahedron*, **22**, 1207 (1966).

[31] A. Yasuda, S. Tanaka, K. Oshima, H. Yamamoto, and H. Nozaki, *J. Am. Chem. Soc.*, **96**, 6513 (1974).

[32] A. Yasuda, H. Yamamoto, and H. Nozaki, *Bull. Chem. Soc. Jpn.*, **52**, 1705 (1979).

[33] A. C. Cope, M. Brown, and H. H. Lee, *J. Am. Chem. Soc.*, **80**, 2855 (1958).

[34] J. Sicher, *Angew. Chem., Int. Ed. Engl.*, **11**, 200 (1972).

[35] A. Nickon and N. H. Werstiuk, *J. Am. Chem. Soc.*, **89**, 3914 (1967).

[36] W. Kirmse, *Carbene Chemistry*, 2nd Ed., Academic Press, New York, 1971, Chapter 7.

[37] G. Köbrich, *Angew. Chem., Int. Ed. Engl.*, **6**, 41 (1967).

[38] R. A. Moss, in *Carbenes*, Vol. 1, M. Jones and R. A. Moss, Eds., Wiley-Interscience, New York, 1973, Chapter 2.

[39] J. K. Crandall and L. H. C. Lin, *J. Am. Chem. Soc.*, **89**, 4526 (1967).

[40] J. K. Crandall and L. H. C. Lin, *J. Am. Chem. Soc.*, **89**, 4527 (1967).

[41] R. W. Thies and R. H. Chiarello, *J. Org. Chem.*, **44**, 1342 (1979).

[42] R. K. Boeckman, Jr., *Tetrahedron Lett.*, **1977**, 4281.

[43] R. P. Thummel and B. Rickborn, *J. Org. Chem.*, **37**, 3919 (1972).

[44] R. P. Thummel and B. Rickborn, *J. Org. Chem.*, **37**, 4250 (1972).

[45] J. K. Crandall, R. D. Huntington, and G. L. Brunner, *J. Org. Chem.*, **37**, 2911 (1972).

[46] C. C. Price and D. D. Carmelite, *J. Am. Chem. Soc.*, **88**, 4039 (1966).

[47] A. C. Cope, P. A. Trumbull, and E. R. Trumbull, *J. Am. Chem. Soc.*, **80**, 2844 (1958).

[48] R. K. Hill, J. W. Morgan, R. V. Shetty, and M. E. Synerholm, *J. Am. Chem. Soc.*, **96**, 4201 (1974).

[49] T. J. Brocksom and J. T. B. Ferreira, *Synth. Commun.*, **11**, 105 (1981).

[50] L. C. Lin, Ph.D. Thesis, Indiana University, 1967 [*Diss. Abstr. B*, **28**, 3650 (1968)].

[51] B. D. Mookherjee, R. W. Trenkle, and R. R. Patel, *J. Org. Chem.*, **36**, 3266 (1971).

[52] J. K. Crandall and L. H. Chang, *J. Org. Chem.*, **32**, 435 (1967).

[53] C. Kissel and B. Rickborn, *J. Org. Chem.*, **37**, 2060 (1972).

[54] P. Warner, W. Boulanger, T. Schleis, S. L. Lu, Z. Le, and S. C. Chang, *J. Org. Chem.*, **43**, 4388 (1978).

[55] L. A. Paquette, W. E. Fristad, C. A. Schuman, M. A. Beno, and G. G. Christoph, *J. Am. Chem. Soc.*, **101**, 4645 (1979).

[56] J. A. Marshall and R. A. Ruden, *J. Org. Chem.*, **37**, 659 (1972).

[57] J. K. Whitesell and S. W. Felman, *J. Org. Chem.*, **45**, 755 (1980).

[58] J. K. Whitesell and P. D. White, *Synthesis*, **1975**, 602.

[59] M. N. Sheng, *Synthesis*, **1972**, 194.

[60] J. K. Crandall, *J. Org. Chem.*, **29**, 2830 (1964).

[61] J. R. Neff and J. E. Nordlander, *Tetrahedron Lett.*, **1977**, 499.

[62] L. C. Crawley, M. S. Thesis, Indiana University, 1969.

[63] S. Terao, K. Kato, M. Shiraishi, and H. Morimoto, *J. Chem. Soc., Perkin Trans. 1*, **1978**, 1101.

[64] O. P. Vig, S. D. Sharma, R. Vig, and S. D. Kumar, *Indian J. Chem.*, **18B**, 31 (1979).

[65] E. E. van Tamelen and J. P. McCormick, *J. Am. Chem. Soc.*, **92**, 737 (1970).

[66] B. M. Trost and T. R. Verhoeven, *J. Am. Chem. Soc.*, **100**, 3435 (1978).

[67] J. K. Crandall and L. H. C. Lin, *J. Org. Chem.*, **33**, 2375 (1968).

[68] J. K. Whitesell, R. S. Matthews, and P. K. S. Wang, *Synth. Commun.*, **7**, 355 (1977).

[69] J. S. Dutcher, J. G. Macmillan, and C. H. Heathcock, *J. Org. Chem.*, **41**, 2663 (1976).

[70] Y. Bessière and R. Derguini-Bouméchal, *J. Chem. Res. (M)*, **1977**, 3519.

[71] K. H. Schulte-Elte and G. Ohloff, *Helv. Chim. Acta*, **51**, 494 (1968).

[72] D. Mrozinska, A. Siemieniuk, K. Piatkowski, and H. Kuczynski, *Pol. J. Chem.*, **53**, 2213 (1979).

[73] S. C. Welch and A. S. C. Prakasa Rao, *J. Org. Chem.*, **43**, 1957 (1978).

[74] C. G. Pitt, M. S. Fowler, S. Sathe, S. C. Srivastava, and D. L. Williams, *J. Am. Chem. Soc.*, **97**, 3798 (1975).

[75] D. R. Williams and J. G. Phillips, *J. Org. Chem.*, **46**, 5452 (1981).

[76] B. A. Pawson, H. C. Cheung, S. Gurbaxani, and G. Saucy, *J. Am. Chem. Soc.*, **92**, 336 (1970).

[77] J. E. McMurry, J. H. Musser, M. S. Ahmad, and L. C. Blaszczak, *J. Org. Chem.*, **40**, 1829 (1975).

[78] R. C. Cambie, R. A. Franich, and T. J. Fullerton, *Aust. J. Chem.*, **24**, 593 (1971).

[79] B. M. Trost and M. J. Bogdanowicz, *J. Am. Chem. Soc.*, **95**, 5311 (1973).

[80] H. Kuczynski and K. Marks, *Rocz. Chem.*, **43**, 943 (1969) [*C.A.*, **71**, 61558 (1970)].

[81] R. C. Cambie and R. A. Franich, *Aust. J. Chem.*, **23**, 93 (1970).

[82] A. R. Lepley, W. G. Khan, A. B. Giumanini, and A. G. Giumanini, *J. Org. Chem.*, **31**, 2047 (1966).

[83] L. E. Overman and L. A. Flippin, *Tetrahedron Lett.*, **22**, 195 (1981).

[84] M. Apparu and M. Barrelle, *Bull. Soc. Chim. Fr.*, **1977**, 947.

[85] S. G. Traynor, B. J. Kane, J. B. Coleman, and C. G. Cardenas, *J. Org. Chem.*, **45**, 900 (1980).

[86] Z. Rykowski, K. Burak, and Z. Chabudzinski, *Rocz. Chem.*, **48**, 1619 (1974) [*C. A.*, **82**, 98160 (1975)].

[87] E. H. Eschinasi, *Isr. J. Chem.*, **6**, 713 (1968).

[88] E. H. Eschinasi, *J. Org. Chem.*, **35**, 1598 (1970).

[89] S. Terao, M. Shiraishi, and K. Kato, *Synthesis*, **1979**, 467.

[90] E. J. Corey, A. Marfat, J. R. Falck, and J. O. Albright, *J. Am. Chem. Soc.*, **102**, 1433 (1980).

[91] R. G. Riley, R. M. Silverstein, J. A. Katzenellenbogen, and R. S. Lenox, *J. Org. Chem.*, **39**, 1957 (1974).

[92] O. P. Vig, S. S. Bari, S. D. Sharma, and S. S. Rana, *Indian J. Chem.*, **15B**, 1076 (1977).

[93] J. Staroscik and B. Rickborn, *J. Am. Chem. Soc.*, **93**, 3046 (1971).

[94] P. H. M. Schreurs, A. J. de Jong, and L. Brandsma, *Recl. Trav. Chim. Pays-Bas*, **95**, 75 (1976).

[95] A. C. Cope and B. D. Tiffany, *J. Am. Chem. Soc.*, **73**, 4158 (1951).

[96] J. K. Crandall and L. H. Chang, *J. Org. Chem.*, **32**, 532 (1967).

[97] P. Miginiac and G. Zamlouty, *Bull. Soc. Chim. Fr.*, **1975**, 1740.

[98] B. M. Trost and S. Kurozumi, *Tetrahedron Lett.*, **1974**, 1929.

[99] S. K. Taylor and C. B. Rose, *J. Org. Chem.*, **42**, 2175 (1977).

[100] M. Apparu, Thesis, Université Scientifique et Médicale de Grenoble, Grenoble, France, 1977.

[101] M. Apparu and M. Barrelle, *Tetrahedron Lett.*, **1976**, 2837.

[102] U. P. Singh and R. K. Brown, *Can. J. Chem.*, **49**, 3342 (1971).

[103] M. Miyano, *J. Org. Chem.*, **46**, 1846 (1981).

[104] K. H. Schulte-Elte and G. Ohloff, *Helv. Chim. Acta*, **50**, 153 (1967).

[105] S. Tanaka, A. Yasuda, H. Yamamoto, and H. Nozaki, *J. Am. Chem. Soc.*, **97**, 3252 (1975).

[106] D. J. Morgans, K. B. Sharpless, and S. G. Traynor, *J. Am. Chem. Soc.*, **103**, 462 (1981).

[107] W. C. Still, A. J. Lewis, and D. Goldsmith, *Tetrahedron Lett.*, **1971**, 1421.

[108] G. Teutsch and R. Bucourt, *J. Chem. Soc., Chem. Commun.*, **1974**, 763.

[109] J. K. Crandall and D. R. Paulson, *J. Org. Chem.*, **33**, 3291 (1968).

[110] R. N. McDonald, R. N. Steppel, and R. C. Cousins, *J. Org. Chem.*, **40**, 1694 (1975).

[111] R. R. Kurtz and D. J. Houser, *J. Org. Chem.*, **46**, 202 (1981).

[112] M. T. Reetz and W. F. Maier, *Justus Liebigs Ann. Chem.*, **1980**, 1471.

[113] H. Normant, T. Cuvigny, and D. Reisdorf, *C. R. Acad. Sci. Paris, Ser. C*, **268**, 521 (1969).

[114] P. Hullot and T. Cuvigny, *Bull. Soc. Chim. Fr.*, **1973**, 2985.

[115] O. F. Beumel and R. F. Harris, *J. Org. Chem.*, **28**, 2775 (1963).

[116] J. J. Eisch and J. E. Galle, *J. Org. Chem.*, **44**, 3277 (1979).

[117] H. W. Gschwend and H. R. Rodriguez, *Org. React.*, **26**, 93 (1979).

[118] H. Gilman and J. W. Morton, *Org. React.*, **8**, 258 (1954).

[119] J. M. Mallan and R. L. Bebb, *Chem. Rev.*, **69**, 693 (1969).

[120] B. J. Wakefield, *The Chemistry of Organolithium Compounds*, Pergamon Press, New York, 1974.

[121] M. F. Lipton, C. M. Sorensen, A. C. Sadler, and R. H. Shapiro, *J. Organomet. Chem.*, **186**, 155 (1980).

[122] D. E. Bergbreiter and E. Pendergrass, *J. Org. Chem.*, **46**, 219 (1981).

[123] M. R. Winkle, J. M. Lansinger, and R. C. Ronald, *J. Chem. Soc., Chem. Commun.*, **1980**, 87.

[124] S. C. Watson and J. F. Eastham, *J. Organomet. Chem.*, **9**, 165 (1967).

[125] Y. Kitagawa, A. Itoh, S. Hashimoto, H. Yamamoto, and H. Nozaki, *J. Am. Chem. Soc.*, **99**, 3864 (1977).

[126] A. Basha, M. Lipton, and S. M. Weinreb, *Tetrahedron Lett.*, **1977**, 4171.

[127] W. Nagata and M. Yoshioka, *Org. React.*, **25**, 354 (1977).

[128] W. Nagata and M. Yoshioka, *Org. Synth.*, **52**, 90 (1972).

[129] R. W. Thies, M. Gasic, D. Whalen, J. B. Grutzner, M. Sakai, B. Johnson, and S. Winstein, *J. Am. Chem. Soc.*, **94**, 2262 (1972).

[130] E. J. Corey and R. S. Glass, *J. Am. Chem. Soc.*, **89**, 2600 (1967).

[131] R. A. Olofson and C. M. Dougherty, *J. Am. Chem. Soc.*, **95**, 581, 582 (1973).

[132] J. P. Monthéard and Y. Chrétien-Bessière, *Bull. Soc. Chim. Fr.*, **1968**, 336.

[133] R. M. Kellogg and J. K. Kaiser, *J. Org. Chem.*, **40**, 2575 (1975).

[134] Z. Rykowski and K. Burak, *Rocz. Chem.*, **50**, 1709 (1976) [*C. A.*, **86**, 140258r (1977)].

[135] A. J. Bridges and G. H. Whitham, *J. Chem. Soc., Perkin Trans. 1*, **1975**, 2264.

[136] L. A. Paquette, G. D. Crouse, and A. K. Sharma, *J. Am. Chem. Soc.*, **104**, 4411 (1982).

[137] P. A. Wender and S. L. Eck, *Tetrahedron Lett.*, **23**, 1871 (1982).

[138] H. Kuczynski and K. Marks, *Rocz. Chem.*, **42**, 647 (1968) [*C. A.*, **71**, 39190 (1970)].

[139] H. Shirahama, K. Hayano, Y. Kanemoto, S. Misumi, T. Ohtsuka, N. Hashiba, A. Furusaki, S. Murata, R. Noyori, and T. Matsumoto, *Tetrahedron Lett.*, **21**, 4835 (1980).

[140] L. A. Paquette and G. Kretschmer, *J. Am. Chem. Soc.*, **101**, 4655 (1979).

[141] M. Gasic, D. Whalen, B. Johnson, and S. Winstein, *J. Am. Chem. Soc.*, **89**, 6382 (1967).

[142] J. K. Crandall and D. R. Paulson, *J. Org. Chem.*, **36**, 1184 (1971).

[143] L. I. Zakharkin, *Bull. Acad. Sci. USSR, Engl. Transl.*, **1961**, 2103.

[144] A. Ogiso, E. Kitazawa, M. Kurabayashi, A. Sato, S. Takahashi, H. Noguchi, H. Kuwano, S. Kobayashi, and H. Mishima, *Chem. Pharm. Bull.*, **26**, 3117 (1978).

[145] H. Shirahama, B. R. Chhabra, and T. Matsumoto, *Chem. Lett.*, **1981**, 717.

[146] Y. Machida and S. Nozoe, *Tetrahedron Lett.*, **1972**, 1969.

[147] F. Y. Perveev and L. N. Gonoboblev, *J. Org. Chem. USSR*, **5**, 987 (1969).

[148] R. P. Thummel, W. E. Cravey, and W. Nutakul, *J. Org. Chem.*, **43**, 2473 (1978).

[149] D. F. Hoeg, J. E. Forrette, and D. I. Lusk, *Tetrahedron Lett.*, **1964**, 2059.

[150] H. J. Fabris, *J. Org. Chem.*, **32**, 2031 (1967).

[151] K. Ranganayakulu, U. P. Singh, T. P. Murray, and R. K. Brown, *Can. J. Chem.*, **52**, 988 (1974).

[152] R. B. Miller and E. S. Behare, *J. Am. Chem. Soc.*, **96**, 8102 (1974).

[153] F. Bellesia, U. M. Pagnoni, and R. Trave, *Tetrahedron Lett.*, **1974**, 1245.

[154] H. B. S. Conacher and F. D. Gunstone, *J. Chem. Soc., Chem. Commun.*, **1968**, 281.

[155] C. E. Slemon and P. Yates, *Can. J. Chem.*, **57**, 304 (1979).

[156] S. Terao, K. Kato, M. Shiraishi, and H. Morimoto, *J. Org. Chem.*, **44**, 868 (1979).

AUTHOR INDEX, VOLUMES 1-29

CHAPTER AND TOPIC INDEX, VOLUMES 1-29

Many chapters contain brief discussions of reactions and comparisons of alternative synthetic methods which are related to the reaction that is the subject of the chapter. These related reactions and alternative methods are not usually listed in this index. In this index the volume number is in BOLDFACE, the chapter number in ordinary type.

SUBJECT INDEX, VOLUME 29

Since the table of contents provides a quite complex index, only those items not readily found from the contents page are listed here. Numbers in **BOLDFACE** refer to experimental procedures.

α-Acetylenic epoxides, 1,4 elimination of, 381
8-Acetyl-2,2,4,4-tetramethyl-8-
    azabicyclo[3.2.1]oct-6-en-3-one, **214**
Aci-nitro ethers, 18
N-Alkoxyimides, 17
Alkoxy(trisdimethylamino)phosphonium
    salts, 34, 36, 39
N-Alkylaziridines, 28, 32
N-Alkyl-*N*,*N*'-dicarbethoxyhydrazine, 9
O-Alkylhydroxylamines, 17
N-Alkylphthalimides, 10
Allylic alcohols, from epoxide isomerization,
    345-443
    further isomerization of, 370, 373-376
    stereochemistry of double bond, 353, 361,
        362, 367, 384
Antibiotic A26771B, 13
Aryl glycosides, 18
Aryl *N*-aryloxycarbonylanthranilates, 17
Arylthio glycosides, 38
Azides, synthesis from alcohols, 11
3-α-Azidocholestane, 11
1-(3'-Azido-3'-desoxy-2',5'-di-*O*-trityl-β-D-
    xylofuranosyl)uracil, **41**

Benzene, from 3,4-epoxycyclohexene, 380
1,2-Benzocyclohepten-4-one, 358-359, **397**
endo-2-Bicyclo[3.1.0]hexanol, 356
endo,*cis*-Bicyclo[3.3.0]octan-2-ol, 347, 364,
    396
endo-Bicyclo[5.1.0]oct-5-en-2-ol, 348, **397**
π-Bromo-(+)-camphor, **210**
3-Bromo-2-cyclohexen-1-one, **42**

Camphenic acid, 190
(+)-Camphor, **210**
Carbocamphenilone, 190
Carbonates of sugar alcohols, 16
Caryophyllenic acid, 21

3-Chloro-2,2-dimethylpropanol, **43**
(*S*)-(−)-5-Chloromethyl-2-pyrrolidinone, **43**
Cord-Factor, 40
α-Cuparenone, single-step synthesis of, 199
3α-Cyanocholestane, **45**
(*E*)-2-Cyclododecenol, 349, **398**
Cyclohexene oxide, isomerization of, 362,
    363, 376
2-(*N*-Cyclohexylimino)-3,3,4,4-tetra-
    methylcyclobutanone, **219**
2,4,7-Cyclononatrien-1-ol, **396**
Cyclopropane formation, 348, 354, 356, 357,
    359, 360, 364-365, 385-387, 391
3-Cyclopropyl-2,5-dimethyl-3-phenyl-
    cyclopentanone, **218**

3'-Desoxy-3'-iodo-5'-*O*-*p*-nitrobenzoyl-
    thymidine, 26
N,*N*-Dialkylpiperazines, 28
trans-Di-*tert*-butylethylene, from *tert*-
    butyllithium and *tert*-butylethylene
    oxide, 356
Diethylaluminum 2,2,6,6-tetramethyl-
    piperidide, for isomerization of
    epoxides, 353, 360, 361, 367, 368, 371,
    376, 377, 388, 389, 394, **398**
1,3-Diiodo-2,2-dimethylpropane, 21
2,3:5,6-Di-*O*-isopropylidene-3-iodoallose, 22
2,3:5,6-Di-*O*-isopropylidene-β-D-
    mannofuranosyl azide, **45**
N-(2,3:5,6-Di-*O*-isopropylidene-β-D-
    mannofuranosyl)phthalimide, **44**
2,4-Diisopropyl-5-methyl-3(2H)-furanone,
    219
trans-2,3-Di-(*o*-methoxyphenyl)oxirane, **46**
2,2-Dimethylbicyclo[3.2.1]oct-6-en-3-one, **215**
3,5-Dimethylbicyclo[5.4.0]undec-1(7)-en-4-
    one, **212**
2,12-Dimethylcyclododecanone, **211**

455